AF675409

Advances in Wine Physicochemical Properties, Sensory Attributes, and Health Benefits

Advances in Wine Physicochemical Properties, Sensory Attributes, and Health Benefits

Guest Editors

Irena Budić-Leto

Jasenka Gajdoš Kljusurić

MDPI

Basel • Beijing • Wuhan • Barcelona • Belgrade • Novi Sad • Cluj • Manchester

Guest Editors

Irena Budić-Leto
Department of Plant Sciences
Institute for Adriatic Crops
and Karst Reclamation
Split
Croatia

Jasenka Gajdoš Kljusurić
Faculty of Food Technology
and Biotechnology
University of Zagreb
Zagreb
Croatia

Editorial Office

MDPI AG
Grosspeteranlage 5
4052 Basel, Switzerland

This is a reprint of the Special Issue, published open access by the journal *Foods* (ISSN 2304-8158), freely accessible at: https://www.mdpi.com/journal/foods/special_issues/CQ6D450JBQ.

For citation purposes, cite each article independently as indicated on the article page online and as indicated below:

Lastname, A.A.; Lastname, B.B. Article Title. *Journal Name* **Year**, *Volume Number*, Page Range.

ISBN 978-3-7258-6298-6 (Hbk)
ISBN 978-3-7258-6299-3 (PDF)
https://doi.org/10.3390/books978-3-7258-6299-3

© 2026 by the authors. Articles in this book are Open Access and distributed under the Creative Commons Attribution (CC BY) license. The book as a whole is distributed by MDPI under the terms and conditions of the Creative Commons Attribution-NonCommercial-NoDerivs (CC BY-NC-ND) license (https://creativecommons.org/licenses/by-nc-nd/4.0/).

Contents

foods

MDPI

Editorial

Advances in Wine Physicochemical Properties, Sensory Attributes, and Health Benefits

Irena Budić-Leto [1,*] and Jasenka Gajdoš Kljusurić [2]

[1] Institute for Adriatic Crops and Karst Reclamation, 21000 Split, Croatia
[2] Faculty of Food Technology and Biotechnology, University of Zagreb, Pierottijeva 6, 10000 Zagreb, Croatia; jasenka.gajdos@pbf.unizg.hr
[*] Correspondence: irena.budic-leto@krs.hr

1. Introduction

Wine is one of the most studied fermented beverages in the world, representing a unique intersection of tradition, culture, and modern science [1,2]. Its complexity arises from the interaction of grape composition, *terroir*, microorganisms, and winemaking techniques [3,4], while its potential health effects continue to attract the interest of researchers and consumers alike [5]. Over the past decades, scientific advances have transformed our understanding of wine, moving beyond simple chemical characterisation toward a multi-dimensional approach that integrates viticulture, oenology, food technology, health sciences, and consumer behaviour [6,7].

This book edition brings together the articles published in the *Foods* Special Issue "Advances in Wine Physicochemical Properties, Sensory Attributes, and Health Benefits," conceived as a continuation of the successful first edition, "Grape Wine: Physicochemical Properties, Sensory Attributes, and Health Benefits." Its purpose is to provide readers with an integrated perspective on the latest research into wine composition, sensory quality, and potential health-promoting properties.

The collected contributions reflect the diversity and innovation of contemporary wine science. They explore a wide spectrum of topics—from the influence of vineyard environment and winemaking techniques on the aromatic and phenolic profile of wines, to the application of advanced analytical techniques such as non-destructive spectroscopy and chemometric modelling. Special attention is given to understanding the compounds responsible for physicochemical characteristics, sensory perception, and antioxidant activity, thereby supporting the view of wine as a potential functional food [8]. Additionally, these studies examine the role of microorganisms, the impact of technological innovations, and consumer perception, thereby opening up new opportunities for the development of wines tailored to evolving market demands and health-conscious lifestyles.

2. Overview of the Contributions

The opening chapter examines how different alcoholic bases affect the flavour and sensory balance of soaked greengage wine, combining chromatographic analyses with trained sensory evaluation to optimize fruit wine production (contribution 1). The study concludes that the choice of base liquor significantly influences the flavour profile of soaked greengage wine, with the 50% edible alcohol treatment performing best overall, providing practical guidance for fruit wine production and optimisation.

Furthermore, many factors influence grape and wine quality. The second study provides a detailed *terroir* and vintage analysis of Cabernet Sauvignon, linking soil composition, climatic variation, and vineyard location with key aroma compounds and wine style

https://doi.org/10.3390/foods14223889

(contribution 2). The findings indicate that within a single slope, *terroir* (soil and plot) plus vintage interact to produce noticeable differences in aroma profile and wine style, and that understanding these relationships can help refine vineyard management to improve wine aroma quality.

Natural additives are increasingly used in winemaking, with many commercial products claiming diverse technological and sensory benefits. Another contribution investigates the influence of oenological additives and individual taster phenotypes on flavour persistence, demonstrating how compounds such as tannins and mannoproteins can modulate astringency and retronasal aroma (contribution 3). The results of this research highlight that both individual taste phenotype (PROP taster vs. non-taster) and choice of oenological additive have measurable impacts on how long flavour (especially astringency and certain aroma attributes) lingers in wine, suggesting these tools can be used strategically to modulate wine sensory profile.

Wine contains a wide array of chemical compounds, attracting researchers eager to explore and understand its intricate composition. Subsequent research has highlighted innovative vinification techniques in Teran wine, demonstrating how extended maceration and thermal pre-treatments can enhance phenolic complexity and sensory quality (contribution 4). Overall, the results indicate that using non-standard vinification techniques can significantly boost the bioactive composition and improve the sensory quality of Teran wine, adding potential market value.

Several chapters focus on enhancing the bioactive and antioxidant potential of wines, including studies on melatonin addition during fermentation (contribution 5) and the selection of indigenous *Saccharomyces cerevisiae* strains to preserve polyphenols and improve functional value (contribution 6). Eremia and coworkers examined the impact of melatonin addition and different maceration techniques (punch-down and pump-over) on the polyphenolic profile and antioxidant activity of Feteasca Neagra and Cabernet Sauvignon wines. Early addition of melatonin during fermentation proved to be an effective tool for enriching wines with antioxidant compounds, though further studies are needed to confirm the stability and reproducibility of these effects during ageing. Romano and collaborators aimed to identify yeast strains that minimise the adsorption of polyphenolic compounds during fermentation, thereby preserving the antioxidant capacity of the wine. A total of 136 yeast strains were analysed through micro-scale fermentations, assessing their impact on total phenolic content and antioxidant capacity. These findings suggest that the selected indigenous yeast strains can be utilized as functional fermentation starters to produce red wines with enriched polyphenolic content, aligning with the growing consumer interest in wines with potential health benefits. This research represents a significant step towards the development of "functional" fermentation starters, offering a biotechnological approach to enhance the nutritional and sensory quality of red wines.

Sustainable approaches to winemaking are also represented through the valorization of Malbec and Torrontés pomace, revealing their rich phenolic content and potential in food, cosmetic, and pharmaceutical industries (contribution 7). Results showed that Malbec pomace had much higher total phenolics and flavonoids compared to Torrontés, and superior antioxidant, reducing, and radical-scavenging activities. The extracts showed no acute toxicity in several models, but Torrontés extract at high concentration reduced the viability of certain intestinal cell lines. Both extracts inhibited lipoxygenase, and Malbec also inhibited tyrosinase effectively.

Novel yeast strategies are presented in research on non-*Saccharomyces* species such as *Lachancea thermotolerans* and *Torulaspora delbrueckii*, showing their capacity to modulate acidity, aromatic complexity, and sensory appeal in red wines (contribution 8). Overall,

Torulaspora delbrueckii × *S. cerevisiae* proved most promising for improving the aromatic complexity and sensory appeal of Babić wines, though further large-scale studies are needed.

In response to climate-driven challenges, Canonico and coworkers (contribution 9) explore ethanol reduction through sequential fermentation with *Starmerella bombicola*, demonstrating improved balance and mouthfeel. Compared to pure *S. cerevisiae* fermentation, the sequential approach lowered ethanol by about 0.8–1% v/v and increased glycerol by 50%, improving wine body and structure. It also enhanced lactic acid and specific aromatic notes, giving wines superior sensory characteristics.

Food safety and quality are addressed through the evaluation of grape pomace as a natural alternative for removing ochratoxin A (OTA) while preserving beneficial metabolites, compared to conventional activated carbon treatment (contribution 10). Importantly, pomace repassage effectively reduced OTA without significantly compromising wine quality. These results suggest grape pomace represents a promising natural adsorbent for OTA decontamination in winemaking.

Finally, the collection concludes with an analysis of consumer perception of moderate wine consumption, highlighting the role of health-related label information and providing insights for clearer communication and the development of functional wine products (contribution 11). Findings suggest that wine labels play a critical role in shaping consumer awareness, highlighting the need for clearer communication about moderation, as well as opportunities for developing functional wines with added health value.

3. Conclusions

The contributions assembled in this volume illustrate the remarkable breadth and dynamism of current wine research. Collectively, they show how advances in viticulture, oenology, and analytical science are converging to improve our understanding of wine's chemical complexity, sensory diversity, and potential health benefits. From exploring *terroir*-driven aroma expression and innovative vinification techniques, to selecting tailored yeast strains and employing sustainable by-product valorization, the studies presented here point toward a more knowledge-based, innovative, and responsible wine industry.

Taken together, these studies highlight the potential to produce wines that are not only sensorially refined and authentic to their origin, but also technologically optimized and aligned with modern nutritional and sustainability goals. We hope that this collection will inspire further interdisciplinary collaboration and innovative research, supporting the continued evolution of wine science and contributing to a future where wine is better understood, appreciated, and responsibly enjoyed.

Author Contributions: Conceptualization, I.B.-L.; writing—original draft preparation, I.B.-L.; writing—review and editing, J.G.K. All authors have read and agreed to the published version of the manuscript.

Data Availability Statement: Not applicable.

Conflicts of Interest: The authors declare no conflict of interest.

List of Contributions

1. Zhao, P.; Liu, C.; Qiu, S.; Chen, K.; Wang, Y.; Hou, C.; Huang, R.; Li, J. Flavor Profile Evaluation of Soaked Greengage Wine with Different Base Liquor Treatments Using Principal Component Analysis and Heatmap Analysis. *Foods* **2023**, *12*, 2016. https://doi.org/10.3390/foods12102016.
2. Chen, Z.; Wu, Y.-P.; Lan, Y.-B.; Cui, Y.-Z.; Shi, T.-H.; Hua, Y.-B.; Duan, C.-Q.; Pan, Q.-H. Differences in Aroma Profile of Cabernet Sauvignon Grapes and Wines from Four Plots in Jieshi Mountain Region of Eastern China. *Foods* **2023**, *12*, 2668. https://doi.org/10.3390/foods12142668.

3. Velázquez-Martínez, R.I.; Criado, C.; Muñoz-González, C.; Crespo, J.; Pozo-Bayón, M.Á. Evaluation of the Long-Lasting Flavour Perception after the Consumption of Wines Treated with Different Types of Oenological Additives Considering Individual 6-n-Propylthiouracil Taster Status. *Foods* **2023**, *12*, 2835. https://doi.org/10.3390/foods12152835.
4. Orbanić, F.; Rossi, S.; Bestulić, E.; Budić-Leto, I.; Kovačević Ganić, K.; Horvat, I.; Plavša, T.; Bubola, M.; Lukić, I.; Jeromel, A.; et al. Applying Different Vinification Techniques in Teran Red Wine Production: Impact on Bioactive Compounds and Sensory Attributes. *Foods* **2023**, *12*, 3838. https://doi.org/10.3390/foods12203838.
5. Eremia, S.A.V.; Albu, C.; Radu, G.-L.; Alecu, A.; Stoica, A.G.; Brinduse, E. Winemaking Technologies for the Production of Cabernet Sauvignon and Feteasca Neagra Wines Enriched with Antioxidant Active Principles Due to the Addition of Melatonin. *Foods* **2024**, *13*, 884. https://doi.org/10.3390/foods13060884.
6. Romano, G.; Taurino, M.; Gerardi, C.; Tufariello, M.; Lenucci, M.; Grieco, F. Yeast Starter Culture Identification to Produce Red Wines with Enhanced Antioxidant Content. *Foods* **2024**, *13*, 312. https://doi.org/10.3390/foods13020312.
7. Tapia, P.E.; Silva, A.M.; Delerue-Matos, C.; Moreira, M.; Rodrigues, F.; Torres Carro, R.; Santi, M.D.; Ortega, M.G.; Blázquez, M.A.; Arena, M.E.; et al. Exploring the Phytochemical Composition and the Bioactive Properties of Malbec and Torrontés Wine Pomaces from the Calchaquíes Valleys (Argentina) for Their Sustainable Exploitation. *Foods* **2024**, *13*, 1795.
8. Ivić, S.; Jeromel, A.; Kozina, B.; Prusina, T.; Budić-Leto, I.; Boban, A.; Vasilj, V.; Jagatić Korenika, A.-M. Sequential Fermentation in Red Wine cv. Babić Production: The Influence of *Torulaspora delbrueckii* and *Lachancea thermotolerans* Yeasts on the Aromatic and Sensory Profile. *Foods* **2024**, *13*, 2000. https://doi.org/10.3390/foods13132000.
9. Canonico, L.; Gattucci, S.; Moretti, L.; Agarbati, A.; Comitini, F.; Ciani, M. Ethanol Reduction in Montepulciano Wine: *Starmerella bombicola* Sequential Fermentation at Pilot Scale under Aeration Conditions. *Foods* **2025**, *14*, 618. https://doi.org/10.3390/foods14040618.
10. Del Coco, L.; Girelli, C.R.; Gambacorta, L.; Solfrizzo, M.; Fanizzi, F.P. ^{1}H NMR Spectroscopy Primitivo Red Wine Screening after Grape Pomace Repassage for Possible Toxin Contamination Removal. *Foods* **2025**, *14*, 734. https://doi.org/10.3390/foods14050734.
11. Ilak Peršurić, A.S.; Težak Damijanić, A.; Radeka, S. Relationship Between Health Benefit Perception, Moderate Wine Consumption, Wine Label and Healthy Behaviour. *Foods* **2025**, *14*, 1937. https://doi.org/10.3390/foods14111937.

References

1. Ribéreau-Gayon, P.; Glories, Y.; Maujean, A.; Dubourdieu, D. The Microbiology of Wine and Vinifications. In *Handbook of Enology*; John Wiley & Sons, Ltd.: West Sussex, UK, 2006; Volume 1.
2. Jackson, R.S. *Wine Science: Principles and Applications*, 3rd ed.; Academic Press: London, UK, 2008; p. 776.
3. Chalvantzi, I.; Banilas, G.; Tassou, C.; Nisiotou, A. Biogeographical Regionalization of Wine Yeast Communities in Greece and Environmental Drivers of Species Distribution at a Local Scale. *Front. Microbiol.* **2021**, *12*, 705001. [CrossRef] [PubMed]
4. Škrab, D.; Sivilotti, P.; Comuzzo, P.; Voce, S.; Cisilino, D.; Carlin, S.; Arapitsas, P.; Masuero, D.; Vrhovsek, U. Influence of Harvest Date on Multi-Targeted Metabolomic Profile and Sensory Attributes of Ribolla Gialla Base and Sparkling Wines. *OENO One* **2024**, *58*, 1–17. [CrossRef]
5. Ćorković, I.; Pichler, A.; Šimunović, J.; Kopjar, M. A Comprehensive Review on Polyphenols of White Wine: Impact on Wine Quality and Potential Health Benefits. *Molecules* **2024**, *29*, 5074. [CrossRef] [PubMed]
6. Boban, D.; Grković, I.; Dželalija, A.M.; Gujinović, D.; Mudnić, I.; Boban, M. The Effects of White Wine and Ethanol Consumption on the Proliferative Phase of Repair after a Surgically Induced Myocardial Infarction in Rats. *Nutrients* **2025**, *17*, 699. [CrossRef] [PubMed]
7. de Sousa Dias, A.L.; Guittin-Leignadier, C.; Roy, A.; Sachot, S.; Maçna, F.; Flores, D.; Meudec, E.; Boulet, J.-C.; Sommerer, N.; Roland, A.; et al. Oenological Performances of New White Grape Varieties. *Beverages* **2025**, *11*, 90. [CrossRef]
8. Giovinazzo, G.; Grieco, F. Tapping into Health: Wine as Functional Beverage. In *Alcoholic Beverages*; Grumezescu, A.M., Holban, A.M., Eds.; Woodhead Publishing: Cambridge, UK, 2019; pp. 279–302, ISBN 978-0-12-815269-0. [CrossRef]

Disclaimer/Publisher's Note: The statements, opinions and data contained in all publications are solely those of the individual author(s) and contributor(s) and not of MDPI and/or the editor(s). MDPI and/or the editor(s) disclaim responsibility for any injury to people or property resulting from any ideas, methods, instructions or products referred to in the content.

foods

MDPI

Article

Flavor Profile Evaluation of Soaked Greengage Wine with Different Base Liquor Treatments Using Principal Component Analysis and Heatmap Analysis

Peipei Zhao [1,2], Chang Liu [2], Shuang Qiu [2], Kai Chen [2,3], Yingxiang Wang [4], Caiyun Hou [2], Rui Huang [5] and Jingming Li [1,2,*]

1 CAU Sichuan Chengdu Advanced Agricultural Industrial Institute, Chengdu 611430, China; z15355952943@163.com
2 College of Food Science & Nutritional Engineering, China Agricultural University, No. 17 Tsinghua Dong Road, Beijing 100083, China
3 College of Food Science & Technology, Henan Agricultural University, No. 63 Nongye Road, Zhengzhou 450002, China
4 Sichuan Mehe Wine Industry Co., Ltd., No. 551 Xiling Avenue, Jinyuan Town, Dayi County, Chengdu 611330, China
5 Luzhou Laojiao Co., Ltd., Luzhou 646000, China
* Correspondence: lijingming@cau.edu.cn

Abstract: The selection of base liquor plays a crucial role in the flavor of soaked greengage wine. This study aimed to investigate the effects of different base liquor treatments on the physicochemical characteristics and aroma composition of greengage wine. We carried out a comprehensive analysis using HPLC for the determination of organic acids and GC-MS for the determination of volatile aroma compounds, combined with sensory evaluation. The results showed that the red and yellow colors were the darkest in the high-alcohol group, while the citric acid content was the highest in the sake group (21.95 $\pm$ 2.19 g/L). In addition, the greengage wine steeped in 50% edible alcohol had more terpenes, a significantly higher concentration of acid–lipid compounds, and a more intense aroma compared to that of the low-alcohol group, whose typical aroma compounds were greatly reduced. The sensory results showed that the greengage wine treated with baijiu had a distinct alcoholic flavor, while almond flavors were more intense in the greengage wine treated with 15% edible alcohol. In this study, base liquor was used as the main influencing factor to provide new research ideas for the flavor optimization of soaked greengage wine.

Keywords: greengage wine; gas chromatography–mass spectrometry; aroma-active compounds; sensory evaluation; correlational analysis

1. Introduction

Greengage (*Prunus mume*) is a plant belonging to the Rosaceae family that is rich in bioactive substances and possesses several healthcare benefits [1]. According to the literature, fresh greengage fruit contains large amounts of protein, fat, many essential amino acids, organic acids, and minerals [2]. In addition to these essential nutrients, greengage contains phenolic substances such as squalene, epicatechin, chlorogenic acid, and rutin, which have antibacterial effects [3,4]. Processed greengage fruit can also be used for the treatment of cough, vomiting, diarrhea, and fever [5]. Mature greengage is mainly used for the production of alcoholic beverages and dried fruit [4]. Currently, greengage wine is popular in Asia due to its unique flavor and nutritional properties, including a low alcohol content [6].

The greengage wine on the market at present can be mainly divided into two types: fermentation and soaking [7]. The former is brewed by adding yeast, while the latter is brewed by soaking the greengage in a base liquor. Related studies show that soaking greengage wine improves fruit flavor more than fermentation [6,8]. Compared with soaked greengage wine, the flavor of fermented greengage wine tends to be more diverse due to the complex fermentation process and byproducts of microbial metabolism, which cause it to lose its unique characteristics and style [9,10]. In most industrial production practices, soaking is a very common method as it is the most economical and easy way to maintain the flavor of fruit wine [11]. After the soaking process, the wine will form a unique flavor and body [6]. South Korea and Japan mainly use 15–18%*v/v* sake as the base liquor for soaking. In China, however, Baijiu (38–65%*v/v*) or edible alcohol (75–95%*v/v*), which is brewed locally, is more commonly used [12].

According to the International Organization of Vine and Wine, wine is generally defined as the product obtained from the fermentation of grape juice by yeast followed by aging; however, the term also applies to fermented beverages made from other fruits or vegetables, generally with an alcohol content of 5–13%*v/v* [13]. Flavor is an important attribute of fruit wine, including aroma, taste, and sense, and is also the main factor determining consumer preference and acceptance [14,15]. Many factors affect the quality of greengage wine. In addition to the main external factors, such as geographical environment, climate, and cultivation technology [16], the conditions selected for processing will also have a considerable impact on the content and quality of flavor compounds in the wine [17]. Gas chromatography–mass spectrometry (GC-MS) and meteorological chromatography-olfaction (GC-O) are commonly used for the analysis of flavor compounds, as both are efficient and highly sensitive techniques [18]. In addition, the contribution of flavor compounds to the olfactory characteristics of wine can be evaluated using odor activity value (OAV) measurements [19].

In recent years, research on the flavor of greengage wine has increased. Tiantian Tian et al. optimized the fermentation conditions of three different greengage wines using the response surface method and central composite experimental design. The results showed that there were significant differences in the flavor compounds and concentrations between the three fermented greengage wines. A total of 53, 30, and 32 flavor substances were identified from the optimized samples [20]. In several studies, researchers have found that when using local acid-resistant non-saccharomyces cerevisiae to ferment greengage, the fermented greengage wine has more esters and higher levels of polyphenols, which can improve the flavor quality and nutritional value of the wine [11,21]. In addition, the content of polyphenols in greengage at different maturity stages also varies greatly. Chang Liu et al. analyzed polyphenols at different maturity stages using ultra-high-performance liquid chromatography-electrospray ionization triple quadrupole mass spectrometry (UPLC-QTRAP-MS/MS) technology and widely targeted metabolomics technology. The results showed that polyphenols first increased and then decreased under the influence of temperature, sunshine duration, humidity, and radiation during the greengage ripening process [22].

Thus far, most research has focused on fermented greengage wine, and studies on soaked greengage wine are less common. The existing research shows that by using ultrasonic-assisted treatment, the aging speed of soaked greengage wine can be greatly accelerated and the content of fusel oil and alcohol compounds in the wine can be significantly reduced, while the content of acid ester compounds can be significantly increased [6]. Tiantian Tian et al. used partial least squares regression analysis to model and analyze the relationship between the flavor-active compounds, aromatic compounds, and sensory properties of 20 greengage wines sold on the market, revealing the relationship between chemical components and sensory properties [20]. However, there have been no reports on the effect of the base liquor on the flavor quality of soaked greengage wine. Therefore, in this study, odor activity value (OAV) combined with PLS-DA was used to evaluate the effect of the key volatile compounds of the base liquor on the aroma of greengage wine. In addition, the interaction between volatile compounds and the sensory evaluation of greengage wine was also determined. The results of this study will provide a useful guide for improving the functional characteristics and quality of soaked greengage wine.

2. Material and Methods

2.1. Preparation of Greengage

Nangao greengage (*Prunus mume*) was collected from the Xiling Greengage Valley (30°25′ N, 102°50′ E), Chujiang Town, Dayi County, Chengdu, Sichuan Province, China in June 2021. Fruits with intact skins were selected for the study, and whole fruits were transported to the laboratory 48 h after fresh picking. Subsequently, the samples were immediately cleaned, pedicled, soaked, and bottled. Chrysanthemum authentic sake (15%*v/v*), manufactured by the chrysanthemum authentic wine factory club, was purchased from Jingdong's self-operated flagship store. Edible alcohol (75%*v/v*) was purchased from the Jingdong Naihui Medical Device Store.

2.2. Soaked Greengage Wine Brew

The prepared greengage fruit was poured into the fermentation tank, followed by the addition of green greengage, rock sugar, and base liquor in a ratio of 1:1:1. According to the experimental requirements, 15%*v/v* of edible alcohol (FS), 50%*v/v* of edible alcohol (TS), sake (SK), and baijiu (BJ) were selected as base liquors, and samples were collected every 15 days and shaken thoroughly. The changes that occurred in the greengage wine during soaking are shown in Figure S1.

2.3. Physicochemical Analyses

The total sugar content (g/L) was determined using the Fehling reagent [13] and the direct titration method. The pH value was measured with a pH meter in the laboratory. According to the Compilation of Chinese Food Industry Standards (2000), the total acid content in greengage wine was neutralized and titrated with 0.1 mol/L of sodium hydroxide and expressed as citric acid content (g/L) as determined by acid–base titration. The alcohol content was determined via distillation using a hydrometer.

2.4. CIELAB Analysis

A description of the CIELAB analysis method can be found in the existing literature [23]. The wine sample was filtered through a 0.45 μm membrane (Syringe Filter, PTFE, Superco, Bellefonte, PA, USA) and transferred to a 1 mm quartz cuvette; then, its absorbances were measured at 440, 530, and 600 nm. Subsequently, the corresponding color characteristic parameters were calculated. Distilled water was used as the blank control during determination.

2.5. Organic Acids in Greengage Wine

Organic acids were analyzed using a Shimadzu Prominence LC-20A HPLC system (Shimadzu, Kyoto, Japan) equipped with a DAD detector and a TechMate C18-ST analytical column (4.6 × 250 mm, 5 μm) maintained at 21 °C, where the mobile phase A was 0.02 M diammonium dihydrogen phosphate solution. The pH of the sample was adjusted to 2.5 with phosphoric acid. The flow rate, detection wavelength, and injection volume were 1.0 mL/min, 210 nm, and 10 μL, respectively. The samples were filtered through a 0.22 μm filter prior to injection. The calibration curves of seven organic acid standards were plotted for quantitative analysis (Table S1).

2.6. Headspace Solid-Phase Microextraction (HS-SPME)

A DVB/CAR/PDMS extraction head (Supelco, Bellefonte, PA, USA) was aged at the gas-phase sampling port (270 °C) for 30 min before use. Aliquots of 5 mL of clarified greengage wine sample, 10 μL of 4-methyl-2 amyl alcohol (1024 mg/L) dissolved in ethanol, and 2.02 g of NaCl were placed in a 10 mL headspace sample injection bottle before the rotor was added. The polyethylene bottle cap with a polytetrafluoroethylene (PTEE) silicon spacer was tightened and the sample bottle was placed on a magnetic stirring table. The sample bottle was maintained at 45 °C for 30 min until the gas and liquid phases reached equilibrium. A pre-treated SPME fiber was inserted into the headspace bottle, placed

1 cm above the liquid level, and extracted for 30 min at a speed of 400 r/min. When the gas, liquid, and solid phases in the headspace bottle reached equilibrium, the extraction fiber was removed from the sample bottle and inserted into the gas chromatography–mass spectrometry (GC-MS) sample inlet to conduct thermal analysis for 8 min in non-split flow mode. Three replicates were conducted for each sample.

2.7. Determination of Volatile Organic Compounds (VOCs) Using GC-MS

An Agilent 7890B gas chromatograph and an Agilent 5977B mass spectrometer were used to analyze the aroma substances. The gas chromatograph was loaded with an ultra-fine capillary column (HP-INNOWAX with dimensions of 60 m × 0.25 mm × 0.25 μm, Agilent, J&W Scientific, Folsom, CA, USA). High-purity helium (He > 99.999%) was used as the carrier gas with a flow rate of 1 mL/min, and SPME was used for manual injection without split flow. The injection port temperature was 250 °C and the thermal analysis time was 8 min. The temperature increase procedure for the column temperature box was as follows: the temperature was maintained at 40 °C for 5 min, then increased to 200 °C at a rate of 3 °C/min, and finally maintained for 2 min. The mass spectrometry ionization mode was electron ionization; the ion source temperature, ionization energy, fourth-stage rod temperature, mass spectrometry interface temperature, and mass scanning range were 230 °C, 70 eV, 150 °C, 280 °C, and 30–350 u, respectively.

The analysis was conducted using the offline software Agilent Chemical Workstation (Agilent Technologies, Inc., J&W Scientific, Folsom, CA, USA). For the qualitative analysis, the detected substances were compared to the retention index in the NIST17 database and the mass spectrum ion fragment information. The retention index of each substance was calculated using normal alkanes (C8-C40), and the values obtained were compared with those in the literature.

2.8. Odor Activity Value

In fruit wine brewing, not all aroma components affect the aroma and sensory properties of the wine. To further identify differences in the aroma substances of greengage wines treated with different base liquors, in combination with the threshold values of aroma compounds reported in the literature, the concept of OAV was used to determine the components contributing to the aroma. The calculation formula for OAV is OAV = c/t, where "c" is the total concentration of each flavor compound in the sample and "t" is the odor threshold of the compound in the 11%*v/v* water/ethanol solution. If OAV $\geq$ 1.0, the substance has an actual contribution to the aroma; otherwise, it has no contribution. The OAV of each substance was calculated according to the qualitative and quantitative results of the GC-MS analysis.

2.9. Sensory Analysis

Forty sensory personnel from the China Agricultural University were trained and assessed, and 25 sensory evaluators were selected to form a sensory evaluation team (15 women and 10 men, aged 20–25 years) to conduct a quantitative descriptive sensory analysis of the greengage wine samples. The greengage wine samples were scored on a 10-point scale (0–9) on the three aspects of appearance, aroma, and taste. The ranges were defined as follows: 0–2 indicates very weak variety, 3–5 indicates medium variety, and 6–9 indicates very strong variety. The evaluation was conducted in a well-ventilated, odorless, and noiseless sensory evaluation room, and mouthwash was provided to the sensory evaluation team during the evaluation to avoid the influence of aftertaste.

2.10. Statistical Analysis

Excel 2019 (Microsoft, Washington, DC, USA) was used for data calculation and Duncan's test and t-test in IBM SPSS (version 25.0; SPSS Inc., Chicago, IL, USA) were used for one-way ANOVA for the analysis of significant differences, $p < 0.05$. The R language and MetaboAnalyst 5.0 (http://www.metaboanalyst.ca (accessed on 22 September 2022.))

software were utilized for principal component analysis (PCA) and mapping, and the remaining figures were drawn using Origin Pro 2018.

3. Results and Discussion

3.1. Basic Physicochemical Characteristics

Table 1 presents the relevant physical and chemical parameters for the four treatment methods. After soaking, significant differences were observed in the total sugar, total acid, alcohol, and pH among the four treatment groups. This is noteworthy because the balance of sweet and sour is essential for fermented drinks. In the low-alcohol treatment group, compared to SK, edible alcohol had a higher total acid content, which has an important impact on the flavor of the fruit wine [24] and may be related to the brown sediment formed after the polymerization of phenolic compounds [25]. The TA also reflects to some extent the quality of fruit wine and is one of the most important indicators [26]. No significant differences were observed in the total sugar, alcohol content, and pH. In the high-alcohol treatment group, significant differences were observed in the total sugar, total acid, and alcohol content between edible alcohol and BJ; however, no significant difference was observed in the pH. Overall, the total sugar content of the high-alcohol treatment group was lower than that of the low-alcohol treatment group, whereas the total acid content was higher. FS had the lowest pH, followed by SK and BJ; however, a significant difference in pH was observed between TS and BJ. As presented in Table 1, the total acid content of the four treatment groups was about 13.30–15.50 g/L, which is consistent with the relevant literature [6].

Table 1. Physiochemical characteristics of soaked greengage beverage.

Treatment	FS	SK	TS	BJ
Total sugar (g/L)	360.09 ± 2.88 a	358.47 ± 6.85 a	345.21 ± 6.85 b	333.09 ± 7.88 c
Total acid (g/L)	15.50 ± 0.50 a	13.90 ± 1.10 b	13.30 ± 0.50 b	15.50 ± 0.50 a
Alcohol degree. (%*v*/*v*)	6.71 ± 0.13 c	7.44 ± 0.16 c	22.94 ± 0.64 b	25.74 ± 0.59 a
pH	2.70 ± 0.01 b	3.40 ± 0.73 a	2.98 ± 0.01 ab	3.05 ± 0.01 ab
CIE*L*	95.02 ± 0.57 a	93.97 ± 0.29 c	92.41 ± 1.31 b	92.14 ± 0.89 b
CIE*a*	1.58 ± 0.32 b	1.98 ± 0.07 a	2.67 ± 0.26 a	2.88 ± 0.3 a
CIE*b*	10.06 ± 0.71 c	15.14 ± 0.04 b	16.59 ± 0.16 a	17.14 ± 0.11 a
Chroma (C)	10.18 ± 0.75 c	15.27 ± 0.05 b	16.81 ± 0.12 a	17.38 ± 0.05 a
Hue angle (*h*)	81.16 ± 1.22 b	82.56 ± 0.24 a	80.85 ± 0.97 b	80.44 ± 1.02 b
Oxalic acid (g/L)	0.52 ± 0.01 b	0.54 ± 0.01 b	0.65 ± 0.01 a	0.46 ± 0.00 c
Tartaric acid (g/L)	0.81 ± 0.01 a	0.08 ± 0.00 d	0.19 ± 0.00 b	0.10 ± 0.00 c
Malic acid (g/L)	2.31 ± 0.00 b	2.05 ± 0.06 d	2.84 ± 0.02 a	2.17 ± 0.05 c
Lactic acid (g/L)	0.86 ± 0.16 b	0.65 ± 0.08 b	1.21 ± 0.18 a	1.20 ± 0.01 a
Acetic acid (g/L)	1.33 ± 0.43 b	1.72 ± 0.05 ab	1.42 ± 0.16 ab	1.91 ± 0.19 a
Critic acid (g/L)	17.5 ± 0.57 b	9.05 ± 0.39 d	21.95 ± 2.19 a	12.55 ± 0.96 c
Succinic acid(g/L)	0.19 ± 0.03 b	7.51 ± 1.37 a	0.50 ± 0.01 b	0.06 ± 0.00 b

Note: Different letters in the same row indicate significant differences ($p \leq 0.05$).

3.2. Analysis of Color Change during Soaking

The CIELAB color coordinate system is an objective color perception tool [24]. The results of this analysis are presented in Table 1. A statistical difference was observed between the CIELAB parameters of soaked greengage wine and wine. As presented in Table 1, FS-soaked greengage wine had the highest CIEL value. This indicated that FS-soaked greengage wine had the highest brightness but also the lowest CIEa and CIEb values, suggesting that FS has the lightest red and yellow colors. Additionally, the alcohol and liquor treatments in the high-alcohol group resulted in the highest CIEa and CIEb values, indicating that their red and yellow colors were the deepest; no significant difference was observed between them. With many foods and beverages, color is an important parameter for consumers [18] as it is usually the first sensory impression perceived. The color of wine provides information about style, maturity, production method, grape variety, growing conditions, etc. [27]. However, the influence of the base liquor on the color of soaked greengage wine has not been extensively studied and the mechanism of its action is not fully understood.

3.3. Organic Acids in Greengage Wine

The levels of organic acids in the different treatment groups after soaking are shown in Figure 1. The organic acids in fruit wine have an important influence on the flavor [28], chemical stability, pH value, and thus the quality of the wine. The coordination of sour tastes depends on the composition of different organic acids and the perceived concentration level [29]. The organic acids in soaked greengage wine mainly originate from the base liquor extracted from the greengage fruit under the action of high osmotic pressure. Seven types of organic acids were detected in the soaked greengage wine. The main organic acid was citric acid, with levels up to 21.95 g/L, representing approximately 81.29% of the total organic acid content (Table 1). Citric acid was the main reason for the refreshing taste of the soaked greengage wine [30]. Malic acid is the second most abundant acid in green wines. It has a strong sour, spicy [31], and bitter taste. It is an important indicator of fruit freshness. Malic acid can increase the freshness of fruit wine by reducing the release of aroma substances. Based on previous reports, it is known that the addition of citric and malic acid enhances the intensity of flavor perception for fruit-flavored beverages and improves participants' flavor recognition ability [32]. In this study, SK contained the highest content of citric, malic, lactic, and oxalic acids, indicating that the SK treatment was the most conducive to promoting the transformation of malic acid to organic acid during the soaking process, followed by FS, BJ, and TS. In addition, BJ had a much higher content of acetic acid than the other three groups. Acetic acid usually reacts with alcohol to form esters, conferring a negative effect on fruit wines [20]. In the present study, SK had significantly higher levels of succinic acid than the other three groups, which is also rare in previous reports.

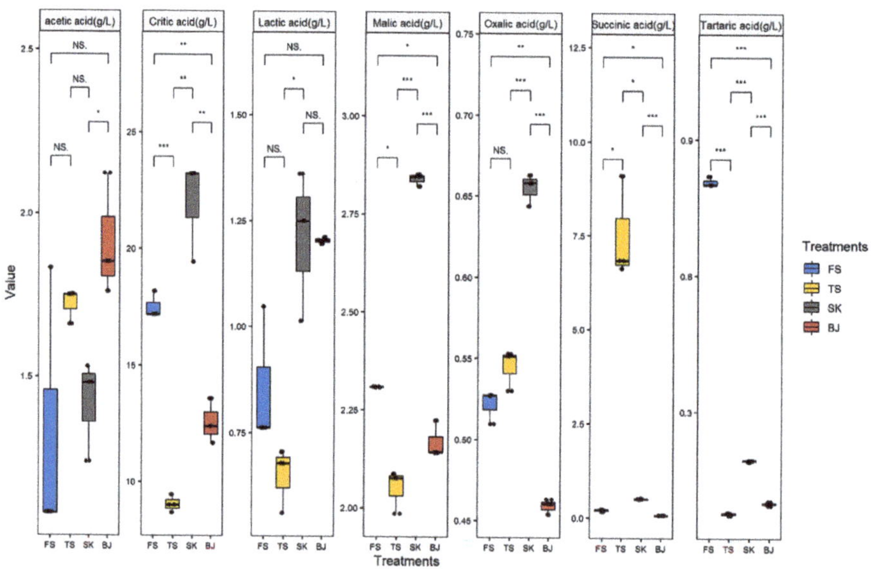

Figure 1. Concentration of organic acids with different base liquor soaking treatments; significance correlations are marked as $p \leq 0.05$ (*), $p \leq 0.01$ (**) and $p \leq 0.001$ (***), respectively.

3.4. Flavor Compounds

As presented in Table S2, 53 flavor compounds were identified in the four types of greengage wine. In addition, the aroma compounds in the four types of greengage wine were analyzed, and 51, 50, 49, and 50 aroma compounds were detected in FS, SK, TS, and BJ, respectively. There were twenty-eight esters, thirteen alcohols, seven aldehydes and ketones, three terpenes, and two other aromatic compounds [33]. To further analyze the significant differences among the four types of soaked greengage wine, we used thermographic cluster analysis (Figure 2). From Figure 2, it can be seen that the BJ treatment was conducive to the production of some alcohols and esters, such as 2-methyl-1-propanol, phenylethyl acetate, 2-methyl-1-butanol, ethyl lactate, isobutyl acetate, phenyl ethanol, 1-nonanol, 1-octanol, ethyl hexanoate, and ethyl isovalerate, which are VOCs that are highly related to the BJ treatment. In contrast, the esters produced by the TS treatment greatly improved the aroma of the fruit wine. The base liquors with a high alcohol content promoted the production of esters, which may be related to the release of aroma compounds in fruit wine due to the concentration of ethanol [34], while the VOCs responsible for the distinctive aroma and flavor of the fruit are mainly accumulated during the ripening period [35]. VOCs are biosynthesized from amino acid derivatives, fatty acid derivatives, and sugar derivatives [36]. Ethyl caproate, ethyl lactate, isobutyl acetate, phenylethyl acetate, and ethyl isovalerate, which represent sweet and fruity aromas, were high in the BJ treatment group and were also considered characteristic compounds of this group. Compared to those in the high-alcohol group, the contents of typical aroma compounds were greatly reduced in the low-alcohol group.

The relationships and differences between the treatment groups during the soaking process were studied using PCA (Figure 2B). PC1 and PC2 accounted for 70.6% and 27.3% of the variation, respectively, indicating a clear separation between the four processed samples of greengage wine. The score chart shows that components 1 and 2 accounted for 97.9% of the total variation. To better analyze the differences in the typical aroma of greengage wine treated with four different base liquors, we used PLS-DA to qualitatively characterize the greengage wine, and the coefficient of each characteristic was used to represent the overall importance (Figure 2C). Fifteen key VOCs were screened, most of which were alcohols and esters, such as 2-methyl-1-butanol, ethyl lactate, ethyl acetate, and 2,3-butanediol. These esters were highly correlated with the BJ treatment group, which was consistent with the results of the thermal polymerization diagram. Esters are often considered to be the most important component of fruit [37] and floral aromas. One source of esters is the esterification of aldehydes, alcohols, ketones, and fatty acids, while another is the metabolic synthesis of higher alcohols by microorganisms in the presence of acetyltransferases [35]. Alcohols are mainly formed during the fermentation of the original wine, while the degradation of amino acids, carbohydrates, and esters may produce fusel oils as well as floral and herbal aromas [38]. The BJ treatment group was rich in fruit alcohol and fruit ester, which is consistent with previous reports [39]. Benzaldehyde has been reported to be a characteristic aroma compound in soaked greengage wine [40]. However, in this study, its content was high only in the FS treatment group, and no significant differences were observed between the other three groups.

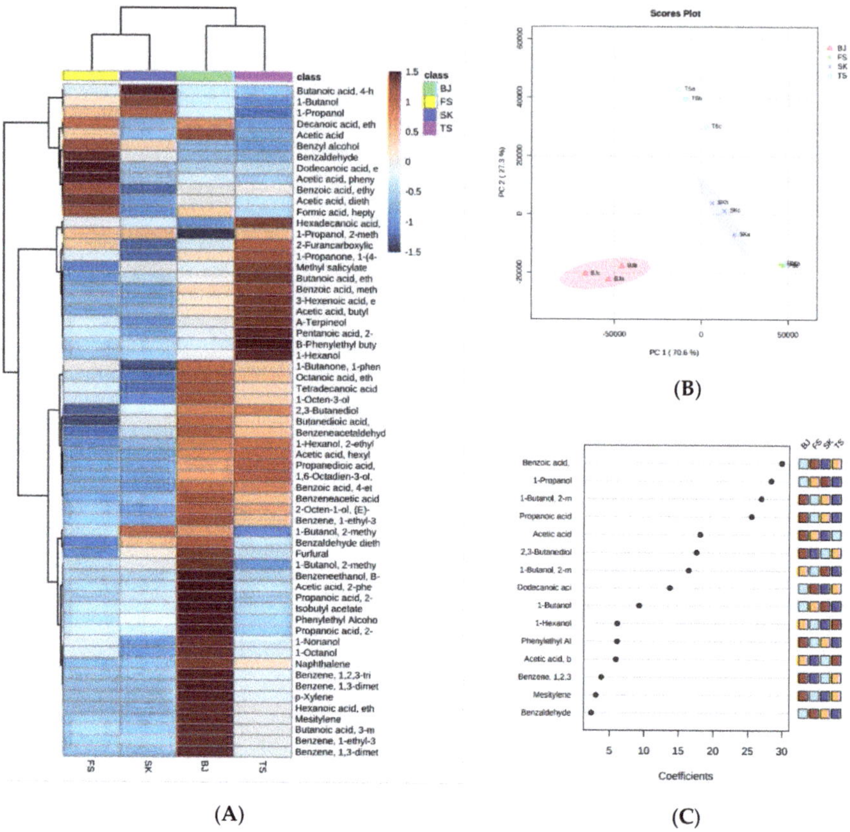

Figure 2. (**A**) Heatmap of flavor compounds in each treatment group during soaking. The Euclidean methodology of distance measure and the Ward clustering algorithm were selected. (**B**) Score plot of PCA for greengage wine from different treatment groups after soaking. (**C**) Ranking of characteristic esters, terpenes, and benzenes, calculated by weighting the sum of absolute regression coefficients in PLS-DA. The colored boxes on the right denote the correlations between the characteristic esters, terpenes, and benzenes and the different treatments at the end of soaking.

Not all volatile compounds affect the overall aroma of greengage wine. We used the OAV to characterize the contribution of certain aroma components to the overall aroma characteristics. When the aroma value of a certain component is greater than or equal to 1, this aroma component contributes to the aroma of the fruit wine. The higher the aroma value of a component, the greater its contribution [41]. Therefore, according to the results of PLS-DA and the components with high OAVs (Table 2), we screened 14 VOCs, including six esters, six alcohols, and two aldehydes. These were not only the key VOCs that distinguished the aroma processed by yeast but were also important contributors to the aroma characteristics of soaked greengage wine. Ethyl benzoate, isobutanol, and 2-methyl-1-butanol were the top three highly correlated VOCs; therefore, these VOCs made an outstanding contribution to the aroma of the wine, being mainly responsible for the "fruity", "fatty", and "banana" flavors, with the highest concentrations in the FS treatment. Although the olfactory threshold of ethyl butyrate is only 0.90 μg/L, its OAV was high, giving the greengage wine a strong apple flavor and sweet smell. Furthermore, although the threshold value of benzaldehyde is high (750.89 μg/L), it was the second richest aroma compound and had an obvious almond aroma, which is also the typical aroma of soaked greengage wine. Notably, as presented in Table 2, the almond flavor of the SK treatment group was the strongest, followed by that of the FS treatment group. Sake has been reported to contain high levels of benzaldehyde, which can increase during storage. Benzaldehyde is a benzene derivative that is mainly formed from terpenes, polyketones, and shikimate in fruits [17]. Benzaldehyde has a typically almond flavor and is one of the most important compounds affecting the overall organoleptic characteristics and consumer acceptability of greengage wine [20], generally found at a concentration of about 300 μg/L [42]. In our results, however, the level of benzaldehyde in the SK treatment group was seven times higher than that in the other groups.

Table 2. VOCs in soaked greengage beverages with OAVs higher than 1.

Category	Odor Description	Threshold (μg/L)	Treatment			
			FS	SK	TS	BJ
Phenylethyl alcohol	Floral, sweet, rosy	564.23	<1	1.52	n.d.	1.92
Benzoic acid, ethyl ester	Sweet, green, fruity, birch	55.56	7.75	9.15	5.57	6.70
Benzeneacetaldehyde	Honey, sweet, floral, chocolate	6.3	<1	1.07	n.d.	n.d.
2,3-Butanediol	Fruity, creamy	>100	2.91	1.19	n.d.	n.d.
Benzaldehyde	Sweet, oily, almond, cherry, nutty	750.89	8.67	9.48	4.72	4.91
1-Octen-3-ol	Mushroom, vegetative,	1.5	2.32	1.53	2.13	3.83
Octanoic acid, ethyl ester	Sweet, fruity, pineapple, creamy	19.3	2.20	<1	1.16	1.81
Propanoic acid, 2-hydroxy-, ethyl ester	Sweet, fruity, creamy, pineapple	50	2.40	25.94	n.d.	54.30
Hexanoic acid, ethyl ester	Sweet, pineapple, fruity, banana	5	1.54	2.34	2.34	6.47
1-Butanol, 2-methyl-	Alcoholic, fatty, cocoa	15.9	92.83	190.44	n.d.	131.66
1-Butanol	Banana	459.2	1.34	2.77	n.d.	n.d.
1-Propanol	Earthy, peanut, nutty, apple, pear	8505.6	1.24	2.15	n.d.	n.d.
Butanoic acid, ethyl ester	Fruity, sweet, apple	0.9	<1	104.73	n.d.	35.48
Isobutyl acetate	Sweet, fruity, banana	25	<1	1.42	n.d.	1.45

Note: "n.d." means not detected.

3.5. Sensory Evaluation

A flavor contour map drawn according to the sensory evaluation results is shown in Figure 3. The score comprised five parts: appearance, aroma, aroma sensory descriptors, overall evaluation, and taste. In terms of appearance, the clarification of greengage wine treated with FS and TS was better. In terms of aroma performance, BJ-treated greengage wine had the highest score for intensity, FS-treated greengage wine had the highest score for intensity, and TS-treated greengage wine had the highest score for coordination. In terms of clarity, there were no significant differences among the four greengage wines. In terms of taste, the greengage wine treated with SK was given a high score for persistence. The greengage wine treated with BJ had the highest score for taste alcohol thickness, that soaked with 50%*v/v* alcohol had the highest score for sour and sweet palatability, and that soaked with 15% alcohol had high scores for sour and sweet palatability, taste alcohol thickness, and persistence. In terms of taste, the score for the alcohol group was much higher than that for the commercial alcohol group, and the overall score order was TS (5.8), FS (5.7), BJ (5), and SK (4.8).

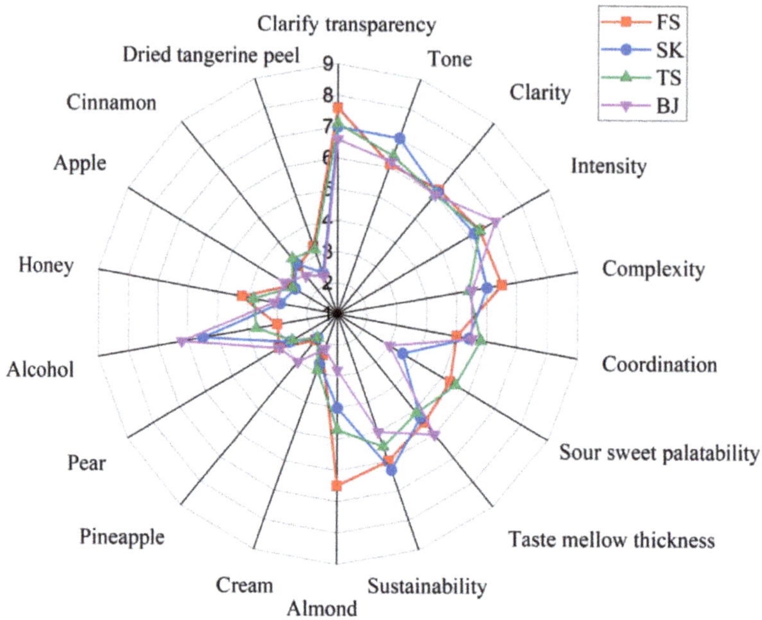

Figure 3. Sensory evaluation results for the four different treatment groups.

The sensory evaluation team screened the aroma of the greengage wine, and nine sensory descriptors were obtained: orange peel, cinnamon, apple, honey, alcohol, pear, pineapple, cream, and almond. Among the sensory evaluation results for the greengage wines soaked in the four base liquors, the greengage wine treated with BJ had more prominent alcohol and pineapple flavors, that treated with FS had more prominent almond and honey flavors, and that treated with TS had more prominent cinnamon and cream flavors.

3.6. Pearson Correlation Analysis of Aroma Sensory Attributes

From Figure 4, it can be seen that the correlation coefficient between the butter and cinnamon flavors (r = 0.95) was large, indicating a strong correlation. The correlation coefficients between the almond, honey (r = 0.87), and orange peel (r = 0.88) flavors were large, indicating that the almond flavor had the same influence on the honey and orange peel flavors. The correlation coefficient between the honey and dried tangerine peel flavors (r = 0.87) was large, indicating a strong correlation. Notably, the three aroma sensory attributes of the soaked greengage wine, namely, the honey, almond, and orange peel flavors, are interrelated, indicating a certain degree of association. Additionally, the correlation coefficients between the apple, pineapple (r = 0.91), and pear flavors (r = 0.80) were also large.

In Figure 4, the flatter the ellipse, the greater the absolute value of the correlation coefficient. A more circular ellipse indicates that the absolute value of the correlation coefficient is small. The direction of the major axis of the ellipse represents the direction of the correlation, where the upper right-lower left direction corresponds to a negative value and the upper left-lower right direction corresponds to a positive value.

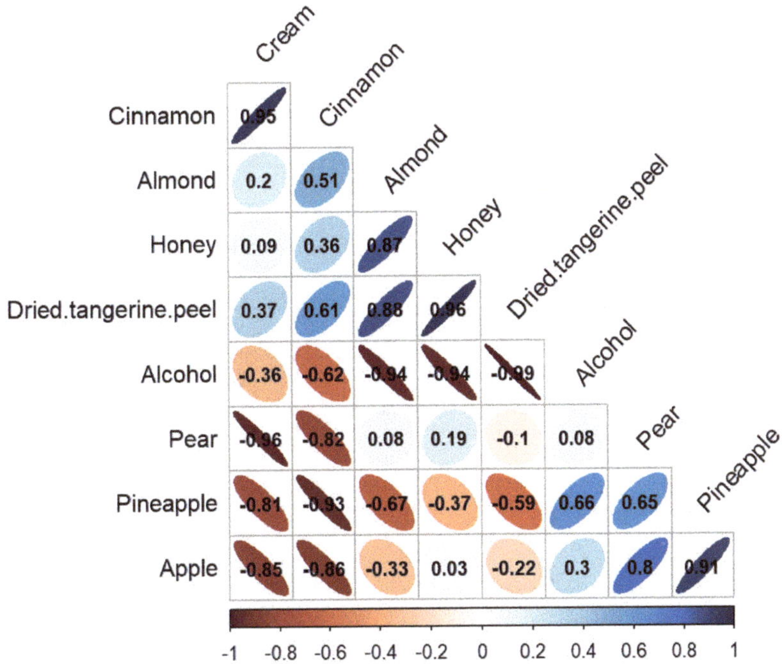

Figure 4. Scatter diagram showing correlation coefficients between sensory aroma attributes. The flatter the ellipse is, the larger the absolute value of the correlation coefficient is. The rounder the ellipse, the smaller the absolute value of the correlation coefficient. The direction of the ellipse's major axis represents the positive and negative of the correlation coefficient: the upper right-lower left direction corresponds to the negative value, and the upper left-lower right direction corresponds to the positive value; The color depth indicates the correlation coefficient.

3.7. Correlation between Aroma-Active Compounds and Sensory Characteristics

Many studies have shown that some of the distinctive flavor characteristics of the wine are determined by its chemical composition; however, the exact part of the chemical composition that affects the senses of the consumer still needs to be further investigated [43]. Therefore, this study explored the association between VOCs and sensory attributes by conducting a Pearson correlation analysis between the GC-MS results and the aroma sensory evaluation results and creating a correlation heat map. From Figure 5, it can be seen that a total of 26 substances were significantly related to the aroma sensory attributes ($p < 0.05$). The aroma sensory evaluation results showed that the pineapple taste was the main aroma sensory attribute of soaked greengage wine. The substances that were significantly positively related to the pineapple taste included isobutyric acid, 2-phenyl-1-propanol, phenylethyl acetate, and isobutyl acetate, while butyl lactate was found to be negatively correlated with the pineapple taste. Acetic acid was found to have a significant positive correlation with the apple taste, ethyl decanoate had a significant positive correlation with the pear taste, and several aroma substances had a significant positive correlation with the alcohol taste, including 2,3-butanediol, ethyl phenylacetate, trans-2-octene-1-ol, ethyl p-ethoxybenzoate, and phenylacetaldehyde. Additionally, phenylethyl alcohol was positively correlated with the tastes of honey and tangerine peels.

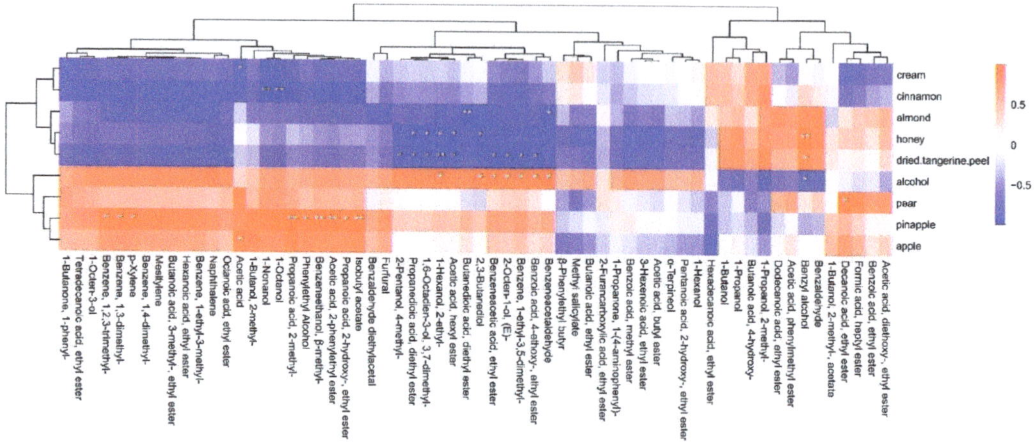

Figure 5. Correlation heat map between the sensory aroma attributes and GC-MS analysis results. Significant correlations are marked as $p \leq 0.05$ (*) and $p \leq 0.01$ (**), respectively.

Overall, fruit aroma was mainly positively correlated with higher alcohols and esters. Xizhen et al. [44] also found that fruit aroma had a good correlation with some esters and alcohols in strong Chinese wines.

4. Conclusions

In this study, the effects of different base liquors on the quality of soaked greengage wine were investigated. The influence of the base liquor on the flavor of soaked greengage wine was comprehensively evaluated via GC-MS and HPLC combined with a sensory evaluation. Our results showed that the basic physical and chemical indices of greengage wine were not significantly different after being soaked in different base liquors. The alcohol content, total acid, total sugar, and pH of the low-alcohol group were 7.44%*v*/*v*, 15.50 g/L, 360.09 g/L, and 3.4, respectively. Compared to the FS treatment, SK was found to significantly improve the red and yellow blending saturation of the greengage wine. In the high-alcohol group, the alcohol content, total acid, total sugar, and pH were approximately 25.74%*v*/*v*, 15.50 g/L, 345.21 g/L, and 3.05, respectively. There was no significant difference in chromaticity. In the low-alcohol group, malic, citric, and tartaric acids were significantly higher in the greengage wine with sake as the base liquor than in the greengage wine soaked in FS. In the high-alcohol group, the liquor significantly accelerated the leaching of citric and malic acids compared to alcohol of the same degree, and alcohol was found to soak out succinic, tartaric, and oxalic acids in greengage more than liquor. Additionally, the results of GC-MS showed that the concentrations of characteristic aroma compounds in the four kinds of soaking greengage wine treated with different base liquors were significantly different. Among them, the greengage wine treated with SK had a stronger nut flavor. In addition, the greengage wine treated with TS contained more terpenes, which can induce a floral and fruity aroma.

Additionally, a sensory evaluation experiment and Pearson coefficient correlation analysis were conducted on the greengage wines soaked in the four base liquors. The results showed that the greengage wines soaked in the four different base liquors had certain differences, among which the greengage wine soaked in TS was superior to the wine in the other three treatment groups in terms of clarity, aroma intensity, and coordination, acidity and sweetness, and overall evaluation. The fruit aroma was found to have a strong positive correlation with most esters and higher alcohols. However, the mechanism of action of the base liquor must be further verified; this is a potential direction for future research on the flavor of soaked fruit wines.

Supplementary Materials: The following supporting information can be downloaded at: https://www.mdpi.com/article/10.3390/foods12102016/s1, Figure S1: Soaked process and changes in greengage fruits, including changes in fruit appearance (SK: soaked).; Table S1: Calibration curves for quantification of organic acids in the experiment. Table S2: Flavor compounds of the four soaked greengage wines.

Author Contributions: Conceptualization, methodology, formal analysis, visualization, writing—original draft, writing—review and editing, P.Z.; Validation, C.L.; Methodology, S.Q.; Data curation, software, K.C.; Resources, Y.W.; Supervision, C.H.; Funding, R.H.; Funding acquisition, project administration, supervision, J.L.; All authors have read and agreed to the published version of the manuscript.

Funding: This research was funded by Sichuan Financial Assistance Program for Agricultural Innovation and Technology Demonstration (SCB-ZNCY-2022003).

Informed Consent Statement: Informed consent was obtained from all subjects involved in the study.

Data Availability Statement: Data is contained within the article or Supplementary Material.

Conflicts of Interest: Author Yingxiang Wang was employed by the conpany Sichuan Mehe Wine Industry. Author Rui Huang was employed by the company luzhou laojiao. The remaining authors declare that the research was conducted in the absence of any commercial or financial relationships that could be construed as a potential conflict of interest.

References

1. Gong, X.P.; Tang, Y.; Song, Y.Y.; Du, G.; Li, J. Comprehensive Review of Phytochemical Constituents, Pharmacological Properties, and Clinical Applications of Prunus mume. *Front. Pharmacol.* **2021**, *12*, 679378. [CrossRef] [PubMed]
2. Li, H.; Huang, Z.; Addo, K.A.; Yu, Y. Evaluation of postharvest quality of plum (*Prunus salicina* L. cv. 'French') treated with layer-by-layer edible coating during storage. *Sci. Hortic.* **2022**, *304*, 111310. [CrossRef]
3. Wang, X.; Du, J.; Zhou, J. Antibiotic activities of extracts from *Prunus mume* fruit against food-borne bacterial pathogens and its active components. *Ind. Crops Prod.* **2019**, *133*, 409–413. [CrossRef]
4. Xia, D.; Wu, X.; Shi, J.; Yang, Q.; Zhang, Y. Phenolic compounds from the edible seeds extract of Chinese Mei (*Prunus mume* Sieb. et Zucc) and their antimicrobial activity. *LWT-Food Sci. Technol.* **2011**, *44*, 347–349. [CrossRef]
5. Yan, X.T.; Li, W.; Sun, Y.N.; Yang, S.Y.; Lee, S.H.; Chen, J.B.; Jang, H.D.; Kim, Y.H. Identification and biological evaluation of flavonoids from the fruits of Prunus mume. *Bioorg. Med. Chem. Lett.* **2014**, *24*, 1397–1402. [CrossRef] [PubMed]
6. Zheng, X.; Zhang, M.; Fang, Z.; Liu, Y. Effects of low frequency ultrasonic treatment on the maturation of steeped greengage wine. *Food Chem.* **2014**, *162*, 264–269. [CrossRef]
7. Jian-jian, Z. Production technology and healthy function of plum wine. *China Brew.* **2005**, *2*, 46–48.
8. Xin-an, Z. Determination of Benzaldehyde Content in Plum Fruit Wine by High Performance Liquid Chromatography (HPLC). *Liquor-Mak. Sci. Technol.* **2009**, *5*, 110–112.
9. Hoang, N.X.; Ferng, S.; Ting, C.-H.; Huang, W.-H.; Chiou, R.Y.-Y.; Hsu, C.-K. Optimizing the initial moromi fermentation conditions to improve the quality of soy sauce. *LWT-Food Sci. Technol.* **2016**, *74*, 242–250. [CrossRef]
10. Franitza, L.; Granvogl, M.; Schieberle, P. Characterization of the Key Aroma Compounds in Two Commercial Rums by Means of the Sensomics Approach. *J. Agric. Food Chem.* **2016**, *64*, 637–645. [CrossRef]
11. Chen, K.; Liu, C.; Wang, Y.; Wang, Z.; Li, F.; Ma, L.; Li, J. Predominance of indigenous non-*Saccharomyces* yeasts in the traditional fermentation of greengage wine and their significant contribution to the evolution of terpenes and ethyl esters. *Food Res. Int.* **2021**, *143*, 110253. [CrossRef] [PubMed]
12. Tian, T.; Sun, J.; Wu, D.; Xiao, J.; Lu, J. Objective measures of greengage wine quality: From taste-active compound and aroma-active compound to sensory profiles. *Food Chem.* **2021**, *340*, 128179. [CrossRef] [PubMed]
13. Kesa, A.L.; Pop, C.R.; Mudura, E.; Salanta, L.C.; Pasqualone, A.; Darab, C.; Burja-Udrea, C.; Zhao, H.; Coldea, T.E. Strategies to Improve the Potential Functionality of Fruit-Based Fermented Beverages. *Plants* **2021**, *10*, 2263. [CrossRef] [PubMed]
14. Yu, M.; Yang, P.; Song, H.; Guan, X. Research progress in comprehensive two-dimensional gas chromatography-mass spectrometry and its combination with olfactometry systems in the flavor analysis field. *J. Food Compos. Anal.* **2022**, *114*, 104790. [CrossRef]
15. Song, H.; Liu, J. GC-O-MS technique and its applications in food flavor analysis. *Food Res. Int.* **2018**, *114*, 187–198. [CrossRef]
16. Bokulich, N.A.; Thorngate, J.H.; Richardson, P.M.; Mills, D.A. Microbial biogeography of wine grapes is conditioned by cultivar, vintage, and climate. *Proc. Natl. Acad. Sci. USA* **2014**, *111*, E139–E148. [CrossRef]
17. Zhao, Y.P.; Tian, T.T.; Li, J.M.; Zhang, B.C.; Yu, Y.; Wang, Y.Y.; Niu, H. Variations in Main Flavor Compounds of Freshly Distilled Brandy during the Second Distillation. *Int. J. Food Eng.* **2014**, *10*, 809–820. [CrossRef]

18. Xu, M.; Jin, Z.; Lan, Y.; Rao, J.; Chen, B. HS-SPME-GC-MS/olfactometry combined with chemometrics to assess the impact of germination on flavor attributes of chickpea, lentil, and yellow pea flours. *Food Chem.* **2019**, *280*, 83–95. [CrossRef]
19. Zhu, J.C.; Chen, F.; Wang, L.Y.; Niu, Y.W.; Chen, H.X.; Wang, H.L.; Xiao, Z.B. Characterization of the Key Aroma Volatile Compounds in Cranberry (*Vaccinium macrocarpon* Ait.) Using Gas Chromatography-Olfactometry (GC-O) and Odor Activity Value (OAV). *J. Agric. Food Chem.* **2016**, *64*, 4990–4999. [CrossRef]
20. Tian, T.; Yang, H.; Yang, F.; Li, B.; Sun, J.; Wu, D.; Lu, J. Optimization of fermentation conditions and comparison of flavor compounds for three fermented greengage wines. *LWT-Food Sci. Technol.* **2018**, *89*, 542–550. [CrossRef]
21. Qiu, S.; Chen, K.; Liu, C.; Wang, Y.; Chen, T.; Yan, G.; Li, J. Non-Saccharomyces yeasts highly contribute to characterisation of flavour profiles in greengage fermentation. *Food Res. Int.* **2022**, *157*, 111391. [CrossRef] [PubMed]
22. Liu, C.; Liu, M.; Yang, L.; Zhang, X. Influence of ripening stage and meteorological parameters on the accumulation pattern of polyphenols in greengages (*Prunus mume* Sieb. Et Zucc) by widely targeted metabolomic. *Curr. Res. Food Sci.* **2022**, *5*, 1837–1844. [CrossRef] [PubMed]
23. Zhang, X.M.; Silverstein, D.A.; Farrell, J.E.; Wandell, B.A. Color image quality metric S-CIELAB and its application on halftone texture visibility. In Proceedings of the IEEE COMPCON 97, San Jose, CA, USA, 23–26 February 1997; pp. 44–48. [CrossRef]
24. Berenguer, M.; Vegara, S.; Barrajon, E.; Saura, D.; Valero, M.; Marti, N. Physicochemical characterization of pomegranate wines fermented with three different Saccharomyces cerevisiae yeast strains. *Food Chem.* **2016**, *190*, 848–855. [CrossRef] [PubMed]
25. Darias-Martin, J.J.; Rodriguez, O.; Diaz, E.; Lamuela-Raventos, R.M. Effect of skin contact on the antioxidant phenolics in white wine. *Food Chem.* **2000**, *71*, 483–487. [CrossRef]
26. Han, X.; Peng, Q.; Yang, H.; Hu, B.; Shen, C.; Tian, R. Influence of different carbohydrate sources on physicochemical properties and metabolites of fermented greengage (*Prunus mume*) wines. *LWT-Food Sci. Technol.* **2020**, *121*, 108929. [CrossRef]
27. Spence, C. Wine psychology: Basic & applied. *Cogn. Res. Princ. Implic.* **2020**, *5*, 22. [CrossRef]
28. Rajkovic, M.B.; Perić, L. The determination of furaldehyde and benzaldehyde in plum brandy. *Hem. Ind.* **2005**, *59*, 78–83. [CrossRef]
29. Guo, L.; Luo, Y.; Zhou, Y.; Bianba, C.; Guo, H.; Zhao, Y.; Fu, H. Exploring microbial dynamics associated with flavours production during highland barley wine fermentation. *Food Res. Int.* **2020**, *130*, 108971. [CrossRef]
30. Coelho, E.M.; Da Silva Padilha, C.V.; Miskinis, G.A.; De Sá, A.G.B.; Pereira, G.E.; De Azevêdo, L.C.; Dos Santos Lima, M. Simultaneous analysis of sugars and organic acids in wine and grape juices by HPLC: Method validation and characterization of products from northeast Brazil. *J. Food Compos. Anal.* **2018**, *66*, 160–167. [CrossRef]
31. Liu, A.; Yang, X.; Guo, Q.; Li, B.; Zheng, Y.; Shi, Y.; Zhu, L. Microbial Communities and Flavor Compounds during the Fermentation of Traditional Hong Qu Glutinous Rice Wine. *Foods* **2022**, *11*, 1097. [CrossRef]
32. Zampini, M.; Wantling, E.; Phillips, N.; Spence, C. Multisensory flavor perception: Assessing the influence of fruit acids and color cues on the perception of fruit-flavored beverages. *Food Qual. Prefer.* **2008**, *19*, 335–343. [CrossRef]
33. Niu, Y.; Wang, P.; Xiao, Z.; Zhu, J.; Sun, X.; Wang, R. Evaluation of the perceptual interaction among ester aroma compounds in cherry wines by GC-MS, GC-O, odor threshold and sensory analysis: An insight at the molecular level. *Food Chem.* **2019**, *275*, 143–153. [CrossRef] [PubMed]
34. Ickes, C.M.; Cadwallader, K.R. Effects of Ethanol on Flavor Perception in Alcoholic Beverages. *Chemosens. Percept.* **2017**, *10*, 119–134. [CrossRef]
35. Li, X.; Qi, L.; Zang, N.; Zhao, L.; Sun, Y.; Huang, X.; Wang, H.; Yin, Z.; Wang, A. Integrated metabolome and transcriptome analysis of the regulatory network of volatile ester formation during fruit ripening in pear. *Plant Physiol. Biochem.* **2022**, *185*, 80–90. [CrossRef] [PubMed]
36. Mertens, N.; Mai, F.; Glomb, M.A. Influence of Nucleophilic Amino Acids on Enzymatic Browning Systems. *J. Agric. Food Chem.* **2019**, *67*, 1719–1725. [CrossRef]
37. Imahori, Y.; Yamamoto, K.; Tanaka, H.; Bai, J. Residual effects of low oxygen storage of mature green fruit on ripening processes and ester biosynthesis during ripening in bananas. *Postharvest. Biol. Technol.* **2013**, *77*, 19–27. [CrossRef]
38. Xie, S.; Hu, F.; Song, C.-z.; Xi, Z.; Zhang, Z.-w. Aromatic profiles of young wines from berries at different heights on grapevines. *Food Sci. Technol. Int.* **2016**, *36*, 248–258. [CrossRef]
39. Xu, Y.; Zhao, J.; Liu, X.; Zhang, C.; Zhao, Z.; Li, X.; Sun, B. Flavor mystery of Chinese traditional fermented baijiu: The great contribution of ester compounds. *Food Chem.* **2021**, *369*, 130920. [CrossRef]
40. Yang, H.Y.; Shao-Hua, W.U.; Wang, X.H.; Peng, Q. Analysis of Flavoring Compositions in Green Plum Fruit Fermenting Wine and in Green Plum Fruit Steeping Wine by GC-MS. *Liquor-Mak. Sci. Technol.* **2005**, *9*, 80–83.
41. Jiang, H.; Zhang, M.; Ye, J.; Qian, M.; Li, X.; Zhao, W.; Bai, W. HS-SPME-GC-MS and OAV analyses of characteristic volatile flavour compounds in salt-baked drumstick. *LWT-Food Sci. Technol.* **2022**, *170*, 114041. [CrossRef]
42. Jiang, B.; Zhang, Z. Volatile Compounds of Young Wines from Cabernet Sauvignon, Cabernet Gernischet and Chardonnay Varieties Grown in the Loess Plateau Region of China. *Molecules* **2010**, *15*, 9184–9196. [CrossRef] [PubMed]

43. Niimi, J.; Tomic, O.; Næs, T.; Jeffery, D.W.; Bastian, S.E.P.; Boss, P.K. Application of sequential and orthogonalised-partial least squaRes. (SO-PLS) regression to predict sensory properties of Cabernet Sauvignon wines from grape chemical composition. *Food Chem.* **2018**, *256*, 195–202. [CrossRef] [PubMed]
44. Sun, X.; Du, J.; Xiong, Y.; Cao, Q.; Wang, Z.; Li, H.; Zhang, F.; Chen, Y.; Liu, Y. Characterization of the key aroma compounds in Chinese JingJiu by quantitative measurements, aroma recombination, and omission experiment. *Food Chem.* **2021**, *352*, 129450. [CrossRef] [PubMed]

Disclaimer/Publisher's Note: The statements, opinions and data contained in all publications are solely those of the individual author(s) and contributor(s) and not of MDPI and/or the editor(s). MDPI and/or the editor(s) disclaim responsibility for any injury to people or property resulting from any ideas, methods, instructions or products referred to in the content.

foods

MDPI

Article

Differences in Aroma Profile of Cabernet Sauvignon Grapes and Wines from Four Plots in Jieshi Mountain Region of Eastern China

Zhuo Chen [1], Yang-Peng Wu [1,2], Yi-Bin Lan [1,2], Yan-Zhi Cui [3], Tong-Hua Shi [3], Yu-Bo Hua [3], Chang-Qing Duan [1,2] and Qiu-Hong Pan [1,2,*]

[1] College of Food Science and Nutritional Engineering, China Agricultural University, Beijing 100083, China
[2] Key Laboratory of Viticulture and Enology, Ministry of Agriculture and Rural Affairs, Beijing 100083, China
[3] Bodega Longes Co., Ltd., Qinghuangdao 066600, China
[*] Correspondence: panqh@cau.edu.cn

Abstract: The Bohai Bay region is a famous wine-growing area in China, where the rainfall is concentrated in the summer due to the influence of the temperate semi-humid monsoon climate. As such, the vineyard terrain has a significant impact on the flavor quality of the grapes and the resulting wines. To explore the relationship between the 'Cabernet Sauvignon' wine style and terrain, this study takes four different plots in the Jieshi Mountain region to investigate the differences in the aroma profile of Cabernet Sauvignon grapes and wines of two consecutive vintages. Based on two-way ANOVA, there were 25 free and 8 glycosylated aroma compounds in the grapes and 21 and 10 aroma compounds with an odor activity value greater than 0.1 in the wines at the end of alcohol fermentation (AF) and malolactic fermentation (MLF), respectively, that varied among the four plots. Wines from the four plots showed a significant difference in floral and fruity aroma attributes, which were mainly related to esters with high odor activity values. The difference in concentration of these compounds between plots was more pronounced in 2021 than in 2020, and a similar result was shown on the Shannon–Wiener index, which represents wine aroma diversity. It has been suggested that high rainfall makes the plot effect more pronounced. Pearson's correlation analysis indicated that concentrations of *(E)*-3-hexen-1-ol in grapes and ethyl 3-methylbutanoate, ethyl hexanoate, isoamyl acetate, isopentanoic acid, and phenethyl acetate in wines were strongly positively correlated with the concentrations of N, P, K, Fe, and electrical conductivity in soil but negatively correlated with soil pH. This study laid a theoretical foundation for further improving the level of vineyard management and grape and wine quality in the Jieshi Mountain region.

Keywords: Cabernet Sauvignon; aroma; wines; plots; differences

1. Introduction

Wine quality is closely linked to the quality of the grape, which largely determines the color, aroma, and flavor of the wine and influences the formation of the fermented aroma profile. Grape quality is influenced by many factors. Studies have shown that climate has the greatest influence on the composition and quality of grapes and wines, followed by soil characteristics, which are able to buffer unfavorable vintage effects even within a small wine region [1–3]. In the same region, under the same climate conditions, the variation in the quality of wine produced from different plots is largely due to the variation in soil, including soil texture [4], soil nutrients [5], etc.

Different styles of wine can be produced from different parts of the same vineyard when under uniform management [6]. Bramley et al. studied the terroir conditions and wine quality of different plots in a vineyard in the Murray Valley region and demonstrated a strong correlation between the two [7]. The study of terroir between different plots of the same vineyard can help to improve understanding of the factors that affect grape and wine quality, which is of great importance for accurate vineyard management and improvement of grape and wine quality.

The quality of wine is mainly judged by indicators such as color, aroma, mouthfeel, and aftertaste [8]. Among these indicators, aroma is very important in evaluating the flavor quality of wine and determining the differences between wine styles around the world, thereby influencing consumer preferences [9]. Many factors influence wine aroma, including the growing environment (climate, soil, and light), raw and auxiliary materials (grape varieties and yeast strains), and the winemaking process (fermentation and ageing) [10]. Reynolds et al. investigated the correlation between the spatial distribution of terroir and grape aroma in a Canadian 'Riesling' vineyard in Ontario and found that soil texture and nutrients were related to the berry weight and grape terpenes [11]. 'Pinot noir' grapes grown in different regions but under standardized winemaking conditions produced wines with unique chemical and sensory profiles, which generally persisted through ageing, and soil pH may be one of the important factors [12].

Research on terroir in China is mainly conducted in the wine regions of the eastern foothills of the Helan Mountains in Ningxia and the northern foothills of the Tianshan Mountains in Xinjiang, focusing on the influence of soil conditions on grape fruit quality [13,14]. Peng et al. analyzed grape aroma compounds from different plots in the eastern foothills of the Helan Mountains in China and found that the contents of C6/C9 compounds, esters, C13-norisoprenoid and terpene were the significant compounds between different plots [14]. Zhang et al. investigated the influence of environmental factors on the physical and chemical parameters of wine produced in the Jieshi Mountain region [15]. Ling et al. identified the styles of white wines produced in the Jieshi Mountain region [16]. The Jieshi Mountain region is located in Changli County, northeastern Hebei Province (39°43′–39°83′ north latitude), bordered by Bohai Bay to the east, Yanshan Mountains to the north, and the Luan River to the southwest, forming a unique regional climate characterized by mountains, seas, and rivers. The grape-growing region has a temperate semi-humid monsoon climate, with an average annual rainfall of 600–650 mm mainly concentrated in July, August, and September, which makes the terrain of the vineyard have a significant impact on the grape berry quality. As an old producing area in China, research on plot influence on grape and wine flavor in Jieshi Mountain region remains limited.

In this study, four plots of a winery in the Jieshi Mountain region, Qinhuangdao, China, were all located on the same slope. Due to the different aroma qualities of the wines made from the grapes of the four plots, it is speculated that the geographical location and soil characteristics may be related, but the specific relationship between the two remains unclear. Based on aroma compound data from 2020 to 2021, this paper investigates the relationship between aroma differences of grapes and wine among four plots and terroir conditions (soil conditions and meteorological conditions), and studies the influence of terroir conditions on the formation of aroma profiles of grapes and wine. The aim of this study is to provide the wine region with a theoretical basis for carrying out fine management of vineyards according to the characteristics of the plots in order to improve wine quality.

2. Materials and Methods

2.1. Materials

2.1.1. Experimental Site

Four plots of *Vitis vinifera* L. cv. Cabernet Sauvignon at Bodega Langes in the Jieshi Mountain region were evaluated over two vintages (2020 and 2021), namely CS1, CS2, CS3, and CS4, respectively (Figure S1). Plot CS1 was located in the northwestern area with an area of 8.54 ha planted in 2001. Plot CS2 was located in the northeast and east area with an area of 5.69 ha planted in 2004. Plot CS3 was located in the central area with an area of 4.94 ha planted in 2006. Plot CS4 was located in the eastern area with an area of 7.14 ha planted in 2011. All plots were located on the same slope with a gradient of 0.3%, with CS1 at the top of the slope and CS2 at the bottom. All the vines were trained to a sloping trunk with a vertical shoot-positioning trellis system, with spacing of 2 m × 1 m and rows planted in a north–south orientation.

2.1.2. Soil Sampling and Analysis

Soil samples were collected using a 9-point sampling method, with these points distributed in a Z-shape in each plot, each point being collected at a distance of 50–70 cm away from the vine in the row, at three depths: 0–30 cm, 30–60 cm, and 60–90 cm. Soils collected at the same depth from the nine sites were completely mixed and then divided into three replicates for analysis of particle content, organic matter, electrical conductivity, pH, and cation exchange capacity (CEC). Soil from the 30–60 cm depth, which is the root enrichment zone, was used for analysis of soil mineral elements. The determination of basic soil physico-chemical properties was based on Han [17]: Soil pH was measured in KCl solution with a soil/solution ratio of 1:2.5 *v*:*v*; organic matter was determined via sulfochromic oxidation; electrical conductivity (EC) was measured with a conductivity meter; and CEC was determined via the ammonium acetate method [18]. N was determined via the Kjeldahl method, which consists of three steps: sample digestion, distillation, and ammonia determination [19]. P, Fe, Ca, K, and Mg contents were determined via inductively coupled plasma mass spectrometry (ICP-MS) (Agilent 7800, Santa Clara, CA, USA). First, 0.25 g of soil was taken, and 5 mL of HNO_3 was added for digestion. Secondly, the solution was heated at 100 °C for 30 min before cooling. After cooling and heating until nearly dry, 1 mL of H_2O_2 was added. After cooling, the double-distilled water volume was reduced to 50 mL and analyzed. The ICP-MS was equipped with an autosampler, a Burgener nebulizer, nickel cones, and a peristaltic sample delivery pump. Detection parameters were as follows: 15 L/min plasma gas flow, 4.3 mL/min helium and reaction gas flow, 0.90 L/min carrier gas flow (>99.99% argon purity), 0.3 r/s sample lift rate, and an atomization chamber temperature at 2 °C. An external standard method was used for quantification, which was prepared with a multi-element standard solution (ICP-MS-CAL2-1, AccuStandard, New Haven, CT, USA) in 0.5% HNO_3 (chromatographically pure) as described by Wu [20].

2.1.3. Grape Berry Sampling

The 5-point sampling method was used for berry sampling. The Cabernet Sauvignon grapes were commercially harvested on 6 October 2020 and 2 October 2021, respectively. During sampling, six berries were randomly selected from the upper, lower, left, right, front, and back positions of different clusters, and the shade and sunny sides of each row were uniformly and randomly sampled. A total of 600 grapes were collected, of which 100 were used for the analysis of physico-chemical analysis, and the rest were frozen in liquid nitrogen and stored at −80 °C for the analysis of aroma compounds.

2.2. Reagents and Equipment

Chromatographically pure grade dichloromethane, methanol, and ethanol were purchased from Honeywell, USA. Analytical grade glucose, sodium hydroxide, sodium chloride, citric acid, malic acid, tartaric acid, and sodium dihydrogen phosphate were purchased from Beijing Chemical Reagent Company. Volatile standards and N-Alkanes (C6-C24) were purchased from Sigma-Aldrich. Yeast Zymaflore FX10, pectinase LAFASE HE GRAND CRU, and Lactobacillus B7 DIRECT 25HL were purchased from LAFFOTA, France.

2.3. Methods

2.3.1. Acquisition of Meteorological Data

Meteorological data were obtained from the self-built weather station in the vineyard of Bodega Langes. Temperature, humidity, rainfall, wind direction, and weather were recorded every 8 h (8:00, 16:00, 24:00). Meteorological data at the vineyard were recorded from 2020 to 2021 (Table S1). It was found that the effective accumulated temperature (from April to September and from August to September) for grapevines was significantly higher in 2020 than in 2021. By contrast, rainfall (from June to August) in 2021 was significantly higher than in 2020. The rainfall in September was only 71.6 mm in 2020, while it was 193.1 mm in 2021. The number of rainy days in September 2021 was 8 more than in the same month in 2021. In terms of monthly indicators (Table S2), the average monthly temperature (from June to August) in 2020 was significantly higher than in 2021, and the average monthly humidity in most months in 2021 was significantly higher than in 2020. Rainfall and number of rainy days in September (harvest period) in 2021 were 2.7 and 3.7 times higher than in 2020, respectively. The number of sunny days per month from April to September in 2020 was larger than in 2021. In general, the weather conditions in 2020 were more suitable for the growth of grapes than those in 2021.

2.3.2. Small-Scale Winemaking Procedure

Grapes from four plots were fermented separately, and two 300-liter stainless steel fermenters were prepared for each plot. A standard winemaking procedure was followed for all wines. For each fermenter, 240 kg of grapes were destemmed and crushed, with the addition of 6% sulfite to give a final concentration of 55 mg/L sulfur dioxide, and then stirred evenly. Zymaflore FX10 yeast (200 mg/L) and LAFASE LE GRAND CRU pectinase (40 mg/L) were added. Fermentation was carried out at 22–25 °C. Three 200-mL bottles of grape juice were collected before the addition of yeast and pectinase and stored at −20 °C for later analysis. During fermentation, the must was stirred with a cap press every 8 h, and the specific gravity and temperature of the must were monitored. At the end of alcoholic fermentation, the residue was separated from the wine, and three bottles of wine (750 mL each) were collected from each fermenter. The separated free-run wine was transferred in its entirety to a 50-liter vessel and inoculated with B7 DIRECT 25HL lactic acid bacteria. After malolactic fermentation, six bottles of wine sample (750 mL each bottle) were collected for each vessel. In 2020, alcohol fermentation and malolactic fermentation took 11 days and 13 days, respectively. In 2021, alcohol fermentation and malolactic fermentation took 10 days and 12 days, respectively.

2.3.3. Determination of Basic Physico-Chemical Properties of Grapes and Wine

One hundred berries were randomly selected and weighed for hundred-grain weight, then the berries were squeezed for their juice for detection. The soluble solids of the juice were measured with a saccharometer and the pH with a pH meter. Titratable acid was titrated with NaOH and measured as tartaric acid (g/L). Titratable acid was determined according to GB/T 15038-2006 'Analytical methods of wine and fruit wine'. The physico-chemical properties of the wine, including alcohol, reducing sugars, titratable acid, pH, and volatile acid, were detected using an OeneFoss wine analyzer (Foss Ltd., Hilleroed, Denmark).

2.3.4. Extraction and Detection of Grape Aroma Compounds

The extraction of free and bound aroma compounds was conducted according to the method of He et al. [21], 60–70 g of de-seeded grape berries were ground with 0.5 g D-gluconolactone and 1 g polyethylpyrrolidone (PVPP) in liquid nitrogen to prevent oxidation of the sample, then were macerated for 4 h at 4 °C and centrifuged at 8000 rpm for 10 min at 4 °C to obtain clear must. Bound aroma compounds were isolated using Cleanert PEP-SPE resins, and enzymatic hydrolysis of glycosidic precursors was conducted at 40 °C for 16 h with the addition of 100 μL AR 2000 (Rapidase, 100 g/L).

Headspace solid-phase microextraction-gas chromatography-tandem mass spectrometry (HS-SPME-GC-MS) was used to analyze the aroma compounds of grapes as described by Wen et al. [22]. Samples were prepared, each consisting of 5 mL of grape juice with an addition of 1 g of sodium chloride and 10 μL of 4-methyl-2-pentanol solution (internal standard). Samples were placed in a CTC-Combi PAL autosampler (CTC Analytics, Zwingen, Switzerland) equipped with a 2 cm DVB/CAR/PDMS 50/30 μm SPME fiber (Supelco, Bellefonete, PA, USA) and agitated at 500 rpm for 30 min at 40 °C. The SPME fiber was then inserted into the headspace to absorb aroma compounds at 40 °C for 30 min and was instantly desorbed into the GC injector to desorb the aroma compounds. Aroma compounds were analyzed with an Agilent gas chromatography-mass spectrometer (Agilent 6890 GC-5975C MS, Santa Clara, CA, USA). fitted with an Agilent 19091N-136hp-InnoWaxPolythyleneol capillary column (60.0 m × 0.25 mm × 0.25 μm). A 1 μL splitless automatic injection procedure was used, with helium as the carrier gas, at 1 mL/min, and the inlet temperature was 250 °C. Oven temperature began with 50 °C for 1 min and then increased to 220 °C at a rate of 3 °C/min and held for 5 min. The ion source (EI) temperature and mass spectrum interface temperature were set at 230 °C and 280 °C, respectively. The ionization energy was set at 70 eV, and the mass scan range was 30–350 u.

The qualitative and quantitative method of measuring aroma compounds followed the method described in our previous report [23]. The aroma compounds were qualified via comparison of the retention indices, mass spectrometry, and the NIST11MS database of analytes and standards. The concentrations of volatile compounds were expressed as μg/L in wines and μg/kg of fresh berry weight of grapes.

2.3.5. Analysis of Aroma Compounds in Wine

HS-SPME-GC-MS was used to analyze aroma compounds in wine according to the method of Lan et al. [24]. The pretreatment and detection methods were almost the same as for grapes, except for the injection mode. The sample was injected in split mode, which was different from that used for grapes.

The qualitative and quantitative characteristics of wine aroma were the same as mentioned above.

2.3.6. Sensory Evaluation of Wine

The quantitative description analysis method (QDA) was adopted for sensory evaluation as described by Lan et al. [25]. The evaluation team (19 persons, including 7 males and 12 females, aged from 22 to 30 years old) was composed of long-term trained evaluators. First, the descriptors were determined. The members of the evaluation team checked the descriptors of wine samples from the vocabulary of red wine descriptors and finally determined the unified descriptors with high check frequency after statistics, discussion, and analysis. The above descriptors were applied to the intensity evaluation of wine samples by using the 10-point system. The analysis results were displayed by the radar image after statistical analysis.

2.3.7. Data Processing

The odor active value (OAV) of aroma compounds was calculated by dividing the concentration of the aroma compound in the sample (μg/L) by the olfactory threshold of the analyte detected in the water medium or simulated wine solution (μg/L). Microsoft Excel 2019 was used for data pre-processing. SPSS Statistic 20.0 was used for statistical analysis, Duncan's method for one-way analysis of variance (ANOVA) with significance level $p < 0.05$, and Pearson's method for linear relationship analysis. GraphPad was used for two-way ANOVA. Alpha diversity is often used for the analysis of biodiversity in systems biology [26]. In this study, the Shannon–Wiener index was used to evaluate the diversity of aroma compounds and the formula was modified as follows [27].

$$\text{Diversity} = -\sum_{i=1}^{N} \mathrm{P_i} \ln \mathrm{P_i} \quad (1)$$

In Formula (1), where Pi represents the relative concentration of VOCi and N represents the total number of VOC in a particular grape sample, Pi was calculated as follows: Pi = ni/N, where ni represents the concentration of VOCi and N represent the total concentration of VOCs in a particular grape sample, respectively.

Triangular diagrams, radar images, and correlation plots were generated using Origin2019b. Box plots were generated using Microsoft Excel 2019. Network charts and heat maps were generated using Gephi 0.10 and TBtools-II v.1.120, respectively.

3. Results

3.1. Soil Characteristics of Four Plots

In accordance with the international system [28], the soil was classified into sand (0.02~2 mm), silt (0.002~0.02 mm), and clay (<0.002 mm). This study showed that the soil of CS1 and CS2 at a depth of 0–30 cm was sandy clay loam, and the soil of CS3 and CS4 at a depth of 0–30 cm was loamy clay based on the soil particle composition (Table S3, Figure 1). The surface soil permeability of sites 1 and 2 was better than that of sites 3 and 4. The organic matter content of CS1 was higher than that of the other three plots. According to the nutrient classification standard formulated by the second general soil survey of China, the organic matter content of all plots was at the lower level of the national soil classification (grades 4–6). CS1 and CS3 were neutral and slightly acidic soils, and CS4 was neutral soil. The CEC of CS4 was higher than the other plots, and the fertilizer holding capacity was higher. Soil mineral elements at a depth of 30–60 cm were measured (Table S4). The results showed that the concentrations of total nitrogen, available phosphorus and available iron were significantly higher in CS1 than those in the other three plots. The concentration of available calcium in CS3 was higher than that in the other three plots, and the available magnesium in CS3 and CS4 was higher than that in CS1 and CS2. In general, the organic matter and total nitrogen contents of the four plots were at a low level, the contents of CEC and available potassium were at a medium to low level, and the contents of available phosphorus, available calcium, available magnesium, and available iron were rich.

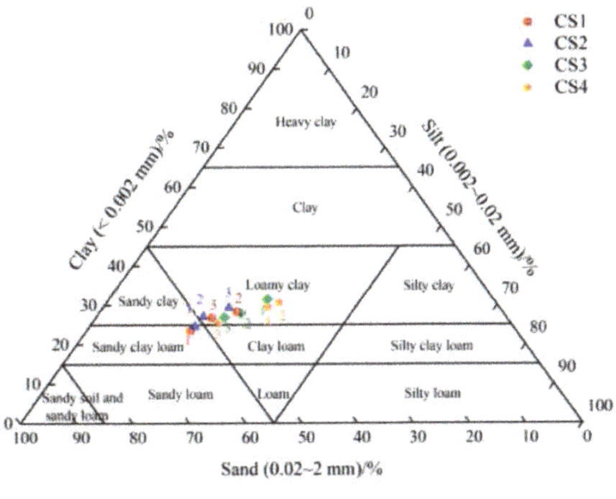

Figure 1. The soil particle composition of four plots: 1, 2, and 3 represent depths of 0–30 cm, 30–60 cm, and 60–90 cm, respectively.

3.2. Physico-Chemical Characteristics of Grapes

The physico-chemical characteristics of the grapes in the 2020 and 2021 vintages were analyzed via one-way and two-way ANOVA to investigate the influence of vintage and plot on characteristics (Table 1). Hundred-grain weight, soluble solid content, titratable acid, and pH showed significant differences between vintages. Hundred-grain weight and pH showed a significant difference between plots. In 2020, the hundred-grain weight of CS2 and CS3 was higher than that of CS1 and CS4. The titratable acid concentration of CS2 was higher than that of the other three plots. The pH of CS1 and CS4 was higher than that of CS2 and CS3. In 2021, the hundred-grain weight of CS4 was higher than that of the other three plots. The pH of CS3 was higher than that of the other three plots. There was no significant difference in soluble solids content between the four plots in any year. The soluble solids content and pH of the four plots was higher in 2020 than that in 2021.

Table 1. Physico-chemical characteristics of grapes from four plots over two vintages.

Characteristics	Vintage	CS1	CS2	CS3	CS4	V	P	V × P
hundred-grain weight (g)	2020	146.8 ± 0.7 b	163.7 ± 4.2 a	166.0 ± 5.5 a	151.6 ± 5.3 b	*	****	****
	2021	147.7 ± 3.5 c	126.3 ± 9.5 d	161.4 ± 1.6 b	172.6 ± 2.8 a			
soluble solids content (Brix)	2020	24.0 ± 0.9 a	23.3 ± 0.1 a	23.4 ± 0.8 a	23.6 ± 0.7 a	***	ns	ns
	2021	22.2 ± 1.8 a	21.2 ± 1.1 a	22.4 ± 0.8 a	21.7 ± 1.0 a			
titratable acid (g tartaric acid/L)	2020	3.7 ± 0.4 b	5.4 ± 1.1 a	3.9 ± 0.1 b	4.1 ± 0.3 b	***	ns	**
	2021	5.0 ± 0.4 ab	4.7 ± 0.2 b	5.2 ± 0.3 a	5.5 ± 0.1 a			
pH	2020	3.87 ± 0.01 a	3.65 ± 0.01 c	3.77 ± 0.02 b	3.84 ± 0.03 a	****	****	****
	2021	3.41 ± 0.01 d	3.43 ± 0.01 c	3.62 ± 0.01 a	3.58 ± 0 b			

Note: Different letters represent significant differences between plots in the same vintage based on one-way ANOVA ($p \leq 0.05$). V: vintages; P: plots; V × P: vintages and plots. *, **, ***, and **** indicate significance at $p < 0.05$, 0.01, 0.001 and 0.0001 based on two-way ANOVA, respectively. ns indicates no significance.

3.3. *Two-Way ANOVA for Differences in Grape Aroma Compounds between the Plots*

A total of 45 free aroma compounds were detected in the 2020 and 2021 grapes. To understand which compounds varied by plot, we performed a two-way ANOVA based on four plots and two vintages. There were 34 compounds that showed a significant difference in concentration between two vintages, 25 compounds that showed a significant difference between the four plots, and 15 compounds that were affected by the combination of vintages and plots (Table S5).

Regarding the glycosylated aroma compounds, a total of 23 compounds were detected in the 2020 and 2021 grapes. Of these, 19 compounds were found to be significantly different between the two vintages and 8 compounds that were significantly different between the four plots. Three compounds varied based on the combination of vintage and plot (Table S6).

These plot-influenced compounds were further compared (Figure 2). It was found that out of the nine free-form norisoprenoids and seven monoterpenoids detected in this study, seven and five compounds varied between the plots, respectively. With the exception of *(Z)*-*β*-damascenone and *p*-cymenene, all other free-form norisoprenoids and monoterpenoids showed higher concentrations in the 2021 grapes than in those from 2020. In contrast, of the eight glycosylated aroma compounds that were affected by plot, seven components showed higher concentrations in 2020.

When CS1 at the top of the slope was compared with CS2 at the bottom, it was observed that many free aroma compounds were higher in CS2 than in CS1 in 2020, of which TDN, *(Z)*-*β*-damascenone, and cadalene showed high levels in CS2 in both vintages. The grapes from CS3 had high concentrations of phenylacetaldehyde, phenylethyl alcohol, hexanal and etc, and the grapes of CS4 had low concentrations of theaspiranes A & B, *p*-cymenene, *p*-cymene and etc.

Free-form (*E*)-2-hexenal, hexanal, and (*E*)-2-hexen-1-ol were the three compounds with the highest concentration among the free-form aroma compounds in this study (Table S5), and the latter two had significant differences between the plots, especially (*E*)-2-hexen-1-ol with the greatest variation. The concentration of (*E*)-2-hexen-1-ol was higher in CS1 than in CS2. Similarly, glycosylated *(E)*-2-hexen-1-ol also had a higher concentration in CS1 compared to CS2 and CS3 in 2020. In addition, the concentration of glycosylated *(Z)*-3-hexen-1-ol was lower in CS4 than in the other three plots in both years.

3.4. *Effect of Soil Characteristics on Aroma Profiles of Grapes*

To investigate the influence of soil characteristics on the aroma profiles of the grapes, Pearson's correlation analysis was performed on soil physico-chemical properties, including 25 free aroma compounds and 8 glycosylated aroma compounds, which were selected via two-way ANOVA. The absolute value of the correlation coefficient between 0.8 and 1.0 was considered a very strong correlation. The absolute value of the correlation coefficient between 0.6 and 0.8 was considered a strong correlation. The absolute value of the correlation coefficient between 0.4 and 0.6 was considered a medium correlation. The result showed that seven free aroma compounds and one bound aroma compound had a strong correlation with soil physico-chemical properties (Figure 3): both free styrene and free *(E)*-3-hexen-1-ol in grapes had a significant positive correlation with soil N, P, and K contents and a negative correlation with pH. Free *(E)*-3-hexen-1-ol was also positively correlated with soil Fe and EC. And free phenylacetaldehyde concentration in grapes was positively influenced by soil Ca and Mg contents. Hexanal in grapes was strongly positively correlated with soil Mg, while *(E)*-2-hexen-1-ol, also a C6 compound, was strongly negatively correlated with Mg. It was found that the bound 3-methyl-1-butanol in grapes was negatively influenced by the soil K and Mg contents.

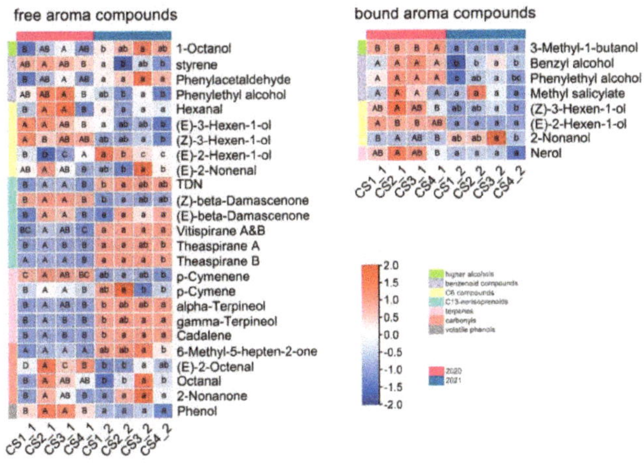

Figure 2. Heat maps for statistical difference in the concentration of aroma compounds between the plots in grapes. The components listed in this figure were shown via two-way ANOVA to be significantly influenced by plot across the two vintages (Tables S5 and S6). The concentrations were converted to log2 fold for standardization. The capital letters represent the differences between the four plots in 2020 determined via one-way ANOVA ($p \leq 0.05$), and the lowercase letters represent the differences in 2021.

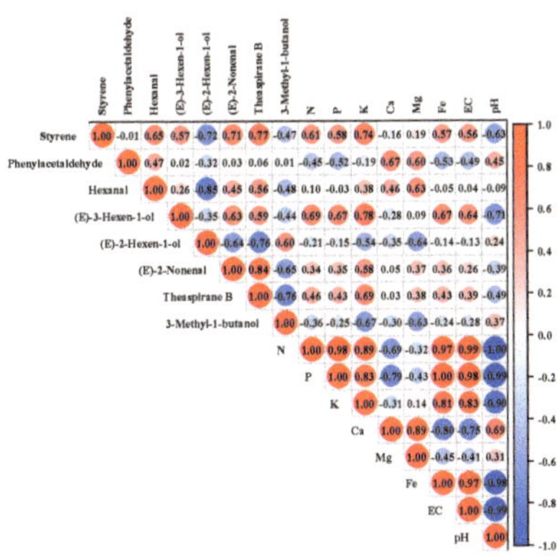

Figure 3. Pearson's correlation coefficients between the concentration of aroma compounds in grapes and the physico-chemical properties of soil. The data from soils at a depth of 30–60 cm was used for analysis. 3-Methyl-1-butanol was the only bound aroma compound.

3.5. Physico-Chemical Characteristics of Wines

Alcohol, reducing sugar, total acidity, pH, and volatile acid of wines showed significant difference between vintage and plot ($p < 0.0001$). Wines produced in 2020 were generally higher in alcohol content than those produced in 2021, which was related to higher soluble solids content. In 2020, CS3 wines had higher residual sugar content. There was not much difference between CS2 and CS4 wines. In 2021, the alcohol content of the CS1 and CS3 wines was higher than that of CS2and CS4. In both years, the CS3 wines had the highest total acidity and the lowest pH (Table 2).

Table 2. Physico-chemical properties of wines at the end of malolactic fermentation.

Characteristics	Vintage	CS1	CS2	CS3	CS4	V	P	V × P
alcohol (%)	2020	13.1 ± 0.02 a	12.7 ± 0.01 c	12.6 ± 0.02 d	12.9 ± 0.0 b	****	****	****
	2021	12.1 ± 0.0 b	11.6 ± 0.02 d	12.1 ± 0.02 a	11.8 ± 0.0 c			
reducing sugar (g/L)	2020	2.9 ± 0.05 c	3.0 ± 0.05 bc	6.5 ± 0.14 a	3.1 ± 0.05 b	****	****	****
	2021	3.5 ± 0.05 a	3.2 ± 0.09 b	3.4 ± 0.08 a	3.2 ± 0.0 b			
total acidity (g tartaric acid/L)	2020	5.9 ± 0.0 c	6.0 ± 0.0 b	7.2 ± 0.05 a	5.7 ± 0.05 d	****	****	****
	2021	5.9 ± 0.0 c	6.0 ± 0.05 b	6.2 ± 0.0 a	5.9 ± 0.0 c			
pH	2020	3.71 ± 0.01 b	3.63 ± 0 c	3.42 ± 0.01 d	3.75 ± 0 a	****	****	****
	2021	3.77 ± 0 a	3.65 ± 0 b	3.59 ± 0.01 c	3.65 ± 0 b			
volatile acid (g acetic acid/L)	2020	0.5 ± 0 a	0.5 ± 0.01 a	0.6 ± 0 a	0.5 ± 0 b	****	****	****
	2021	0.5 ± 0 a	0.5 ± 0.01 b	0.4 ± 0 c	0.4 ± 0 c			

Note: Different letters represent significant differences determined via one-way ANOVA ($p \leq 0.05$). V: vintages; P: plots; V × P: vintages and plots. **** indicates significance at $p < 0.0001$ determined via two-way ANOVA. ns indicates no significance.

3.6. Two-Way ANOVA for Differences in Wine Aroma Compounds between the Plots

A total of 74 aroma compounds were detected in the wines after AF (referred to as AF wine) and the wines after MLF (referred to as MLF) in the 2020 and 2021 vintages. To find out which compounds were influenced by plot, a two-way ANOVA was performed for four plots and two vintages (Tables S7 and S8). There were 66 compounds in the AF wine and 61 compounds in the MLF wine that showed a statistical difference in concentration between two vintages. In comparison, 58 compounds in the AF wine and 44 compounds in the MLF wine showed a significant difference between the plots, of which 26 and 21 compounds, respectively, had an OAV ≥ 0.1. In addition, 23 compounds in the AF wine and 16 compounds in the MLF wines with an OAV ≥ 0.1 were influenced by the combination of vintage and plot.

These plot-influenced compounds in the wines were further compared (Figure 4). Overall, the concentrations of many aromatic compounds in the AF and MLF wines were higher in 2020 than in 2021, especially the esters that contribute to fruity aroma, such as ethyl acetate, ethyl lactate, ethyl 3-methylbutanoate, and isoamyl acetate. In addition, (Z)-β-damascenone and linalool were also higher in the AF wines in 2020 than in 2021.

Comparing the plot-influenced compounds in AF and MLF wines, the most significant compounds were 2-heptanol, 1-octen-3-ol, 1-octanol, and isoamyl acetate. Of these, the concentration of isoamyl acetate far exceeded the olfactory threshold (Tables S8 and S9), indicating a significant contribution by this compound to the fruity aroma of the wine. Moreover, this compound had an overall high concentration in the CS1 wines in both years and a low concentration in the CS3 wines. However, the concentrations of 2-heptanol, 1-octen-3-ol, and 1-octanol were well below their olfactory threshold (Tables S7 and S8) and had no substantial impact on the wine aroma profile, although they showed significant differences between the plots. In the MLF wines, the concentrations of ethyl acetate, ethyl lactate, and ethyl 3-methylbutanoate were all higher than their olfactory threshold (Table S8), and their difference between the plots appeared to be more pronounced in 2021 than in 2020, which may be related to the higher rainfall in September 2021.

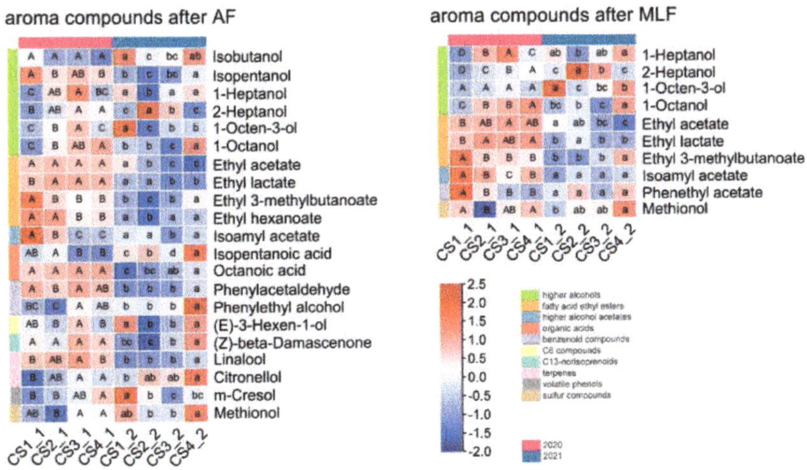

Figure 4. Heat maps for statistical difference in the concentration of aroma compounds between plots in wines. AF: Alcohol fermentation; MLF: Malolactic fermentation. The components listed in this figure were shown via two-way ANOVA to be significantly influenced by plot across two vintages (Tables S7 and S8). The concentrations were converted to log2 fold for standardization. The capital letters represented the differences between four plots in 2020 by one-way ANOVA ($p \leq 0.05$) and the lowercase letters represented the differences in 2021.

3.7. *Effect of Soil Characteristics on Aroma Profiles of Wines*

To investigate the impact of soil characteristics on the aroma profiles of wines, Pearson's correlation analysis was performed on 21 aroma compounds after AF and 10 aroma compounds after MLF in 2020, which were selected by two-way ANOVA (Figure 5). The result showed that 10 and 8 aroma compounds had a strong correlation with soil physico-chemical properties after AF and MLF, respectively. In AF wines, 1-heptanol had a strong–positive correlation with soil Ca and Mg; 2-heptanol and 1-octanol had a strong–negative correlation with N, P, K, Fe, EC, and a strong–positive correlation with soil pH value. 1-octen-3-ol had a strong to extremely strong positive correlation with soil Ca and Mg, respectively. Linalool had a strong positive correlation with Mg, and m-cresol had a strong negative correlation with soil K.

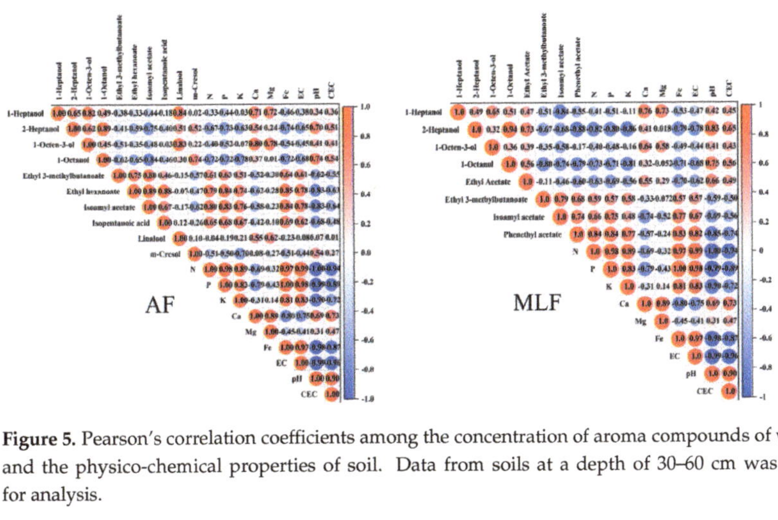

Figure 5. Pearson's correlation coefficients among the concentration of aroma compounds of wines and the physico-chemical properties of soil. Data from soils at a depth of 30–60 cm was used for analysis.

It is worth noting that ethyl 3-methylbutanoate, ethyl hexanoate, isoamyl acetate, and isopentanoic acid in AF wines had a strong to extremely strong negative correlation with N, P, K, Fe, and EC (except for ethyl 3-methylbutanoate and K) and an extremely strong and strong negative correlation with soil pH (Figure 5), and the four compounds had a significant aroma contributionin the wine aroma due to their concentrations far exceeding the thresholds. In addition, the four aroma compounds showed very significant differences in concentration between the two years; the concentration was higher in 2020 than in the 2021 (Table S7).

In the MLF wines, the correlation of several compounds such as higher alcohol was consistent with the compounds in the AF wines. Ethyl acetate had a strong to extremely strong negative correlation with N, P, K, Fe, and EC in soil and a strong–positive correlation with soil pH. Phenethyl acetate had a strong to extremely strong positive correlation with N, P, K, Fe, and EC in soil and an extremely strong and strong–negative correlation with soil pH.

3.8. Aroma Profiles of Wines Based on OAV

The aroma compounds in the MLF wines were grouped into 8 categories according to the odor descriptors (Table S9), including fruity, floral, green, spicy, solvent, fatty, cooked fruit/vegetable, and others. The OAVs of the compounds from the same categories were summed and converted to log10-fold, and radar maps were generated for the two vintages (Figure 6). It was found that the aroma profiles of the wines were characteristic of fruity, floral, and cooked fruit/vegetable odor attributes, and both fruity and floral intensities were generally higher in the 2020 wines than in the 2021 wines. The number of odor descriptors that differed between the plots was higher in 2020 than in 2021. The odor intensities of the fruity, floral, fatty, and cooked fruit/vegetable descriptors differed between plots in both vintages, among which the variation of floral and fruity odor between plots in 2021 was greater than in 2020. In 2020, the CS1 wines had the highest fruity and floral intensities, whereas in 2021, they had the lowest intensities.

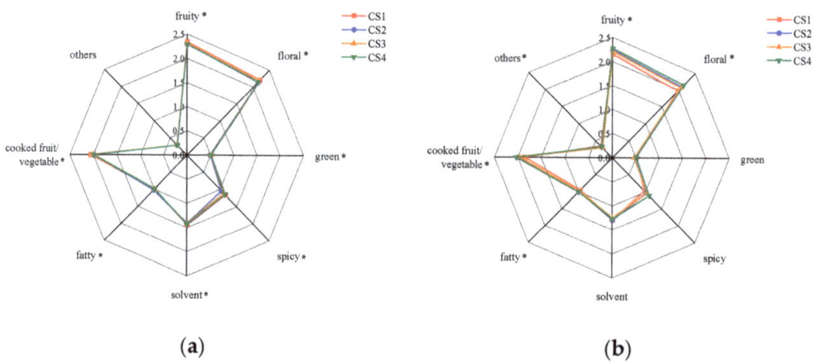

Figure 6. Aromatic categories calculated by adding the odor activity values of the compounds grouped in each one. (**a**) 2020; (**b**) 2021. The values were converted to log10 fold for visualization. * indicates that the values of one-way ANOVA ($p \leq 0.05$) varied significantly between the four plots.

3.9. Aroma Profiles of Wines Based on Sensory Analysis

Sensory analysis was carried out on the post-MLF wines from the four plots in 2020 and 2021. Nine dimensions, including floral, red berry, black berry, fruity intensity, toasted/caramel, wood/oak, vanilla/cream, and smoky/spicy were scored and averaged to produce a radar map (Figure 7).

In general, there was little difference in the sensory characteristics of the post-MLF wines between the two years in the categories of medium fruity intensity, red fruity, and blackberry flavor. In the 2020 wines, the sensory characteristics of floral, black berry, toasted/caramel and vanilla/cream flavor were more pronounced and the intensity of green and red berry were lower compared to the 2021 wines. In 2020, the sensory characteristics varied between the plots, and the wine from CS3 showed a better aroma quality. In 2021, the sensory characteristics did not show any significant difference between the plots.

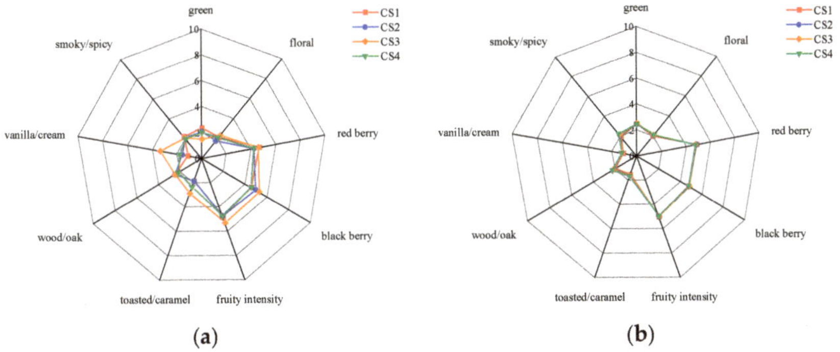

Figure 7. Sensory analysis results of wines after MLF in the year of 2020 and 2021. (**a**) 2020; (**b**) 2021.

3.10. Alpha Diversity Analysis of Aroma Compounds

To quantify the aroma diversity of grapes and wines, the Shannon–Wiener index was used to evaluate grape aroma compounds and wine aroma compounds in two vintages (Figure 8). The diversity of CS1 and CS4 grapes was higher than CS2 and CS3 grapes in 2020, whereas the diversity of CS1 and CS2 grapes was higher than CS3 and CS4 grapes in 2021. The diversity of grape aroma was lower in 2020 than in 2021 in the same plots, except for CS4. For wines, the diversity of those from CS2 was lower than that of those from the other three plots in 2020, while the diversity of CS1 was the highest and CS4 was the lowest of the four plots in 2021. The wine aroma diversity of 2020 was lower than that of 2021 in CS1 and CS2, while the wine aroma diversity of CS3 and CS4 did not show any difference between the two vintages. Overall, the CS1 wines had a rich aroma diversity in both vintages, whereas the CS2 wines had low Shannon–Wiener index. From yearly perspective, the difference in the Shannon–Wiener index was more pronounced in 2021 than in 2020.

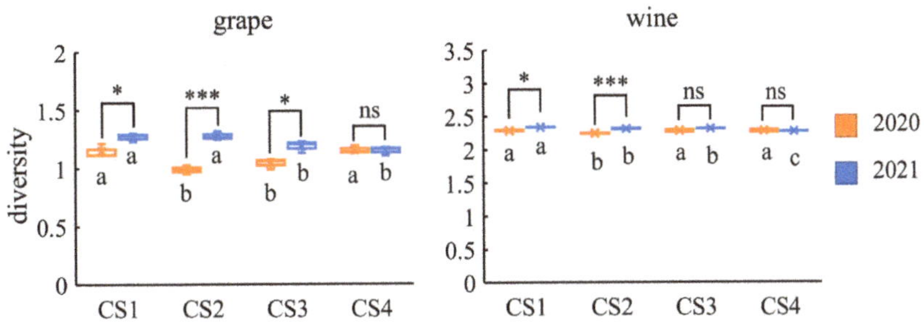

Figure 8. Variation of Shannon–Wiener index of grape and wine aroma compounds in two vintages. Different letters indicate significant differences between plots in the same year ($p < 0.05$). * and *** indicate significant differences between vintages in the same plot at $p < 0.05$ and 0.001 based on one-way ANOVA, respectively. ns indicates no significance.

4. Discussion

The four Cabernet Sauvignon plots studied are located on a gentle slope with a gradient of 0.3%, where the influence of rainfall is relatively greater than that of sunlight and temperature. Practical experience shows that there are certain differences in the aromatic characteristics of the wines from these four plots. This study investigated the soil characteristics and aroma compounds in grapes and wines from these plots. The results clarified the main aroma components that differed between the plots and dissected their correlation with soil physicochemical properties. Based on the differences in rainfall between the two years, it is assumed that high rainfall will make the differences in aroma profiles between the plots more apparent.

4.1. Variation of Aroma Compounds in Grapes and Wines between the Plots

Water deficit in the special zone helps to increase the content of C6 alcohols in grapes [23]. Researchers have found that reducing water supply improved the content of 1-hexanol in grapes, which was associated with the up-expression of two genes, *Vv*LOX and *Vv*HPL, in the biosynthetic pathway of hexanol [29]. The accumulation of (*E*)-*β*-damascenone and *β*-ionone in grapes was strongly influenced by temperature, humidity, sunshine duration, frost-free days, etc. [30]. At the same sugar concentrations, higher temperatures contributed to lower monoterpene levels in white aromatic grape varietals, resulting in reduced aromatic intensity [31]. In this study, the grapes had higher concentrations of free-form C6 compounds and lower norisoprenoids concentrations of monoterpenoids in 2020 than in 2021, which may be due to the lower rainfall and more sunshine in 2020. Interestingly, glycosylated aroma compounds had higher concentrations in 2020, which is different from the concentrations of free-form aroma compounds.

The grape aroma components affected by the plot were selected via two-way ANOVA (Figure 2). Among these components, monoterpenoids and norisoprenoids were the most affected, with most of them exhibiting plot differences. Norisoprenoids and monoterpenoids are synthesized via carotenoid metabolism and the 2-C-methyl-D-erythritol-4-phosphate (MEP) pathway [10,32]. Norisoprenoids and monoterpenoids often contribute to the floral and fruity odor of grapes and wines, especially in non-aromatic grape varieties such as Cabernet Sauvignon. The concentration of norisoprenoids in grapes is regulated by sun exposure [11]. In addition, Yuan et al. observed that low nitrogen status was associated with low β-damascenone content in wines [12]. In our study, most norisoprenoid and terpene compounds, especially theaspirane B, were at lower concentrations in wines from CS4 than in those from the other three plots in both vintages, possibly because of the lower soil N content in CS4. Four green leaf odor components, hexanal, (*E*)-3-hexen-1-ol, (*Z*)-3-hexen-1-ol and (*E*)-2-hexen-1-ol, showed differences between the plots in this study, especially (*E*)-2-hexen-1-ol, which had the greatest variation. Water deficit favors the increase of the content of C6 alcohols in grapes [23]. The concentration of (*E*)-2-hexen-1-ol was higher in CS1 than in CS2. Similarly, the concentration of glycosylated (*E*)-2-hexhen-1-ol in CS1 was higher compared to CS2 and CS3 in 2020, which may be related to the fact that the CS1 plot was located at the top of the slope. The location of CS1 facilitates water permeation.

Glycosically bound aroma compounds are composed of components with free hydroxyl group(s), mainly aliphatic alcohol derivatives (higher alcohols, C6 compounds), terpenoids, norisoprenoids, and benzenoids in grapes [33]. Researchers have found that sunshine is beneficial for the accumulation of glycosylated terpenes [34]. In Agiorgitiko vines, limited water supply can increase the levels of the glycoconjugates of the main aroma compounds [35]. The previous reports explained why most of the bound aroma compounds had a higher concentration in 2020, when there was more abundant sunshine and less rainfall.

The aroma components affected by the plot were selected from AF and MLF wines, respectively, via two-way analysis of variance and consisted mainly of higher alcohols and fatty acid ethyl esters (Figure 4), which was different from previous studies. Slaghenaufia et al. studied the differences in wine aroma of different plots in a vineyard in the Valpolicella wine region and found that the compounds causing the differences between plots were mainly benzenoid compounds, terpenes, and norisoprenoids [36]. In our study, the compounds of most concern are ethyl acetate, ethyl lactate, ethyl 3-methylbutanoate, and isoamyl acetate. Esters are synthesized mainly by yeast but can also be synthesized by lactic acid bacteria. They play a central role in characterization of fruity aromas, with ethyl esters contributing more [37]. It is believed that in non-aromatic grape varieties such as Cabernet Sauvignon, fruity odors are generated mainly by ethyl esters such as ethyl isobutyrate, ethyl butyrate, ethyl 3-methylbutyrate, ethyl hexanoate, ethyl octanoate, etc. [33]. In the MLF wines in 2020, the concentration of higher alcohols such as 1-heptanol, 2-heptanol, 1-octen-3-ol, and 1-octanol in CS1 was at a lower level and the concentration of ethyl 3-methylbutanoate and isoamyl acetate was at a higher level, contributing to more intense floral and fruity odor in CS1 (Figure 6). In MLF wines in 2021, the concentration of 1-octen-3-ol, ethyl acetate and ethyl lactate were at a higher level in CS1 and the concentration of ethyl 3-methylbutanoate and isoamyl acetate was at a higher level in CS4, leading to the higher aroma quality in CS4.

Overall, isoamyl acetate, which has a very high OAV in the wines in this study, was present at a high concentration in the CS1 wines in both years and a low concentration in the CS3 wines. This compound contributes to the banana-like fruity note of the wine. According to the radar map of the added OAV of the aroma categories, the intensity of fruity and floral of wines in 2020 was higher than that in 2021, and the intensity of solvent was lower in 2020. It was found that rainfall can affect the maturity of grapes and excessive rainfall can reduce the sugar content of grapes, increase acidity, and dilute the flavor of grapes [38]. From the content of soluble solids and titratable acid in the grapes from the four plots, the maturity of grapes in 2020 was better than that in 2021 except acidity in CS2, and the CS2 grapes had higher acidity in 2020 and lower acidity in 2021 (Table 1). The higher maturity and aroma quality of the wines in 2020 may be related to its lower rainfall compared to 2021. Studies have shown that as the maturity of Cabernet Sauvignon increases, green (vegetable) flavor decreases and red berry aroma shifts to blackberry aroma in the resulting wines [39]. In addition, in the MLF wines, ethyl esters of acetate, lactate, and 3-methylbutanoate with high OAVs showed a greater difference in concentration between the plots in 2021 than in 2020, suggesting that high rainfall would accentuate the differences in vineyard terrain.

Sensory analysis revealed a significant difference between the two vintages and indicated that the fruity and floral aroma characteristics were better in 2020, which was consistent with the aroma profile based on OAV and may be due to the lower rainfall in 2020. However, there was a contradictory result, namely that the sensory analysis showed greater difference between plots in 2020, while the aroma profile based on OAV showed a greater difference between plots in 2021. Furthermore, the aroma quality of CS3 wines was higher than other plots based on sensory analysis, while CS1 wine was the best based on OAV. It is commonly known that wine aroma is not only the simple complex of individual aroma compounds, but also related to the interaction between aroma compounds and the influence of the wine matrix, such as polyphenols, proteins, carbohydrates, alcohols, etc. Researchers have found that glucose in wine can increase the release of volatile compounds [40–44]. The higher reducing sugars and other nonvolatile components in CS3 wines in 2020 may have influenced the sensory aroma profile.

4.2. Correlation between Some Aroma Compounds in Grapes and Wines and Soil Physico-Chemical Properties

The physical loss of soil through mechanical cultivation and displacement through erosion is likely to be exacerbated by heavy rainfall [45]. Although the four plots in this study are located on the same slope, due to the temperate monsoon climate and concentrated rainfall in the production area, there are certain differences in soil physico-chemical properties between the plots. It is known that the physico-chemical properties of the soil influence the growth of roots of the vine and thus the quality of the grapes. Among all characteristics, nitrogen is the most important nutrient element that restricts plant growth. The nitrogen content of the soil affects the concentration of nitrogenous compounds in the grapes, such as total nitrogen, amino acids, ammonium salts, and assimilable nitrogen [46]. Nitrogen influences the vine vigor, yield, and berry size and has an effect on the major metabolites (sugars, organic acids) and secondary metabolites (phenolic compounds, flavors and aroma precursors) of grapes [47]. In addition, phosphorus is one of the most important elements for plant growth and reproduction. Phosphorus plays an important role in improving the uptake and transformation of nitrogen and can affect flower bud differentiation and fruit development as well as improving the uptake capacity of the root system in the plant [48,49]. Most of the potassium elements that can be absorbed and utilized by grapes come from the soil, and the potassium content in plants is high and similar to that of nitrogen. During the growth period of wine grapes, there is a high demand for potassium, which can enhance photosynthesis, improve nitrogen metabolism and carbohydrate metabolism, improve the rate of water use absorbed by the grapes, and increase grape stress resistance and disease resistance [50]. In this study, there was a positive correlation between free theaspirane B concentration in the grapes and soil K and N contents, suggesting that the lower theaspirane B concentration in the CS4 grapes in both vintages may be associated with the lower soil K and N contents in CS4.

Available iron is one of the elements that make up chlorophyll, which is involved in photosynthesis and respiration, and iron deficiency usually results in yellowing of new shoots and young leaves [51,52]. Electrical conductivity value (EC) is a parameter of water-soluble salts in the soil, which is a factor that determines whether salt ions in the soil will limit plant growth; too high or too low a concentration can hinder plant growth [53]. This study showed that free €-3-hexen-1-ol, a component of green leaf odor, had a strong positive correlation with soil N, P, K, Fe, and EC and a strong negative correlation with soil pH. CS1 soil had higher levels of N, P, K, and Fe, corresponding to a higher concentration of (*E*)-3-hexen-1-ol in the CS1 grapes.

Previous studies have demonstrated the effects of soil physico-chemical properties on wine aroma quality. Nitrogen is the most abundant soil-derived macronutrient in a grapevine and plays an important role in fermentative microorganisms [46]. The present study showed that higher alcohol concentration was negatively correlated with soil N content, which may explain why wines from CS1, with its higher soil N content, contained a lower concentration of higher alcohols. The relationship was also consistent with the research showing that low YAN results in high content of higher alcohols [46].

Compounds that promoted wine aroma quality, such as ethyl 3-methylbutanoate, ethyl hexanoate, isoamyl acetate, and phenethyl acetate, were positively correlated with soil N, P, K, Fe, and EC, but negatively correlated with soil pH. Conversely, compounds with certain negative effects on wine flavor, such as 2-heptanol and 1-octanol, were strongly negatively correlated with soil N, P, K, Fe, and EC and strongly positively correlated with soil pH. The result was consistent with a previous report that N fertilization of a Riesling vineyard increased 1-butanol, trans-3-hexen-1-ol, benzyl alcohol, and most of the esters in wines [54].The content of N, P, and Fe of CS1 was significantly higher than that of the other three plots, which may be the reason why the wine aroma quality of CS1 was better than that of the other three plots. Soil pH may be important in defining the unique ageing characteristics of a particular vineyard [12]. Given the present results, it is suggested that the range of soil pH was from 6.46 to 7.05, which seemed to produce a pleasant odor with lower soil pH in a certain range.

As previously reported, the water status of the grapevine is an important determinant driver of terroir expression. It depends on climatic conditions (rainfall and reference evapotranspiration) and soil type (soil water holding capacity, SWHC). Wine aromatic typicity is strongly influenced by vine water status [55]. Based on the above research, it has been concluded that soil physico-chemical properties play an important role in wine aroma, and different vintages may also affect soil physico-chemical properties and thus wine aroma quality. In 2020, when there was less rainfall than in 2021, the physico-chemical properties of CS1 resulted in a better wine aroma profile based on OAVs. In 2021, when there was a great deal of rain, the soils of CS3 and CS4 with loamy clay texture had higher viscosity and better fertility protection than those of CS1 and CS2 with their sandy soil layers, and the wine produced from CS4 had better aroma quality with higher soil Fe content than CS3. On the other hand, compared to CS2 at the bottom, CS1 at the top of the slope had better soil, so the fertility protection was poor and the wine aroma quality was poor. It was speculated CS1 and CS2 were more likely affected by rainfall.

4.3. Quantifying the Complexity of Aroma Compounds

In systems biology, α-diversity refers to the assessment of the diversity of a single ecosystem or sample [26]. In this study, in order to quantify the complexity of aroma compounds in grapes and wines from different plots, we introduced the α-diversity index commonly used in systems biology research. Here, we use the Shannon–Wiener index to comprehensively evaluate the quantity and concentration of aroma compounds detected in grapes or wines from each plot [27]. Overall, the aroma compound diversity of CS1 and CS2 wines was affected by vintage, while the aroma compound diversity of CS3 and CS4 wines was not found to differ between vintages. Aroma diversity varied between plots from one vintage to another. The amplitude of variation of the Shannon–Wiener index was more pronounced in 2021 than in 2020. As mentioned above, high rainfall in 2021 could have led to a more pronounced vineyard terrain effect. Soil microbial diversity may also lead to more efficient mineralization of soil nutrients [56]. In the vineyards of southern Australia, the soil fungal community plays an important role in wine aroma [57]. It has therefore been speculated that aroma diversity may be related to soil microbial diversity. Liu et al. [58] proved that the functional diversity of microorganisms at the Chihuahuan Desert Ranch in northern New Mexico, USA was lower in the summer drought test site than in the summer and spring controls. In this context, the lower rainfall in 2020 may reduce microbial diversity and thus affect the aroma diversity. As previously reported [59], microbial diversity was higher in soils with higher organic matter content. In this study, the aroma compound diversity of CS1 wines was the highest among the four plots in 2020 and 2021, which may be related to the higher microbial diversity caused by higher organic matter in CS1.

5. Conclusions

In this study, we investigated grape and wine aroma compounds and the terrain of four plots on the same slope so as to find the deep relationship between the two and thus provide a new way to improve wine aroma quality in this region. One-way ANOVA, two-way ANOVA, and Pearson's correlation analysis were used to find the elements that influenced the aroma profiles. In summary, the variation in aroma compounds was greater between vintages than between plots. Based on two-way ANOVA, the most plot-variant aroma compounds were identified, including 25 free and 8 bound components in grapes and 21 and 10 components in AF and MLF wines, respectively. Of these, most of the free norisoprenoids, monoterpenoids, and C6 compounds in grapes varied between plots. The concentration of free and bound (*E*)-2-hexen-1-ol was higher in CS1 than in CS2, which may be related to the location of CS1 at the top of the slope with better water permeability. Higher alcohols and esters in AF and MLF wines were most affected by the plots, including 1-heptanol, 2-heptanol, 1-octen-3-ol, ethyl 3-methylbutanoate, isoamyl acetate, etc. Wines from the four plots showed a significant difference in floral and fruity aroma attributes, which were mainly related to ethyl esters with high odor activity values. In addition, the variation of these compounds between the plots was more pronounced in the 2021 vintage with more rainfall compared to the 2020 vintage. Pearson's correlation analysis showed that concentrations of (*E*)-3-hexen-1-ol in grapes and ethyl 3-methylbutanoate, ethyl hexanoate, isoamyl acetate, isopentanoic acid, and phenethyl acetate in wines were strongly positively correlated with the concentrations of N, P, K, Fe, and EC in the soil, but negatively correlated with soil pH. And the result showed the esters contributing to floral and fruity odor were positively correlated with N, P, K, Fe, and EC in soil and negatively correlated with soil pH, while the situation was exactly reversed for higher alcohols contributing to off-odor in wines. Interestingly, the MLF wines from CS1, with its higher soil N, P, and Fe contents, showed higher aroma quality in 2020 but lower aroma quality in 2021, which may be related to the negative effect of fertility loss caused by more rainfall in 2021. The α-diversity of aroma compounds was also calculated, and it was speculated that the variations in α-diversity between vintages and between plots were caused by the differences in rainfall and soil organic matter. This study was the first to dissect the influence of vineyard terrain on the aroma profile of grapes and wines in the Jieshi Mountain region with a temperate monsoon climate and provides some guidance for improving vineyard management and the quality of grapes and wines. The research can be complemented in the future by expanding the scale of wine making, increasing the number of grape varieties, and monitoring the changes in grapevine ageing.

Supplementary Materials: The following supporting information can be downloaded at: https://www.mdpi.com/article/10.3390/foods12142668/s1, see refs [8,60–92], Figure S1: Map of four plots; Table S1: Meteorological data of Jieshi Mountain region in 2020 and 2021; Table S2: Monthly meteorological data of Jieshi Mountain region in 2020 and 2021; Table S3: Physico-chemical properties of soil of four plots in Bodega Langes; Table S4: Soil mineral elements of four plots in Bodega Langes; Table S5: The concentration and differences of free aroma compounds of grapes of four plots in the vintage of 2020 and 2021; Table S6: The concentration and differences of bound aroma compounds of grapes of four plots in the vintage of 2020 and 2021; Table S7: The concentration and differences of aroma compounds of AF wines of four plots in the vintage of 2020 and 2021; Table S8: The concentration and differences of aroma compounds of MLF wines of four plots in the vintage of 2020 and 2021; Table S9: The qualitative information, threshold, descriptor and series of aroma compounds of MLF wines.

Author Contributions: Conceptualization, Q.-H.P. and C.-Q.D.; formal analysis, Z.C. and Y.-P.W.; investigation, Y.-P.W.; resources, Y.-Z.C., T.-H.S. and Y.-B.H.; data curation, Z.C. and Y.-B.L.; writing—original draft preparation, Z.C.; writing—review and editing, Q.-H.P.; visualization, Z.C.; Project administration, Y.-B.L., C.-Q.D. and Q.-H.P.; funding acquisition, Y.-Z.C. and Q.-H.P. All authors have read and agreed to the published version of the manuscript.

Funding: This research was funded by the Science & Technology Cooperation Project between China Agricultural University and Longus Vineyard Co., Ltd., and the Earmarked Fund for China Agriculture Research System (CARS-29 to C-QD).

Data Availability Statement: Data is contained within the article or Supplementary Materials.

Acknowledgments: The author sincerely thanks Longus Vineyard Co., Ltd., for its help with experimental implementation and grape and wine sampling.

Conflicts of Interest: Yan-Zhi Cui, Tong-Hua Shi and Yu-Bo Hua were employed by Bodega Longes Co. Ltd. The remaining authors declare that the research was conducted in the absence of any commercial or financial relationships that could be construed as a potential conflict of interest.

References

1. Pereira, G.E.; Gaudillere, J.P.; Van Leeuwen, C.; Hilbert, G.; Maucourt, M.; Deborde, C.; Moing, A.; Rolin, D. H-1 NMR metabolite fingerprints of grape berry: Comparison of vintage and soil effects in bordeaux grapevine growing areas. *Anal. Chim. Acta* **2006**, *563*, 346–352. [CrossRef]
2. Zsofi, Z.; Gal, L.; Szilagyi, Z.; Szucs, E.; Marschall, M.; Nagy, Z.; Balo, B. Use of stomatal conductance and pre-dawn water potential to classify terroir for the grape variety kekfrankos. *Aust. J. Grape Wine Res.* **2009**, *15*, 36–47. [CrossRef]
3. Priori, S.; Pellegrini, S.; Perria, R.; Puccioni, S.; Storchi, P.; Valboa, G.; Costantini, E.a.C. Scale effect of terroir under three contrasting vintages in the Chianti Classico area (Tuscany, Italy). *Geoderma* **2019**, *334*, 99–112. [CrossRef]
4. Gomez-Miguez, M.J.; Gomez-Miguez, M.; Vicario, I.M.; Heredia, F.J. Assessment of colour and aroma in white wines vinifications: Effects of grape maturity and soil type. *J. Food Eng.* **2007**, *79*, 758–764. [CrossRef]
5. Wheeler, S.J.P.G.J. *Effects of Soil Management Techniques on Grape and Wine Quality*; WFL Publisher: Helsinki, Finland, 2005; pp. 195–208.
6. Bramley, R.G.V.; Hamilton, R.P. Terroir and precision viticulture: Are they compatible? *J. Int. Des Sci. De La Vigne Et Du Vin* **2007**, *41*, 1–8. [CrossRef]
7. Bramley, R.G.V.; Ouzman, J.; Boss, P.K. Variation in vine vigour, grape yield and vineyard soils and topography as indicators of variation in the chemical composition of grapes, wine and wine sensory attributes. *Aust. J. Grape Wine Res.* **2011**, *17*, 217–229. [CrossRef]
8. Hua, L. *Wine Tasting*; Science Press: Beijing, China, 2006; pp. 83–102.

9. Jiang, B.; Zhang, Z.W. A preliminary study of aroma composition and impact odorants of cabernet franc wines under different terrain conditions of the Loess Plateau region (China). *Molecules* **2018**, *23*, 1096. [CrossRef] [PubMed]
10. He, Y.; Wang, X.Y.; Li, P.H.; Lv, Y.C.; Nan, H.L.; Wen, L.K.; Wang, Z.T. Research progress of wine aroma components: A critical review. *Food Chem.* **2023**, *402*, 134491. [CrossRef]
11. Reynolds, A.G.; Senchuk, I.V.; Van Der Reest, C.; De Savigny, C. Use of GPS and GIS for elucidation of the basis for terroir: Spatial variation in an ontario riesling vineyard. *Am. J. Enol. Vitic.* **2007**, *58*, 145–162. [CrossRef]
12. Cantu, A.; Lafontaine, S.; Frias, I.; Sokolowsky, M.; Yeh, A.; Lestringant, P.; Hjelmeland, A.; Byer, S.; Heymann, H.; Runnebaum, R.C. Investigating the impact of regionality on the sensorial and chemical aging characteristics of pinot noir grown throughout the us west coast. *Food Chem.* **2021**, *337*, 127720. [CrossRef]
13. Li Jun, N.; Ling Li, S.; Ya Shan, L.; Chang Wei, C.; Na, N.; Jing, H.; Ying Li, Z.; Cheng Dong, X. Changes in quality of 'Cabernet Sauvignon' grapevines from three plots in Shihezi region during the ripening. *Sino-Overseas Grapevine Wine* **2018**, *222*, 44–49.
14. Jing, P.; Xiao Tong, R.; Xiao, H.; Jun, W.; Fei, H. Effects of different parcels on the aroma substances of wine grapes from eastern foothill of Helan mountain. *Food Sci.* **2022**, *43*, 291–300.
15. Xiao Yan, Z.; Rong Ping, C.; Peng Bao, S. Environmental conditions of Jieshi mountain production area in Changli and their effects on wine grape Syrah. *Mod. Agric. Sci. Technol.* **2022**, *813*, 49–53.
16. Meng Qi, L.; Shi, W.; Tong Hua, S.; Yu Bo, H.; Ying, S.; Chang Qing, D.; Yan Zhi, C.; Yi Bin, L. Analysis of varietal traits and wine flavor characteristics of different white grape varieties from Jieshi mountain region. *Food Ferment. Ind.* **2023**. [CrossRef]
17. Han, X.; Lu, H.C.; Wang, Y.; Gao, X.T.; Li, H.Q.; Tian, M.B.; Shi, N.; Li, M.Y.; Yang, X.L.; He, F.; et al. Region, vintage, and grape maturity co-shaped the ionomic signatures of the Cabernet Sauvignon wines. *Food Res. Int.* **2023**, *163*, 112165. [CrossRef]
18. Brun, L.A.; Maillet, J.; Hinsinger, P.; Pepin, M. Evaluation of copper availability to plants in copper-contaminated vineyard soils. *Environ. Pollut.* **2001**, *111*, 293–302. [CrossRef]
19. Saez-Plaza, P.; Navas, M.J.; Wybraniec, S.; Michalowski, T.; Asuero, A.G. An overview of the kjeldahl method of nitrogen determination. Part ii. Sample preparation, working scale, instrumental finish, and quality control. *Crit. Rev. Anal. Chem.* **2013**, *43*, 224–272. [CrossRef]
20. Wu, H.; Lin, G.H.; Tian, L.; Yan, Z.; Yi, B.Q.; Bian, X.H.; Jin, B.H.; Xie, L.Q.; Zhou, H.C.; Rogers, K.M. Origin verification of french red wines using isotope and elemental analyses coupled with chemometrics. *Food Chem.* **2021**, *339*, 127760. [CrossRef]
21. Lan, Y.B.; Qian, X.; Yang, Z.J.; Xiang, X.F.; Yang, W.X.; Liu, T.; Zhu, B.Q.; Pan, Q.H.; Duan, C.Q. Striking changes in volatile profiles at sub-zero temperatures during over-ripening of 'Beibinghong' grapes in northeastern china. *Food Chem.* **2016**, *212*, 172–182. [CrossRef]
22. Wen, Y.Q.; He, F.; Zhu, B.Q.; Lan, Y.B.; Pan, Q.H.; Li, C.Y.; Reeves, M.J.; Wang, J. Free and glycosidically bound aroma compounds in cherry (*Prunus avium* L.). *Food Chem.* **2014**, *152*, 29–36. [CrossRef]
23. Xu, X.Q.; Liu, B.; Zhu, B.Q.; Lan, Y.B.; Gao, Y.; Wang, D.; Reeves, M.J.; Duan, C.Q. Differences in volatile profiles of Cabernet Sauvignon grapes grown in two distinct regions of china and their responses to weather conditions. *Plant Physiol. Biochem.* **2015**, *89*, 123–133. [CrossRef] [PubMed]
24. Lan, Y.B.; Xiang, X.F.; Qian, X.; Wang, J.M.; Ling, M.Q.; Zhu, B.Q.; Liu, T.; Sun, L.B.; Shi, Y.; Reynolds, A.G.; et al. Characterization and differentiation of key odor-active compounds of 'Beibinghong' icewine and dry wine by Gas Chromatography-Olfactometry and aroma reconstitution. *Food Chem.* **2019**, *287*, 186–196. [CrossRef] [PubMed]
25. Lan, Y.B.; Guo, J.X.; Qian, X.; Zhu, B.Q.; Shi, Y.; Wu, G.F.; Duan, C.Q. Characterization of key odor-active compounds in sweet Petit Manseng (*Vitis vinifera* L.) wine by gas chromatography-olfactometry, aroma reconstitution, and omission tests. *J. Food Sci.* **2021**, *86*, 1258–1272. [CrossRef] [PubMed]
26. Moreno, C.E.; Calderon-Patron, J.M.; Martin-Regalado, N.; Martinez-Falcon, A.P.; Ortega-Martinez, I.J.; Rios-Diaz, C.L.; Rosas, F. Measuring species diversity in the tropics: A review of methodological approaches and framework for future studies. *Biotropica* **2018**, *50*, 929–941. [CrossRef]
27. Chen, K.; Qiu, S.; Liu, C.P.; Zhang, L.; Wu, X.G.; Ma, L.Y.; Li, J.M. Abiotic factors play important roles in complexity and characterization of aroma precursors in Vidal blanc grape. *Food Res. Int.* **2022**, *162*, 112015. [CrossRef] [PubMed]
28. Ke Ning, W.; Rui, Z. Soil texture classification and its application in China. *Acta Pedol. Sin.* **2019**, *56*, 227–241.
29. Ju, Y.L.; Liu, M.; Tu, T.Y.; Zhao, X.F.; Yue, X.F.; Zhang, J.X.; Fang, Y.L.; Meng, J.F. Effect of regulated deficit irrigation on fatty acids and their derived volatiles in 'Cabernet Sauvignon' grapes and wines of Ningxia, China. *Food Chem.* **2018**, *245*, 667–675. [CrossRef]
30. Xie, S.; Lei, Y.J.; Wang, Y.J.; Wang, X.Q.; Ren, R.H.; Zhang, Z.W. Influence of continental climates on the volatile profile of Cabernet Sauvignon grapes from five Chinese viticulture regions. *Plant Growth Regul.* **2019**, *87*, 83–92. [CrossRef]
31. Karaoglan, S.Y. The expected impact of climate change on grape flavor components - a review. *Ann. Univ. Craiova-Agric. Mont. Cadastre Ser.* **2018**, *48*, 215–220.
32. Gonzalez-Barreiro, C.; Rial-Otero, R.; Cancho-Grande, B.; Simal-Gandara, J. Wine aroma compounds in grapes: A critical review. *Crit. Rev. Food Sci. Nutr.* **2015**, *55*, 202–218. [CrossRef]
33. Ferreira, V.; Lopez, R. The actual and potential aroma of winemaking grapes. *Biomolecules* **2019**, *12*, 818. [CrossRef]
34. Skinkis, P.A.; Bordelon, B.P.; Butz, E.M. Effects of sunlight exposure on berry and wine monoterpenes and sensory characteristics of traminette. *Am. J. Enol. Vitic.* **2010**, *61*, 147–156. [CrossRef]

35. Koundouras, S.; Marinos, V.; Gkoulioti, A.; Kotseridis, Y.; Van Leeuwen, C. Influence of vineyard location and vine water status on fruit maturation of nonirrigated cv. Agiorgitiko (*Vitis vinifera* L.). Effects on wine phenolic and aroma components. *J. Agric. Food Chem.* **2006**, *54*, 5077–5086. [CrossRef]
36. Slaghenaufi, D.; Guardini, S.; Tedeschi, R.; Ugliano, M. Volatile terpenoids, norisoprenoids and benzenoids as markers of fine scale vineyard segmentation for Corvina grapes and wines. *Food Res. Int.* **2019**, *125*, 108507. [CrossRef] [PubMed]
37. Antalick, G.; Perello, M.C.; De Revel, G. Characterization of fruity aroma modifications in red wines during malolactic fermentation. *J. Agric. Food Chem.* **2012**, *60*, 12371–12383. [CrossRef] [PubMed]
38. Yan Er, M.; Li, Y.; Yu Yun, H.; Hua, L.; Wang, H. Effects of rainfall on polyphenols in grapes. *Jiangsu Agric. Sci.* **2016**, *44*, 221–224.
39. Bindon, K.; Varela, C.; Kennedy, J.; Holt, H.; Herderich, M. Relationships between harvest time and wine composition in *vitis vinifera* L. cv. *Cabernet Sauvignon* L. Grape and wine chemistry. *Food Chem.* **2013**, *141*, 147. [CrossRef]
40. Munoz-Gonzalez, C.; Martin-Alvarez, P.J.; Moreno-Arribas, M.V.; Pozo-Bayon, M.A. Impact of the nonvolatile wine matrix composition on the in vivo aroma release from wines. *J. Agric. Food Chem.* **2014**, *62*, 66–73. [CrossRef] [PubMed]
41. Rodriguez-Bencomo, J.J.; Munoz-Gonzalez, C.; Andujar-Ortiz, I.; Martin-Alvarez, P.J.; Moreno-Arribas, M.V.; Pozo-Bayon, M.A. Assessment of the effect of the non-volatile wine matrix on the volatility of typical wine aroma compounds by headspace solid phase microextraction/gas chromatography analysis. *J. Sci. Food Agric.* **2011**, *91*, 2484–2494. [CrossRef]
42. Villamor, R.R.; Evans, M.A.; Mattinson, D.S.; Ross, C.F. Effects of ethanol, tannin and fructose on the headspace concentration and potential sensory significance of odorants in a model wine. *Food Res. Int.* **2013**, *50*, 38–45. [CrossRef]
43. Villamor, R.R.; Ross, C.F. Wine matrix compounds affect perception of wine aromas. *Annu. Rev. Food Sci. Technol.* **2013**, *4*, 1–20. [CrossRef] [PubMed]
44. Robinson, A.L.; Ebeler, S.E.; Heymann, H.; Boss, P.K.; Solomon, P.S.; Trengove, R.D. Interactions between wine volatile compounds and grape and wine matrix components influence aroma compound headspace partitioning. *J. Agric. Food Chem.* **2009**, *57*, 10313–10322. [CrossRef] [PubMed]
45. Curtin, D.; Beare, M.H.; Hernandez-Ramirez, G. Temperature and moisture effects on microbial biomass and soil organic matter mineralization. *Soil Sci. Soc. Am. J.* **2012**, *76*, 2055–2067. [CrossRef]
46. Bell, S.J.; Henschke, P.A. Implications of nitrogen nutrition for grapes, fermentation and wine. *Aust. J. Grape Wine Res.* **2005**, *11*, 242–295. [CrossRef]
47. Van Leeuwen, C.; Roby, J.P.; De Resseguier, L. Soil-related terroir factors: A review. *Oeno One* **2018**, *52*, 173–188. [CrossRef]
48. Crews, T.E.; Brookes, P.C. Changes in soil phosphorus forms through time in perennial versus annual agroecosystems. *Agric. Ecosyst. Environ.* **2014**, *184*, 168–181. [CrossRef]
49. Maltais-Landry, G.; Scow, K.; Brennan, E. Soil phosphorus mobilization in the rhizosphere of cover crops has little effect on phosphorus cycling in california agricultural soils. *Soil Biol. Biochem.* **2014**, *78*, 255–262. [CrossRef]
50. Jalali, M.; Ostovarzadeh, H. Evaluation of phosphorus leaching from contaminated calcareous soils due to the application of sheep manure and ethylenediamine tetraacetic acid. *Environ. Earth Sci.* **2009**, *59*, 441–448. [CrossRef]
51. Hong Bo, R. The Research on the Transference Law of the Mineral Elements in the Soil. Master's Thesis, Ocean University of China, Qingdao, China, 2008.
52. Rui, W. Relationship between Soil Quality with Grape Growth and Composition at the Eastern Foot of Helan Mountain Wine Production Regions. Ph.D. Thesis, Northwest A&F University, Yangling, China, 2016.
53. Singh, B.P.; Cowie, A.L.; Chan, K.Y. *Soil Health and Climate Change*; Springer: Heidelberg, Germany, 2011; pp. 33–34.
54. Webster, D.R.; Edwards, C.G.; Spayd, S.E.; Peterson, J.C.; Seymour, B.J. Influence of vineyard nitrogen-fertilization on the concentrations of monoterpenes, higher alcohols, and esters in aged riesling wines. *Am. J. Enol. Vitic.* **1993**, *44*, 275–284. [CrossRef]
55. Van Leeuwen, C.; Barbe, J.C.; Darriet, P.; Geffroy, O.; Gomes, E.; Guillaumie, S.; Helwi, P.; Laboyrie, J.; Lytra, G.; Le Menn, N. Recent advancements in understanding the terroir effect on aromas in grapes and wines. *OENO One* **2020**, *54*, 985–1006. [CrossRef]
56. Jiang, P.K.; Xu, Q.F.; Zhou, G.M.; Wu, Q.F.; Wu, J.S. Effects of green manure on soil nutrients and bio-properties of Castanea mollissima Blume plantations. *J. Beijing For. Univ.* **2007**, *29*, 120–123.
57. Liu, D.; Chen, Q.L.; Zhang, P.Z.; Chen, D.L.; Howell, K.S. The fungal microbiome is an important component of vineyard ecosystems and correlates with regional distinctiveness of wine. *Msphere* **2020**, *5*, 1–15. [CrossRef] [PubMed]
58. Liu, X.Y.; Lindemann, W.C.; Whitford, W.G.; Steiner, R.L. Microbial diversity and activity of disturbed soil in the northern Chihuahuan desert. *Biol. Fertil. Soils* **2000**, *32*, 243–249. [CrossRef]
59. Ovreas, L.; Torsvik, V. Microbial diversity and community structure in two different agricultural soil communities. *Microb. Ecol.* **1998**, *36*, 303–315. [PubMed]
60. Peinado, R.A.; Moreno, J.; Bueno, J.E.; Moreno, J.A.; Mauricio, J.C. Comparative study of aromatic compounds in two young white wines subjected to pre-fermentative cryomaceration. *Food Chem.* **2004**, *84*, 585–590. [CrossRef]
61. Guth, H. Quantitation and sensory studies of character impact odorants of different white wine varieties. *J. Agric. Food Chem.* **1997**, *45*, 3027–3032. [CrossRef]
62. Gomez-Miguez, M.J.; Cacho, J.F.; Ferreira, V.; Vicario, I.M.; Heredia, F.J. Volatile components of Zalema white wines. *Food Chem.* **2007**, *100*, 1464–1473. [CrossRef]

63. Li, N.; Wang, L.Y.; Yin, J.; Ma, N.; Tao, Y.S. Adjustment of impact odorants in hutai-8 rose wine by co-fermentation of Pichia fermentans and Saccharomyces cerevisiae. *Food Res. Int.* **2022**, *153*, 110959. [CrossRef]
64. Franco, M.; Peinado, R.A.; Medina, M.; Moreno, J. Off-vine grape drying effect on volatile compounds and aromatic series in must from Pedro Ximenez grape variety. *J. Agric. Food Chem.* **2004**, *52*, 3905–3910. [CrossRef]
65. Furdikova, K.; Makysova, K.; Durcanska, K.; Spanik, I.; Malik, F. Influence of yeast strain on aromatic profile of Gewurztraminer wine. *LWT-Food Sci. Technol.* **2014**, *59*, 256–262.
66. Garcia-Carpintero, E.G.; Sanchez-Palomo, E.; Gallego, M.a.G.; Gonzalez-Vinas, M.A. Characterization of impact odorants and sensory profile of Bobal red wines from Spain's la Mancha region. *Flavour Fragr. J.* **2012**, *27*, 60–68. [CrossRef]
67. Moreno, J.A.; Zea, L.; Moyano, L.; Medina, M. Aroma compounds as markers of the changes in sherry wines subjected to biological ageing. *Food Control* **2005**, *16*, 333–338. [CrossRef]
68. Peinado, R.A.; Mauricio, J.C.; Moreno, J. Aromatic series in sherry wines with gluconic acid subjected to different biological aging conditions by Saccharomyces cerevisiae var. Capensis. *Food Chem.* **2006**, *94*, 232–239. [CrossRef]
69. Tao, Y.; Zhang, L. Intensity prediction of typical aroma characters of Cabernet Sauvignon wine in Changli county (China). *LWT - Food Sci. Technol.* **2010**, *43*, 1550–1556. [CrossRef]
70. La Guerche, S.; Dauphin, B.; Pons, M.; Blancard, D.; Darriet, P. Characterization of some mushroom and earthy off-odors microbially induced by the development of rot on grapes. *J. Agric. Food Chem.* **2006**, *54*, 9193–9200. [CrossRef] [PubMed]
71. 1-octen-3-ol. Available online: https://www.vcf-online.nl/VcfCompoundDetails.cfm?volatkey=0205251300 (accessed on 28 April 2023).
72. Callejon, R.M.; Morales, M.L.; Ferreira, A.C.S.; Troncoso, A.M. Defining the typical aroma of sherry vinegar: Sensory and chemical approach. *J. Agric. Food Chem.* **2008**, *56*, 8086–8095. [CrossRef]
73. Gurbuz, O.; Rouseff, J.M.; Rouseff, R.L. Comparison of aroma volatiles in commercial Merlot and Cabernet Sauvignon wines using gas chromatography - olfactometry and gas chromatography - mass spectrometry. *J. Agric. Food Chem.* **2006**, *54*, 3990–3996. [CrossRef]
74. Ferreira, V.; Lopez, R.; Cacho, J.F. Quantitative determination of the odorants of young red wines from different grape varieties. *J. Sci. Food Agric.* **2000**, *80*, 1659–1667. [CrossRef]
75. Zhang, M.; Zhong, T.; Heygi, F.; Wang, Z.; Du, M. Effects of inoculation protocols on aroma profiles and quality of plum wine in mixed culture fermentation of Metschnikowia pulcherrima with Saccharomyces cerevisiae. *LWT* **2022**, *161*, 113338. [CrossRef]
76. Li, H.; Tao, Y.S.; Wang, H.; Zhang, L. Impact odorants of chardonnay dry white wine from Changli county (China). *Eur. Food Res. Technol.* **2008**, *227*, 287–292. [CrossRef]
77. Mayr, C.M.; Geue, J.P.; Holt, H.E.; Pearson, W.P.; Jeffery, D.W.; Francis, I.L. Characterization of the key aroma compounds in shiraz wine by quantitation, aroma reconstitution, and omission studies. *J. Agric. Food Chem.* **2014**, *62*, 4528–4536. [CrossRef] [PubMed]
78. Diez-Ozaeta, I.; Lavilla, M.; Amarita, F. Wine aroma profile modification by Oenococcus oeni strains from Rioja Alavesa region: Selection of potential malolactic starters. *Int. J. Food Microbiol.* **2021**, *356*, 109324. [CrossRef] [PubMed]
79. Takeoka, G.; Buttery, R.G.; Flath, R.A.; Teranishi, R.; Wheeler, E.L.; Wieczorek, R.L.; Guentert, M. Volatile constituents of pineapple (*Ananas-comosus* [L] Merr). *ACS Symp. Ser.* **1989**, *388*, 223–237.
80. Meilgaard, M.C. Flavor chemistry of beer: Part ii: Flavor and threshold of 239 aroma volatiles. *MBAA TQ* **1975**, *12*, 151–168.
81. Yang, F.; Shi, C.; Yan, L.; Xu, Y.; Dai, Y.; Bi, S.; Liu, Y. Low-frequency ultrasonic treatment: A potential strategy to improve the flavor of fresh watermelon juice. *Ultrason. Sonochem.* **2022**, *91*, 106238. [CrossRef]
82. Avalos-Martinez, E.; Pino, J.A.; Sayago-Ayerdi, S.; Sosa-Moguel, O.; Cuevas-Glory, L. Assessment of volatile compounds and sensory characteristics of Mexican hibiscus (*Hibiscus sabdariffa* L.) calyces hot beverages. *J. Food Sci. Technol.* **2019**, *56*, 360–366. [CrossRef]
83. Yu, H.Y.; Guo, W.; Xie, T.; Ai, L.Z.; Tian, H.X.; Chen, C. Aroma characteristics of traditional huangjiu produced around winter solstice revealed by sensory evaluation, gas chromatography-mass spectrometry and gas chromatography-ion mobility spectrometry. *Food Res. Int.* **2021**, *145*. [CrossRef]
84. Noguerol-Pato, R.; Gonzalez-Alvarez, M.; Gonzalez-Barreiro, C.; Cancho-Grande, B.; Simal-Gandara, J. Evolution of the aromatic profile in Garnacha Tintorera grapes during raisining and comparison with that of the naturally sweet wine obtained. *Food Chem* **2013**, *139*, 1052–1061. [CrossRef]
85. Chin, S.T.; Eyres, G.T.; Marriott, P.J. Application of integrated comprehensive/multidimensional gas chromatography with mass spectrometry and olfactometry for aroma analysis in wine and coffee. *Food Chem* **2015**, *185*, 355–361. [CrossRef]
86. Welke, J.E.; Nicolli, K.P.; Hernandes, K.C.; Biasoto, A.C.T.; Zini, C.A. Adaptation of an olfactometric system in a GC-FID in combination with GC×GC/MS to evaluate odor-active compounds of wine. *Food Chem.* **2022**, *370*, 131004. [CrossRef]
87. Du, X.; Finn, C.E.; Qian, M.C. Volatile composition and odour-activity value of thornless 'black diamond' and 'marion' blackberries. *Food Chem.* **2010**, *119*, 1127–1134. [CrossRef]
88. Cullere, L.; Escudero, A.; Cacho, J.; Ferreira, V. Gas chromatography-olfactometry and chemical quantitative study of the aroma of six premium quality Spanish aged red wines. *J. Agric. Food Chem.* **2004**, *52*, 1653–1660. [CrossRef] [PubMed]
89. Kong, C.L.; Li, A.H.; Su, J.; Wang, X.C.; Chen, C.Q.; Tao, Y.S. Flavor modification of dry red wine from Chinese spine grape by mixed fermentation with Pichia fermentans and s. cerevisiae. *LWT-Food Sci. Technol.* **2019**, *109*, 83–92. [CrossRef]

90. Delgado, J.A.; Sánchez-Palomo, E.; Osorio Alises, M.; González Viñas, M.A. Chemical and sensory aroma typicity of la mancha petit verdot wines. *LWT* **2022**, *162*, 113418. [CrossRef]
91. Parker, M.; Osidacz, P.; Baldock, G.A.; Hayasaka, Y.; Black, C.A.; Pardon, K.H.; Jeffery, D.W.; Geue, J.P.; Herderich, M.J.; Francis, I.L. Contribution of several volatile phenols and their glycoconjugates to smoke-related sensory properties of red wine. *J. Agric. Food Chem.* **2012**, *60*, 2629–2637. [CrossRef] [PubMed]
92. Ling, M.Q.; Bai, X.X.; Cui, D.S.; Shi, Y.; Duan, C.Q.; Lan, Y.B. An efficient methodology for modeling to predict wine aroma expression based on quantitative data of volatile compounds: A case study of oak barrel-aged red wines. *Food Res. Int.* **2023**, *164*, 112440. [CrossRef]

Disclaimer/Publisher's Note: The statements, opinions and data contained in all publications are solely those of the individual author(s) and contributor(s) and not of MDPI and/or the editor(s). MDPI and/or the editor(s) disclaim responsibility for any injury to people or property resulting from any ideas, methods, instructions or products referred to in the content.

foods

Article

Evaluation of the Long-Lasting Flavour Perception after the Consumption of Wines Treated with Different Types of Oenological Additives Considering Individual 6-n-Propylthiouracil Taster Status

Rafael I. Velázquez-Martínez [1], Celia Criado [1], Carolina Muñoz-González [1], Julia Crespo [2] and María Ángeles Pozo-Bayón [1,*]

1 Instituto de Investigacion en Ciencias de la Alimentacion, CSIC-UAM, Nicolas Cabrera 9, 28049 Madrid, Spain; rafael.v.m@csic.es (R.I.V.-M.); celia.criado@csic.es (C.C.); c.munoz@csic.es (C.M.-G.)
2 Departamento de Investigación Agroambiental, Instituto Madrileño de Investigación y Desarrollo Rural, Agrario y Alimentario (IMIDRA), El Encín, A-2 km 38.2, 28805 Alcalá de Henares, Spain; julia.crespo.garcia@madrid.org
* Correspondence: m.delpozo@csic.es; Tel.: +34-910-017-961

Abstract: Due to the limited scientific knowledge on the impact of commercial oenological additives on flavour perception, the aim of this work was to evaluate the effect of different types of oenological additives on the long-lasting flavour perception (flavour persistence) during wine tasting, also considering the effect of the individual PROP (6-n-propylthiouracil) taster status (PTS). To do so, white and red wines with two oenotannins (ellagitannin and gallotannin) and a commercial yeast mannoprotein were prepared. A control wine of each type was also made without additives. All the wines were spiked with a mixture of aromatic compounds responsible for the "fruity" and "woody" notes. Retronasal aroma and astringency were evaluated at the same time using time–intensity (TI) methodology and a trained panel (n = 40), including PROP non-tasters (NTs) and tasters (Ts). The results showed a significant effect of PTS on the long-lasting perception of astringency, being Ts who showed higher values than NTs for most TI parameters. However, PTS did not affect aroma persistence. In addition, the three oenological additives had an effect on astringency and retronasal aroma perception. They significantly increased the long-lasting perception of astringency compared to the control, while gallotannin also increased the persistence of the woody aroma.

Keywords: wine; oenotannins; mannoproteins; flavour persistence; PROP taste phenotype; time–intensity sensory analysis

1. Introduction

The use of natural additives to improve the winemaking process is becoming a current practice in the oenological industry, leading to a large variety of commercial additives with different claims related to many technological and sensory properties in wines. Nonetheless, the number of scientific studies supporting the potential benefits of these compounds and their optimal usage during winemaking is still quite scarce. In general, most of these additives come from different sources, being oenotannins and yeast derivatives among the most commonly used.

Oenotannins can come from grapes or other different botanical sources. The International Organization of Vine and Wine (OIV) defines their use for the stabilisation and fining of musts and wines, as well as to increase the antioxidant and antioxidasic capacity of grape juice and to promote colour stability (resolutions OIV-OENO-612–2019 and OIV-OENO-613–2019). Nonetheless, besides these effects, they can also affect some wine sensory properties such as aroma, astringency, and bitterness [1–6]. This mainly depends on their chemical structure and concentration [7,8]. Depending on their chemical characteristics, oenotannins can be generally divided into condensed tannins (or proanthocyanidins) and hydrolysable tannins. Proanthocyanidins can be polymers of flavan-3-ols and flavan-3,4-diols. In grapes, they are flavan-3-ols-based polymers, namely procyanidins, which are composed of (+)-catechin and (−)-epicatechin with different extents of galloylation [9]. Additionally, there are condensed tannins with flavan-3,4-diols subunits extracted from exotic woods (quebracho, acacia, etc.) for oenological use [7,8]. Regarding hydrolysable tannins, they can be classified as gallotannins and ellagitannins. Gallotannins are present in plant gallnuts, and they are composed of gallic acid and D-glucose, with different extents of substitution with galloyl moiety. Meanwhile, ellagitannins are formed by D-glucose and ellagic, gallic, or hexahydroxydiphenic acids, and they are usually extracted from chestnut and oak [10].

Besides oenological tannins, yeast derivative products constitute a second type of oenological additives widely used during winemaking. They include different products such as inactivated yeast, inactivated yeast with glutathione, autolysate, yeast protein extract, and yeast wall mannoproteins [11,12]. Traditionally, they have been used to provide assimilable nitrogen or to stimulate yeast and lactic bacteria growth and prevent stuck fermentations [11,12]. Currently, mannoproteins are also being used for increasing wine colloidal stability, which is an application supported by the OIV (resolution OIV-OENO 417–2011). Nonetheless, the use of yeast derivatives in wines has also been shown to have additional effects on wine sensory characteristics. For instance, they preserve the intensity of the colour of wines, eliminate the excess of tannins related to wine astringency, and also have an impact on wine aroma [13–19].

Interestingly, tannins and mannoproteins are also polymers with the ability to bind the oral surfaces (mucoadhesion properties). For instance, Ginsburg and collaborators [20] showed that tannins can be retained in the oral cavity for long periods despite a constant salivary flow. More recently, it has been proven the existence of intermolecular interactions between tannic acid and mucin [21]. These interactions affect the retention and release of aroma compounds in the oral cavity [22–25].

In the case of mannoproteins, recent works using salivary proteins, protein-rich proline proteins (PRPs), and a cell-based model of the oral epithelium system also show that mannoproteins can interact with some salivary proteins and tannins, with an impact on astringency modulation [26–28]. Besides this, well known is the existence of interactions between aroma compounds and mannoproteins, which depend on the hydrophobicity of the aroma compounds and on the composition of the mannoprotein (glucidic/protein ratio) affecting aroma release, as shown when using headspace analysis [15,16,29].

Given the ability of tannins and mannoproteins to be retained on the oral surface together with the capacity of both types of polymers for binding aroma molecules, the hypothesis arises that they can have a preponderant role in flavour persistence, which is the long-lasting perception of flavour stimuli produced immediately after wine swallowing [30]. This phenomenon is a key factor in the sensory experience of consumers and therefore very much related to wine preference and liking.

In this sense, it is also important to notice that wine flavour perception, including flavour persistence, might be greatly variable depending on many types of genetic, biological, physiological, and psychological factors [31]. Among them, taste phenotype, or PROP taster status (PTS), measured as the sensibility to taste the bitter compound 6-n-propylthiouracil (PROP), has been one of the most studied [32]. It has been described that PROP taste individuals also have a higher acuity to perceive other basic tastes, mouthfeel (astringent), and olfactory stimuli [33]. Although still under study, the higher sensory ability of PROP taste individuals has been related to many different factors, such as differences in TAS2R38 gene polymorphisms, differences in fungiform papillae density, salivary protein composition, or age and gender, among others [34]. Interestingly, in previous studies on wine, the association between PTS and a higher sensory ability to perceive taste, mouthfeel, and olfactory stimuli has been rather controversial, with some studies showing a positive correlation [35] while others do not [36–38].

Considering the lack of specific studies focused on the effect of oenological additives on wine flavour persistence, the objective of this work was to test if three common commercial oenological additives, namely, hydrolysable tannins (gallotannin and ellagitannin) and yeast mannoproteins, might affect the long-lasting flavour perception (astringency and retronasal aroma) of wines (red and white). For this purpose, a dynamic sensory technique (time–intensity) was used, which is based on scoring the evolution of the intensity of the flavour stimulus (fruity and woody aroma and astringency) from the moment it appears (immediately after the wine is tasted) until it is no longer perceived. Additionally, the effect of PTS (ability to perceive the bitter compound 6-n-propylthiouracil) was also tested by using a trained panel formed of tasters (n = 20) and non-tasters (n = 20).

2. Materials and Methods

2.1. Wine Samples

A red wine and a white wine from Tempranillo and Malvar grape varieties were industrially produced at the IMIDRA experimental winery (Alcalá de Henares, Madrid, Spain). These wines were considered the control wines (red, CRW; white, CWW). Their chemical composition is shown in Table S1. From each control wine, three more wines were formulated by adding three types of commercial oenological additives (Table 1). One of them was a mannoprotein from yeast wall and the other two were oenological hydrolysable tannins, specifically a gallotannin and an ellagitannin. Their chemical compositions are shown in Table S2. All of them were provided by Laffort Ibérica S.A. They were added before bottling at the concentration recommended by the manufacturer for wine applications. The wine types and the final concentration of each additive in the wine are shown in Table 1.

Table 1. Wine types and concentration used of each oenological additive.

Oenological Additive	White Wines [a]		Red Wines [a]	
	Wine Type (f/w)	Concentration	Wine Type (f/w)	Concentration
No additive (control)	CWW	--	CRW	--
Gallotannin	GTWW	300 mg/L	GTRW	300 mg/L
Ellagitannin	ETWW	700 mg/L	ETRW	700 mg/L
Mannoprotein	MWW	1.5 mL/L	MRW	0.9 mL/L

[a] All the wines (white and red) with and without oenological additives were aromatised with a fruity and a woody aroma mixture. Therefore, 8 types of white wines (fruity and woody) and 8 types of red wines (fruity and woody) were prepared for this study.

2.2. Wine Aromatisation

To reinforce the aroma profile of the wines in two aroma descriptors of interest (fruity and woody), before the sensory test, all the wines were independently aromatised with two aroma mixtures, responsible for these aroma nuances using food-grade odorant compounds (Table 2) from Merck (Darmstadt, Germany). For the preparation of the aroma solutions, individual solutions of each aroma compound responsible for the fruity and woody aroma notes were weighted and diluted in food-grade ethanol. From this, a working stock solution containing all the aroma compounds of each aroma mixture was prepared. Two hundred microliters of this solution was added to each wine (15 mL contained in a wine glass) before each test to have the final concentrations shown in Table 2. These concentrations were chosen in previous lab trials, considering that they should be easily distinguished by the panel but not unpleasant, which might negatively affect the completion of the test. Likewise, the aromatic concentrations were taken as a reference from previously published works [38]. Each aroma mixture was independently poured and evaluated in all four wine types (reds and whites). Aromatisation was performed 5–10 min before the beginning of the sensory evaluation. During this time, the wine glasses were covered with plastic Petri dishes to prevent volatile loss.

Table 2. Concentrations of aroma compounds included in each aroma mixture (fruity and woody).

Aroma Mixture	Aroma Compounds	CAS Number	Concentration in Wine (μg/L)
Woody	Whiskylactone	80041-00-5	165
	Vainillin	121-33-5	55
	Eugenol	97-53-0	8
	Guaiacol	90-05-1	8
	Furaneol	3658-77-3	55
Fruity	2,3-butanedione	431-03-08	1400
	Isoamyl acetate	123-92-2	550
	Ethyl acetate	141-78-6	5000
	Ethyl cinnamate	103-36-6	12
	B-damascenone	23726-93-4	0.3

Therefore, a total of sixteen wines were produced: (a) eight red wines: control wine without and with the three additives aromatised with fruity and woody aroma mixtures, and (b) eight white wines produced under the same conditions (Table 1).

2.3. Individual Panel

Forty individuals from two different PROP taste phenotypes (PROP tasters and non-tasters) (see Section 2.4) were recruited for this study. Additionally, the inclusion criteria for participation were healthy, non-pregnant, and adult volunteers (over 18 years old). In addition, all volunteers completed a food allergy screening document, which included allergy/intolerance to wine or any of its components. Each volunteer attended one-hour sessions eight times. All the participants were informed of the nature of this study and gave written consent to participate. This work was approved by the Bioethics Committee of the Spanish National Research Council (CSIC, 008/2021). This study was conducted in April and May 2022 at the CIAL (Madrid, Spain).

2.4. Taste PROP Phenotype

Individual taste PROP phenotype was tested using commercial strips impregnated with 6-n-propylthiouracil (3 μg/strip) from Sensonic International (Haddon Heights, New Jersey). Triangular tests were conducted with two blank samples (no impregnated strips) (Sensonic International) and one impregnated sample to test whether consumers were able to recognise the PROP sample. If the volunteers did not perform the triangular test correctly, they were considered to be in the non-taster group. Individuals who positively recognised the PROP sample were retested and they evaluated the perceived intensity, which was assessed using the Generalised Labelled Magnitude Scale (gLMS) scale (0–100; from "no sensation" = 0, to "strongest imaginable sensation of any kind" = 100) [39]. For each sample, they were instructed to swipe the paper strip across the tongue, remove the strip, press the tongue against the roof of the mouth, and swallow. Between samples, they were asked to drink water. Individuals were classified into two groups: non-tasters (NTs; n = 20) and tasters (Ts; n = 20).

2.5. Dynamic Sensory Analyses

2.5.1. Training

The absence of anosmia was previously confirmed by means of a triangular test with flavoured hydroalcoholic solutions. Individuals received specific training for the recognition of fruity and woody aromas as well as astringency using red and white wines. In the following sessions, training sessions for the recognition of the intensity of the different sensory stimuli perceived with the different types of wine were also carried out. Additionally, during these sessions, individuals were instructed in the use of the intensity scale (15 cm unstructured scale delimited at the ends), and in the TI methodology using tablets for sensory data collection.

Samples (15 mL) were served in standard tasting glasses (20 mL) covered with plastic Petri dishes to avoid loss of volatiles. For this purpose, wine glasses were labelled with random 3-digit codes and presented simultaneously in a randomised order using a Balanced Complete Block design [40]. Mineral water (Aliada, Madrid, Spain) and breadsticks (El Corte Inglés, Cádiz, Spain) were offered to cleanse the palate.

Sensory evaluation sessions were conducted using Compusense® Cloud software (Compusense Inc., Guelph, ON, Canada) via tablets, where each consumer had a user profile and could access each of the tests.

2.5.2. Sensory Evaluation

For the evaluation of the sensory stimuli (retronasal aroma and astringency), panellists gently rinsed their mouths with the wine (15 mL) for 30 s, then spat it out. During rinsing, special care was taken to keep the lips closed, not to swallow and not to open the velum–tongue border prior to expectoration. Then, they were instructed to swallow the remaining saliva in their mouth and to start the TI evaluation. For this, panellists moved the cursor along the unstructured scale (15 cm) to evaluate the astringency and aroma intensity perceived (of one single aroma attribute) that lasted for two minutes. The evaluation of the intensity of both stimuli in the same trial avoided the halo-dumping effect [41,42]. Data were recorded at a frequency of 1s.

2.5.3. TI Data Analyses

For each sensory stimulus, TI curves were obtained by averaging the data at each point of time across the two groups of subjects (Ts and NTs). The raw data were obtained by Compusense software. Moreover, four typical time–intensity parameters were extracted from the TI curves using the XLSTAT Sensory software: time to reach the maximum intensity (Tmax), maximum intensity (Imax), duration time of the perceived stimuli (Tend), and area under the curve (AUC).

2.6. Statistical Analyses

Analysis of variance ANOVA (one-way or two-way) and mean comparison tests (Tukey) were applied to check the effect of PROP taste phenotype and the effects of the oenological additives on the different parameters (Imax, Tmax, Tend, AUC) extracted from the astringency and aroma TI curves. A significance level of $p < 0.05$ was always used. Statistical analyses were carried out using XLSTAT (Version 2019.01).

3. Results

3.1. Effect of PROP Taste Phenotype on Flavour Perception in Red and White Wines

PROP taster status (PTS) has been associated with a higher individual ability to perceive sensory stimuli from wines [35]. However, there were no previous studies that tested this using dynamic sensory methods that better represent the flavour perception experienced during wine consumption. Therefore, the effect of PROP phenotype on wine astringency and retronasal aroma perception over time was first checked.

3.1.1. Effect of PROP Taster Status in Wine Astringency Perception over Time

As can be seen in Table 3, in the case of astringency, tasters (Ts) showed significantly higher values of Text and AUC in the case of red wines with a woody aroma. For instance, T phenotypes showed values of Text and AUC that were 9.7% and 18% higher compared to NT. The same results were found in the case of the red wine with a fruity aroma. In this case, Ts also exhibited higher values for most TI parameters. For instance, Tmax, Text, and AUC values were 26.9%, 9.6%, and 21% higher, respectively, compared to the non-taster (NT) phenotype. Figure 1 shows the TI curves (average values from 20 volunteers) for the astringency perception obtained for this wine considering both PROP taste phenotypes.

Table 3. ANOVA and Tukey test results showing the effect of PROP taster status on astringency perception considering the different TI parameters in the red and white wines aromatised with woody and fruity aromas.

Wine Type	PROP Phenotype	Astringency in Red Wines				Astringency in White Wines			
		I max	T max	T ext	AUC	I max	T max	T ext	AUC
Woody aromatised wine	T	14.1 a	4.4 a	**104.5 a**	**34,560 a**	11.8 a	5.0 a	91.4 a	22,299 a
	NT	13.9 a	1.8 a	**94.3 b**	**28,326 b**	13.8 a	4.2 a	81.6 a	19,765 a
Fruity aromatised wine	T	13.7 a	**2.6 a**	**103.4 a**	**33,856 a**	**10.9 b**	**4.0 a**	90.9 a	20,397 a
	NT	13.7 a	**1.9 b**	**93.5 b**	**26,776 b**	**13.1 a**	**2.4 b**	80.6 a	21,222 a

T, taster; NT, non-taster. Different letters in bold (a–b) show significant differences among PROP taster status groups from Tukey test ($p < 0.05$).

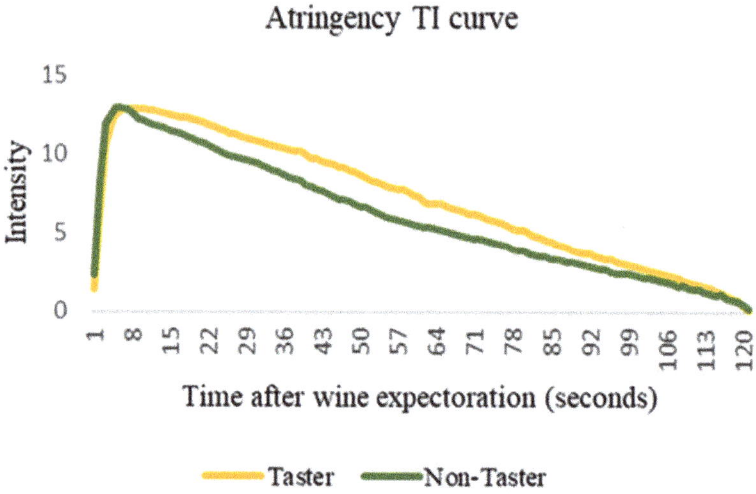

Figure 1. Example of the time–intensity average profile (n = 20) of astringency obtained during the tasting of the fruity flavoured red wine considering both PROP taste phenotypes (taster and non-taster).

In the case of white wines, there were also significant differences in the perception of astringency depending on PROP taster status, but only in white wines aromatised with fruity aroma notes. As occurred in red wines, in white wines, Ts showed higher T max values (40% more) than NTs. On the contrary, Ts showed 16.8% lower values of Imax compared to NTs. There were no differences between PROP phenotypes in the other TI parameters, showing a lower impact of the perception on the astringency in the case of white wines.

3.1.2. Effect of PROP Taster Status on Wine Retronasal Aroma Perception over Time

As shown in Table 4, in the case of both retronasal aroma attributes (woody and fruity aroma), there were no significant differences based on PTS in any of the parameters extracted from the TI curves in neither red nor white wines.

Table 4. ANOVA and Tukey test results showing the effect of PROP taster status on retronasal aroma perception considering the different TI parameters in the red and white wines flavoured with woody and fruity aromas.

Sensory Stimuli	PROP Phenotype	Time–Intensity Parameters in Red Wines				Time–Intensity Parameters in White Wines			
		I max	T max	T ext	AUC	I max	T max	T ext	AUC
Woody aroma	T	14.0 a	4.6 a	100.2 a	28,396 a	12.7 a	3.1 a	97.3 a	24,039 a
	NT	13.7 a	3.5 a	98.3 a	28,310 a	12.4 a	4.7 a	97.2 a	24,544 a
Fruity aroma	T	12.7 a	2.4 a	99.9 a	18,863 a	12.9 a	2.5 a	98.4 a	24,905 a
	NT	13.1 a	2.3 a	99.5 a	26,328 a	13.4 a	2.0 a	96.6 a	29,358 a

T, taster; NT, non-taster. The letter "a" indicates no significant differences among PROP taster status groups from Tukey test ($p > 0.05$).

3.2. *Effect of the Oenological Additives on Flavour Perception in Red and White Wines*

In a further step of the work, the effect of the oenological additives in the perception of the astringency and retronasal aroma (fruity and woody) over time was also checked.

3.2.1. Effect of Oenological Additives in Wine Astringency

Results corresponding to the differences in TI parameters considering the control wine and the wines supplemented with oenological additives are shown in Table 5. Since, as shown before, the evaluation of astringency was affected by the individual PROP taste phenotype, Table 5 shows these results considering both taste phenotypes.

Table 5. ANOVA and Tukey test results showing the effect of the oenological additives on the TI parameters from the astringency evaluation in red and white wines also considering the PROP taster status.

Wine	PROP Phenotype	Wine Type	Astringency							
			Fruity Wine				Woody Wine			
			I max	T max	T ext	AUC	I max	T max	T ext	AUC
Red	T	CRW	13.3 a	**4.8 a**	100.9 a	28,796 a	15.4 a	9.2 a	105.6 a	35,749 a
		GTRW	13.6 a	**2.1 b**	99.0 a	31,371 a	13.4 a	3.4 a	104.1 a	33,543 a
		MRW	13.9 a	**1.9 b**	107.0 a	36,046 a	13.8 a	1.6 a	104.3 a	35,410 a
		ERW	13.9 a	**1.9 b**	106.3 a	38,261 a	13.8 a	2.9 a	104.1 a	33,643 a
	NT	CRW	14.0 a	**2.6 a**	91.7 a	24,492 a	13.9 a	1.8 a	90.6 a	28,702 a
		GTRW	14.0 a	**1.7 b**	101.3 a	31,903 a	14.0 a	2.1 a	95.9 a	25,510 a
		MRW	12.9 a	**1.6 b**	88.7 a	21,175 a	13.8 a	1.8 a	96.6 a	30,489 a
		ERW	13.8 a	**1.8 b**	91.6 a	28,957 a	14.0 a	1.4 a	93.9 a	28,467 a
White	T	CWW	**9.0 b**	4.8 a	82.4 a	16,442 a	12.1 a	5.8 a	89.2 a	20,387 a
		GTWW	**10.4 ab**	6.5 a	83.7 a	18,909 a	13.7 a	2.2 a	96.2 a	25,089 a
		MWW	**12.4 a**	2.7 a	101.1 a	23,279 a	11.2 a	5.8 a	86.3 a	19,368 a
		EWW	**11.7 ab**	2.3 a	95.3 a	22,635 a	10.2 a	6.9 a	92.9 a	23,854 a
	NT	CWW	12.0 a	3.5 a	72.0 a	17,265 a	12.8 a	4.9 a	81.0 a	18,607 a
		GTWW	11.7 a	2.4 a	77.7 a	21,779 a	16.6 a	6.8 a	93.6 a	22,720 a
		MWW	13.3 a	1.9 a	88.6 a	22,720 a	12.7 a	2.3 a	65.0 a	14,221 a
		EWW	15.2 a	1.9 a	83.2 a	22,955 a	12.7 a	2.5 a	82.8 a	22,904 a

T, taster; NT, non-taster. Different letters in bold (a–b) show significant differences from Tukey test ($p < 0.05$) among wine types for each of the PROP taster status groups. Control wine (CW); gallotannin wine (GTW); mannoprotein wine (MW); ellagitannin wine (EW).

As can be seen in Table 5, the effect of the oenological additives on astringency was significant ($p < 0.05$) in the case of red and white fruity wines, but not in the case of wines aromatised with a woody aroma. In the case of red fruity wines, for both PROP taste phenotypes (T and NT), wines with oenological additives exhibited lower Tmax values compared to the control. The lower Tmax was more pronounced for Ts (60% lower compared to the control) than for NTs (34.6% lower compared to the control). A lower Tmax value means that the maximum intensity of perceived astringency will be quicker in wines with additives, and even faster for individuals of the T phenotype.

In the case of fruity white wines, the effect of oenological additives was only significant ($p < 0.05$) for Ts. Interestingly, they also exhibited higher Imax values of astringency for the three wines with additives compared to the control wine. Of the three additives, mannoproteins provoked the largest effect, increasing Imax by 27.4% compared to the control wine. On the contrary, the impact of the additives was not significant for the other TI parameters, nor in the case of wines (red or white) with a woody aroma.

3.2.2. Effect of Oenological Additives on Retronasal Aroma

To check the effect of the oenological additives on TI parameters obtained from the retronasal aroma evaluation, data from both PROP taste phenotypes were considered together, since, as previously shown, this factor did not affect TI retronasal aroma evaluation.

As shown in Table 6, only fruity red wines with additives exhibited significantly lower (above 30%) T max values compared to the control wine. Similar results were found in the case of fruity white wines, although in this case, the effect of the additive was slightly different. Although the wines with additives exhibited lower Tmax values compared to the control, a higher effect was noticed in the wine with mannoprotein, in which a reduction in Tmax of 48% was found. These results mean that the three additives reduced the time to reach the Imax, and therefore, their addition in wines produces a quicker fruity sensation, which is more pronounced in the case of mannoprotein in white wine. Interestingly, white wines with mannoprotein also exhibited higher Text and AUC values, although these results were not significant.

Table 6. ANOVA and Tukey test results showing the effect of oenological additives on retronasal aroma on the TI parameters from the retronasal aroma evaluation of the red and white wines aromatised with fruity and woody aroma mixtures.

Wine	Wine Type	Retronasal Aroma							
		Fruity Aroma				Woody Aroma			
		I max	T max	T ext	AUC	I max	T max	T ext	AUC
Red	CRW	13.1 a	**2.5 a**	100.5a	27,967 a	13.3 a	2.1 a	99.6 a	27,354 a
	GTRW	13.4 a	**1.8 b**	101.0 a	28,541 a	13.1 a	2.1 a	100.2 a	28,167 a
	MRW	13.0 a	**1.7 b**	100 a	2772 a	13.6 a	1.6 a	97.8 a	29,526 a
	ERW	12.1 a	**1.7 b**	100 a	26,411 a	13.4 a	1.7 a	100.7 a	28,336 a
White	CWW	13.1 a	**2.9 a**	90.5 a	25,537 a	12.9 a	2.9 a	102.3 a	**25,203 ab**
	GTWW	13.2 a	**2.5 ab**	98.6a	28,214 a	13.4 a	2.2 a	103.8 a	**29,886 a**
	MWW	13.2 a	**1.5 c**	101.2 a	26,889 a	11.93 a	2.9 a	9.0 a	**20,467 b**
	EWW	13.8 a	**1.8 bc**	99.3 a	27,939 a	12.0 a	3.1 a	91.9 a	**20,999 ab**

Different letters in bold (a–b) show significant differences among wine types from Tukey test ($p < 0.05$). Control wine (CW); gallotannin wine (GTW); mannoprotein wine (MW); ellagitannin wine (EW).

Additionally, the results in Table 6 also show that in white wines with a woody aroma, the addition of gallotannin increased the AUC (18.6% more than the control), which is related to the total aroma perceived, while on the contrary, the addition of mannoprotein reduced this parameter (18.8% less than the control).

4. Discussion

The main objective of this work was to evaluate the effect of three widely used commercial oenological additives of different natures, based on hydrolysable tannins (gallotannin and ellagitannin) or yeast mannoproteins, on the long-lasting flavour (astringency and retronasal aroma) perception (also called wine flavour persistence) of red and white wines. Besides this main objective, the study also focused on testing if individual PROP taste phenotype (or PROP taster status, PTS) might have an effect on flavour persistence. The results of this study show that PTS has a significant impact on astringency perception, mainly in red wines. In this sense, the analysis of the parameters extracted from the TI curves (Table 3) confirmed that taster individuals (individuals able to perceive the bitter compound 3-propylthiouracil, PROP) showed higher values of most TI parameters, such as Tmax, Text, and AUC. Tmax is related to the time necessary for reaching the maximum intensity of astringency (Tmax), while Text is related to the time required until the extinction of this sensation (Tex). Therefore, Text is directly related to the long-lasting perception of astringency. Furthermore, higher AUC values are related to a higher overall astringency perception. Therefore, the results from this work show that the persistence of astringency was higher in individuals classified as PROP tasters. For instance, in red wines, AUC values were significantly higher (about 21%) in PROP taster individuals compared to non-tasters (Figure 1). These results were, as indicated, more relevant in red wines than in white wines. In white wine, PROP taster individuals only exhibited the highest Tmax values of astringency in the fruity aromatised wine, but these differences were not noticed in the woody white wine (Table 3). Even PROP taster individuals showed lower Imax values when tasting this wine type compared to non-taster individuals. These differences could be linked to the higher astringency of red wines, which are generally richer in natural astringent phenolic compounds compared to white wines. Since astringency is lower in white wines, this sensation could have been more difficult to distinguish and to be rated by the panel (independently of the taste phenotype) in white wines compared to red wines.

Interestingly, previous works with wines have related the individual ability to detect the bitterness of PROP with a higher ability to detect other basic tastes (sweet, salty, acid) and trigeminal sensations such as astringency [35]. Nonetheless, pioneer works on the topic [43] did not show an association between PROP phenotypes and the ability to perceive astringency. This question has been somehow quite controversial in the scientific literature. While Pickering and co-workers did confirm this association [35], in more recent works, the same authors and others did not show a relationship between PROP phenotype and wine astringency perception [36–38]. Some of the reasons suggested to explain this lack of agreement were that astringency is a sensation that evolves over time, and therefore, differences among PROP taste phenotypes could be masked when astringency is only rated considering a single time point [36]. The results from the present work seem to confirm this hypothesis, and to the best of our knowledge, this is the first time that the difference in the ability to perceive wine astringency between PROP taster and non-taster individuals has been confirmed using a dynamic sensory approach, evaluating the development of this mouthful sensation from its onset until it is no longer perceived.

Additionally, some previous works also suggested that the higher sensory ability of PROP taster individuals could also include a higher retro-olfactive performance [33]. Results from the present study using two types of aroma mixtures representing congruent aroma notes of red (woody aroma) and white (fruity) wines do not confirm this hypothesis. Neither in white wines with woody or fruity aromas nor in red wines eliciting the above-mentioned aroma nuances were there significant differences in any of the TI parameters when considering PROP taster and non-taster individuals. Although the genetic and physiological mechanisms behind differences in sensitivity among taste PROP phenotypes are not sufficiently understood [34], some positive correlations such as a higher number of taste papillae and greater trigeminal innervation are often positively linked to PROP taster individuals [34,35]. These factors, however, seem to be more related to taste and trigeminal sensations than to olfactive inputs, which could explain the lack of an effect of PROP taste phenotype on retronasal aroma perception. On the contrary, the sensation of astringency involves the activation of mechanoreceptors in the oral cavity, which themselves are innervated by trigeminal fibres, and therefore, the effect of taste PROP phenotype could be much more relevant to taste or mouthfeel sensations than when considering olfactive stimulus [35].

A second objective of the work was to check if the use of typical oenological additives in wines such as oenotannins and mannoproteins might have an impact on the long-lasting flavour perception. Although these types of additives have long been used and represent a common technological practice currently available for winemakers, the scientific studies devoted to understanding their sensory impact are relatively scarce, and to the best of our knowledge, there is no previous study devoted to knowing their impact on flavour persistence.

The results from the present work show an impact of these additives on wine astringency in red and white wines, but only in those wines aromatised with a fruity aroma mixture and not in the case of the wines aromatised with a woody aroma (Table 5). This could be related to the existence of aroma–astringency interactions at a cognitive level, which can modify astringency perception [44,45]. For instance, in previous works [46], authors have shown that the fruity aroma extracted from a Chardonnay wine and added to a dearomatised red wine induced a lower astringency in the latter. However, more recent works [47] did not find an effect of different categories of odours on astringency perception, although they did not specifically evaluate the effect of a woody odour mixture. Another possible explanation is that the task of rating astringency in fruity wines could have been easier for the panel than in woody wines, considering that this odour is more associated with a mouthfeel sensation such as astringency.

In the case of fruity red and white wines, the addition of additives induced small but significant changes in the astringency perception. In general, although the overall astringency perception, intensity, and duration were not modified by these additives, other TI parameters such as Tmax were significantly affected. Nonetheless, the effect was different in red and white wines. For instance, in red fruity wines, all the additives provoked significantly ($p < 0.05$) lower Tmax values compared to the control (Table 5). This means that the time necessary to perceive the maximum astringency is shorter when adding these additives. The drop in Tmax was more pronounced for individuals belonging to the T phenotype (60% lower compared to the control) than in the individuals from the NT phenotype (34.6% lower compared to the control), supporting the idea that the higher sensitivity of taster individuals allowed them to detect the changes induced by the additives to a larger extent compared to the NT group. Nonetheless, in the case of white fruity wines, the additives significantly increased the astringency Imax compared to the control, but this was only observed in the taster phenotype, adding scientific support to the hypothesis that taster phenotypes might be more sensitive to the changes in some sensory stimuli (e.g., astringency) induced by this type of winemaking practice. Additionally, in the case of white fruity wines, the addition of mannoproteins induced the highest Imax values (Table 5). It is interesting to note that although some previous works have shown that commercial mannoprotein-rich yeast extracts reduce wine astringency [48,49], other works did not find any effect, even when they were added at very high concentrations to the wines (6 g/L) [50]. Nonetheless, it is worth noting that these previous works evaluated this mouthfeel sensation through descriptive sensory analysis at a single point after wine tasting, but not the astringency persistence like in the present work, which might lead to very different results. Additionally, as shown before, there are many types of mannoproteins that can vary in their glucidic and protein content, and, therefore, in their chemical and sensory properties [51], which might induce different effects in wines. Even the wine composition can have an influence on the role of mannoproteins in astringency and other wine sensory properties [19]. Although mannoproteins induced the highest Imax values for astringency in white fruity wines, a slightly lower but significant effect on Imax was also observed after the addition of the polyphenol-based additives (quertannin and gallotannin) (Table 3), which could be related to the increase in the phenolic content of the wine, which, in turn, can affect wine astringency, as recently shown [24].

Regarding the effect of the oenological additives on aroma persistence, even though a significant effect was observed, this effect was different depending on the wine type (red or white) but also depending on the aroma considered (fruity or woody). In this sense, the addition of additives in red fruity wines decreased Tmax (above 30%) compared to the control wine (Table 6). This means that the fruity intensity will be perceived earlier or faster in wines with additives. On the contrary, this effect was not observed when the panel considered the same wine but aromatised with a woody aroma. This agrees with the results related to the lack of an impact of additives on astringency in wines (red or whites) aromatised with a woody flavour, supporting the hypothesis of the existence of woody aroma–astringency interactions, which might affect the performance of the panel in the evaluation of either astringency or retronasal woody aroma. Additionally, other explanations, such as physicochemical interactions between the additives and the volatile compounds included in the aroma mixtures, cannot be discarded. The different chemical compositions of the fruity and woody aroma mixture might have induced different types of interactions (aroma–polyphenol, aroma–mannoprotein) with the additives, affecting aroma release, and, therefore, aroma perception [15,16,29,52,53].

In the case of white fruity wines, the effect of the additive or retronasal aroma was similar to that found for the fruity red wine, that is, a reduction in Tmax compared to the control (Table 6). Nonetheless, in this wine, the effect induced by the mannoprotein was significantly higher (48% lower Tmax compared to the control wine) than the effect produced by the polyphenol-type additives. Additionally, white fruity wines with mannoproteins also exhibited the highest Text values from the four tested wines (control wines and wines with additives), although these results were not statically significant (Table 6). Interestingly, in a previous work, an increase in fruity, floral, and balsamic aromas was found in Sangiovese wines aged for six months in contact with three different commercial mannoprotein-rich yeast extracts at a concentration of 20 g/hL [19]. Considering the large diversity of these types of additives, the results found in this work are interesting since, besides the quicker fruity perception that mannoproteins are able to induce, they also could improve and extend the fruity perception in white wines over time, which is an interesting technological feature associated with this type of additive, which will be necessary to investigate in the future.

On the other hand, besides the effect of the additives on the fruity aroma perception of white wines, they also had an effect on the woody aroma persistence. For instance, the addition of gallotannin in white woody wines significantly increased AUC compared to the control (Table 6), while the addition of mannoprotein produced the opposite effect, a decrease in AUC compared to the control. The addition of ellagitannin did not produce a significant effect on woody aroma persistence, which was very similar to that of the control wine. The addition of gallotannins could have incorporated some volatile compounds into the wine, contributing to an increased overall woody aroma. This effect has been recently shown when incorporating ellagitannin extracts into wines [24]. Additionally, the gallotannins could have an effect on wine astringency, which, as previously described, could induce taste–aroma interactions at a cognitive level and a higher perception of woody notes usually associated with astringent wines. This idea could be also valid to explain the reduction in woody aroma persistence (AUC) compared to the control induced by mannoproteins in the white woody wine (Table 6). It is known that yeast polysaccharides modify the aggregation between tannins and proteins, affecting astringency through two possible mechanisms: (i) competition between polysaccharides and salivary proteins towards tannins and (ii) polysaccharides forming a ternary complex, protein–polyphenol–polysaccharide, which enhances solubility in an aqueous medium [51]. As previously explained, the reduction in this mouthfeel sensation could be related to a lower perception of woody aroma, often associated with astringency. This seems a plausible explanation, since, as shown in Table 5, the lowest values of AUC (indicative of the total astringency perception) in white wines with a woody aroma were found when mannoproteins were added.

Besides an effect at the cognitive level, other explanations, involving physicochemical interactions between the additives and the aroma compounds included in the aroma mixture, could also contribute to explaining the opposite effect of gallotannin and mannoproteins on the woody persistence in white wine. In this sense, it has been previously shown that the addition of some kinds of polyphenolic extracts, especially flavan-3-ols, increased oral aroma persistence, therefore prolonging the release of certain volatile compounds by the tertiary interactions among oral mucosa, polyphenols, and aroma compounds [28]. These associations can form aroma reservoirs at the surface of the oral mucosa, ready to be released by the respiratory air flows several minutes after swallowing the wine [25,54]. More recently, it has been proven the existence of intermolecular interactions between tannic acid and mucin, which indeed affect the retention and release of aroma, suggesting that these types of polyphenols have the ability to delay aroma release, prolonging aroma perception [20]. Additionally, Pittari and co-workers [24] also show that hydrolysable tannins (ellagitannin) in non-oxidised wines increased the in-nose release of wine volatiles. However, in this work, the authors did not evaluate aroma persistence by sensory analysis.

The fact that the addition of ellagitannin did not induce a higher aroma persistence like gallotannin did shows the necessity of more studies comparing chemically different types of oenotannins in order to unravel the molecular action mechanisms which might help in the development of additives with better technological properties for improving aroma persistence.

Additionally, the existence of physicochemical interactions between mannoproteins and aroma compounds can also explain the reduction in aroma persistence that these yeast polysaccharides produced in white woody wines. In fact, previous studies showed hydrophobic interactions between mannoproteins and aroma compounds, mainly of hydrophobic nature, able to reduce aroma release [15,16]. Considering that woody aroma is composed of some relatively higher hydrophobic compounds (e.g., vainilline, whiskylactone) (Table 1), a large retention of these compounds on mannoproteins could be expected. Depending on the degree of interaction, oral aroma release could be slowed down compared to the control wine, translating into a lower overall aroma perception. Nonetheless, these results show the necessity of new in vivo analytical and sensory studies including these types of additives from different sources and with different chemical properties in order to confirm their role in aroma persistence and their usefulness to improve wine aroma persistence.

5. Conclusions

Results from this study using time–intensity methodology confirm the significant effect of individual PROP taste phenotype on the long-lasting astringency perception after wine consumption. PROP taster individuals (those able to perceive the bitter compound 3-propylthiouracil, PROP) showed higher values for most TI parameters (Tmax, Text, and AUC) compared to non-taster individuals. The effect of taste phenotype was more evident in red wines than in white wines. Nonetheless, PROP phenotype did not affect the long-lasting aroma perception of the tested aromas (fruity and woody) in either red or white wines. Additionally, the results showed that common oenological additives of different natures, such as hydrolysable tannins (gallotannin and ellagitannin) and yeast polysaccharides (mannoproteins) added at recommended concentrations to red and white wines have an impact on flavour persistence, affecting astringency and retronasal aroma perception. This effect depends on the wine type (red or white) and the type of aroma considered. For instance, they affect the astringency perception of red and white wines, but only in those wines with a fruity flavour. In red wines, they reduce the time to reach the maximum astringency intensity (Tmax), while in white wines, they induce a higher maximum intensity. In the case of aroma persistence, in red and white fruity wines, the addition of these additives allows for reaching the maximum intensity of fruitiness sooner. This was especially relevant in the case of mannoproteins in white fruity wines. On the contrary, none of the additives have a significant effect on the woody aroma in red wines, while in white wines, the addition of gallotannin significantly increased the global woody aroma perception (AUC) compared to the control. Overall, the wide diversity of chemical structures and properties of oenotannins and mannoproteins offers a great opportunity to use them as oenological tools to improve and/or modulate the aromatic persistence of wine. However, further in vivo analytical and sensory studies will be necessary to select the ones that provide the best results.

Supplementary Materials: The following supporting information can be downloaded at: https://www.mdpi.com/article/10.3390/foods12152835/s1, Table S1: Chemical compositions of the control wines without oenological additives (average values); Table S2: Chemical compositions of the oenological additives used in this study.

Author Contributions: R.I.V.-M.: Investigation, Methodology, Formal analysis, Data curation, Writing—original draft; C.C.: Investigation, Methodology, Formal analysis, Data curation, Writing—original draft; C.M.-G.: Investigation, Supervision, Data curation, Writing—review and editing; J.C.: Methodology, Winemaking, Wine analysis, Writing—review and editing; M.Á.P.-B.: Conceptualisation, Funding acquisition, Supervision, Data curation, Writing—review and editing. All authors have read and agreed to the published version of the manuscript.

Funding: This research was funded by the Spanish MCIN through PID 2019-111734RB-I00.

Data Availability Statement: The data presented in this study are available on request from the corresponding author.

Acknowledgments: The authors thank the volunteers for their participation in this study. Juan Mariano Cabellos and Teresa Arroyo are gratefully acknowledged for their technical assistance during the winemaking steps. The authors also thank Laffort Ibérica for donating the oenological additives for this study.

Conflicts of Interest: The authors declare no conflict of interest.

References

1. Bautista-Ortin, A.B.; Martinez-Cutillas, A.; Ros-Garcia, J.M.; Lopez-Roca, J.M.; Gomez-Plaza, E. Improving colour extraction and stability in red wines: The use of maceration enzymes and enological tannins. *Int. J. Food Sci. Technol.* **2005**, *40*, 867–878. [CrossRef]
2. Harbertson, J.F.; Parpinello, G.P.; Heymann, H.; Downey, M.O. Impact of exogenous tannin additions on wine chemistry and wine sensory character. *Food Chem.* **2012**, *131*, 999–1008. [CrossRef]
3. Larcher, R.; Tonidandel, L.; Villegas, T.R.; Nardin, T.; Fedrizzi, B.; Nicolini, G. Pre-fermentation addition of grape tannin increases the varietal thiols content in wine. *Food Chem.* **2015**, *166*, 56–61. [CrossRef]
4. Chen, K.; Escott, C.; Loira, I.; Del Fresno, J.M.; Morata, A.; Tesfaye, W.; Calderon, F.; Benito, S.; Suárez-Lepe, J.A. The Effects of Pre-Fermentative Addition of Oenological Tannins on Wine Components and Sensorial Qualities of Red Wine. *Molecules* **2016**, *21*, 1445. [CrossRef]
5. Li, L.; Li, Z.; Wei, Z.; Yu, W.; Cui, Y. Effect of tannin addition on chromatic characteristics, sensory qualities and antioxidant activities of red wines. *RSC Adv.* **2020**, *10*, 7108–7117. [CrossRef]
6. Corona, O.; Bambina, P.; De Filippi, D.; Cinquanta, L. Influence of pre-fermentative addition of aqueous solution tannins extracted from oak wood (Quercus petraea) on the composition of Grillo wines. *Eur. Food Res. Technol.* **2021**, *247*, 1595–1608. [CrossRef]
7. Ma, W.; Guo, A.; Zhang, Y.; Wang, H.; Liu, Y.; Li, H. A review on astringency and bitterness perception of tannins in wine. *Trends Food Sci. Technol.* **2014**, *40*, 6–19. [CrossRef]
8. Paissoni, M.A.; Bitelli, G.; Vilanova, M.; Montanini, C.; Segade, S.R.; Rolle, L.; Giacosa, S. Relative impact of oenological tannins in model solutions and red wine according to phenolic, antioxidant, and sensory traits. *Food Res. Int.* **2022**, *157*, 111203. [CrossRef]
9. Souquet, J.-M.; Cheynier, V.; Brossaud, F.; Moutounet, M. Polymeric proanthocyanidins from grape skins. *Phytochemistry* **1996**, *43*, 509–512. [CrossRef]
10. Hagerman, A.E. Extraction of Phenolics from Plants, Sephadex LH 20 and Separation of Tannin from Non-Tannin Phenolics. In *The Tannin Handbook*; Miami University: Oxford, OH, USA, 2011.
11. Pozo-Bayón, M.; Andújar-Ortiz, I.; Moreno-Arribas, M.V. Volatile profile and potential of inactive dry yeast-based winemaking additives to modify the volatile composition of wines. *J. Sci. Food Agric.* **2009**, *89*, 1665–1673. [CrossRef]
12. Rigou, P.; Mekoue, J.; Sieczkowski, N.; Doco, T.; Vernhet, A. Impact of industrial yeast derivative products on the modification of wine aroma compounds and sensorial profile. A review. *Food Chem.* **2021**, *358*, 129760. [CrossRef] [PubMed]
13. Andújar-Ortiz, I.; Chaya, C.; Martín-Álvarez, P.J.; Moreno-Arribas, M.V.; Pozo-Bayón, M. Impact of Using New Commercial Glutathione Enriched Inactive Dry Yeast Oenological Preparations on the Aroma and Sensory Properties of Wines. *Int. J. Food Prop.* **2014**, *17*, 987–1001. [CrossRef]
14. Feuillat, M.; Charpentier, C. Autolysis of Yeasts in Champagne. *Am. J. Enol. Vitic.* **1982**, *33*, 6–13. [CrossRef]
15. Lubbers, S.; Voilley, A.; Feuillat, M.; Charpentier, C. Influence of Mannaproteins from Yeast on the Aroma Intensity of a Model Wine. *LWT* **1994**, *27*, 108–114. [CrossRef]

16. Lubbers, S.; Charpentier, C.; Feuillat, M.; Voilley, A. Influence of Yeast Walls on the Behavior of Aroma Compounds in a Model Wine. *Am. J. Enol. Vitic.* **1994**, *45*, 29–33. [CrossRef]
17. Pozo-Bayón, M.; Andújar-Ortiz, I.; Moreno-Arribas, M.V. Scientific evidences beyond the application of inactive dry yeast preparations in winemaking. *Food Res. Int.* **2009**, *42*, 754–761. [CrossRef]
18. Rodríguez-Bencomo, J.J.; Andújar-Ortiz, I.; Moreno-Arribas, M.V.; Simó, C.; González, J.; Chana, A.; Dávalos, J.; Pozo-Bayón, M. Impact of Glutathione-Enriched Inactive Dry Yeast Preparations on the Stability of Terpenes during Model Wine Aging. *J. Agric. Food Chem.* **2014**, *62*, 1373–1383. [CrossRef]
19. Rinaldi, A.; Gonzalez, A.; Moio, L.; Gambuti, A. Commercial Mannoproteins Improve the Mouthfeel and Colour of Wines Obtained by Excessive Tannin Extraction. *Molecules* **2021**, *26*, 4133. [CrossRef]
20. Ginsburg, I.; Koren, E.; Shalish, M.; Kanner, J.; Kohen, R. Saliva increases the availability of lipophilic polyphenols as antioxidants and enhances their retention in the oral cavity. *Arch. Oral Biol.* **2012**, *57*, 1327–1334. [CrossRef]
21. Li, Y.; Gao, Z.; Guo, J.; Wang, J.; Yang, X. Modulating aroma release of flavour oil emulsion based on mucoadhesive property of tannic acid. *Food Chem.* **2022**, *388*, 132970. [CrossRef]
22. Esteban-Fernández, A.; Muñoz-González, C.; Jiménez-Girón, A.; Pérez-Jiménez, M.; Pozo-Bayón, M. Aroma release in the oral cavity after wine intake is influenced by wine matrix composition. *Food Chem.* **2018**, *243*, 125–133. [CrossRef]
23. Esteban-Fernández, A.; Rocha-Alcubilla, N.; Muñoz-González, C.; Moreno-Arribas, M.V.; Pozo-Bayón, M. Intra-oral adsorption and release of aroma compounds following in-mouth wine exposure. *Food Chem.* **2016**, *205*, 280–288. [CrossRef] [PubMed]
24. Pittari, E.; Piombino, P.; Andriot, I.; Cheynier, V.; Cordelle, S.; Feron, G.; Gourrat, K.; Le Quéré, J.-L.; Meudec, E.; Moio, L.; et al. Effects of oenological tannins on aroma release and perception of oxidized and non-oxidized red wine: A dynamic real-time in-vivo study coupling sensory evaluation and analytical chemistry. *Food Chem.* **2021**, *372*, 131229. [CrossRef] [PubMed]
25. Muñoz-González, C.; Criado, C.; Pérez-Jiménez, M.; Pozo-Bayón, M. Evaluation of the Effect of a Grape Seed Tannin Extract on Wine Ester Release and Perception Using In Vitro and In Vivo Instrumental and Sensory Approaches. *Foods* **2021**, *10*, 93. [CrossRef] [PubMed]
26. Manjón, E.; Brás, N.F.; García-Estévez, I.; Escribano-Bailon, M.T. Cell Wall Mannoproteins from Yeast Affect Salivary Protein–Flavanol Interactions through Different Molecular Mechanisms. *J. Agric. Food Chem.* **2020**, *68*, 13459–13468. [CrossRef]
27. Manjón, E.; Recio-Torrado, A.; Ramos-Pineda, A.M.; García-Estévez, I.; Escribano-Bailón, M.T. Effect of different yeast mannoproteins on the interaction between wine flavanols and salivary proteins. *Food Res. Int.* **2021**, *143*, 110279. [CrossRef]
28. Ramos-Pineda, A.M.; Manjón, E.; Macías, R.I.R.; García-Estévez, I.; Escribano-Bailón, M.T. Role of Yeast Mannoproteins in the Interaction between Salivary Proteins and Flavan-3-ols in a Cell-Based Model of the Oral Epithelium. *J. Agric. Food Chem.* **2022**, *70*, 13027–13035. [CrossRef]
29. Dufour, C.; Bayonove, C.L. Interactions between Wine Polyphenols and Aroma Substances. An Insight at the Molecular Level. *J. Agric. Food Chem.* **1999**, *47*, 678–684. [CrossRef]
30. Muñoz-González, C.; Brule, M.; Martin, C.; Feron, G.; Canon, F. Molecular mechanisms of aroma persistence: From noncovalent interactions between aroma compounds and the oral mucosa to metabolization of aroma compounds by saliva and oral cells. *Food Chem.* **2021**, *373*, 131467. [CrossRef]
31. Francis, I.; Williamson, P. Application of consumer sensory science in wine research. *Aust. J. Grape Wine Res.* **2015**, *21*, 554–567. [CrossRef]
32. Yang, Q.; Hollowood, T.; Hort, J. Phenotypic variation in oronasal perception and the relative effects of PROP and Thermal Taster Status. *Food Qual. Preference* **2014**, *38*, 83–91. [CrossRef]
33. Hayes, J.E.; Keast, R.S.J. Two decades of supertasting: Where do we stand? *Physiol. Behav.* **2011**, *104*, 1072–1074. [CrossRef]
34. Robino, A.; Concas, M.P.; Spinelli, S.; Pierguidi, L.; Tepper, B.J.; Gasparini, P.; Prescott, J.; Monteleone, E.; Toschi, T.G.; Torri, L.; et al. Combined influence of TAS2R38 genotype and PROP phenotype on the intensity of basic tastes, astringency and pungency in the Italian taste project. Food Qual. *Preference* **2021**, *95*, 104361. [CrossRef]
35. Pickering, G.J.; Simunkova, K.; DiBattista, D. Intensity of taste and astringency sensations elicited by red wines is associated with sensitivity to PROP (6-n-propylthiouracil). Food Qual. *Preference* **2004**, *15*, 147–154. [CrossRef]
36. Pickering, G.J.; Robert, G. Perception of mouthfeel sensations elicited by red wine are associated with sensitivity to 6-n-propylthiouracil. *J. Sens. Stud.* **2006**, *21*, 249–265. [CrossRef]
37. Pickering, G.J.; Moyes, A.; Bajec, M.R.; DeCourville, N. Thermal taster status associates with oral sensations elicited by wine. *Aust. J. Grape Wine Res.* **2010**, *16*, 361–367. [CrossRef]
38. Vitorino, G.; Mota, M.; Malfeito-Ferreira, M. Characterization of sensory perceptions elicited by white wine spiked with different aroma, taste and mouth-feel active molecules. *Ciência E Técnica Vitivinícola* **2021**, *36*, 139–150. [CrossRef]
39. Bartoshuk, L.M.; Duffy, V.B.; Green, B.G.; Hoffman, H.J.; Ko, C.-W.; Lucchina, L.A.; Marks, L.E.; Snyder, D.J.; Weiffenbach, J.M. Valid across-group comparisons with labeled scales: The gLMS versus magnitude matching. *Physiol. Behav.* **2004**, *82*, 109–114. [CrossRef]
40. Criado, C.; Chaya, C.; Fernández-Ruíz, V.; Álvarez, M.D.; Herranz, B.; Pozo-Bayón, M. Effect of saliva composition and flow on inter-individual differences in the temporal perception of retronasal aroma during wine tasting. *Food Res. Int.* **2019**, *126*, 108677. [CrossRef]
41. Lawless, H.T.; Corrigan, C.J.; Lee, C.B. Interactions of astringent substances. *Chem. Senses* **1994**, *19*, 141–154. [CrossRef]

42. Ayya, N.; Lawless, H.T. Quantitative and qualitative evaluation of high-intensity sweeteners and sweetener mixtures. *Chem. Senses* **1992**, *17*, 245–259. [CrossRef]
43. Lshikawa, T.; Noble, A.C. Temporal perception of astringency and sweetnessin red wine. *Food Qual. Prefer.* **1995**, *6*, 27–33. [CrossRef]
44. Ferrer-Gallego, R.; Hernández-Hierro, J.M.; Rivas-Gonzalo, J.C.; Escribano-Bailón, M.T. Sensory evaluation of bitterness and astringency sub-qualities of wine phenolic compounds: Synergistic effect and modulation by aromas. *Food Res. Int.* **2014**, *62*, 1100–1107. [CrossRef]
45. Sáenz-Navajas, M.-P.; Campo, E.; Fernández-Zurbano, P.; Valentin, D.; Ferreira, V. An assessment of the effects of wine volatiles on the perception of taste and astringency in wine. *Food Chem.* **2010**, *121*, 1139–1149. [CrossRef]
46. Sáenz-Navajas, M.-P.; Campo, E.; Avizcuri, J.M.; Valentin, D.; Fernández-Zurbano, P.; Ferreira, V. Contribution of non-volatile and aroma fractions to in-mouth sensory properties of red wines: Wine reconstitution strategies and sensory sorting task. *Anal. Chim. Acta* **2012**, *732*, 64–72. [CrossRef] [PubMed]
47. De-La-Fuente-Blanco, A.; Fernández-Zurbano, P.; Valentin, D.; Ferreira, V.; Sáenz-Navajas, M.-P. Cross-modal interactions and effects of the level of expertise on the perception of bitterness and astringency of red wines. Food Qual. *Preference* **2017**, *62*, 155–161. [CrossRef]
48. Rinaldi, A.; Coppola, M.; Moio, L. Aging of Aglianico and Sangiovese wine on mannoproteins: Effect on astringency and colour. *LWT* **2019**, *105*, 233–241. [CrossRef]
49. Wang, S.; Wang, X.; Zhao, P.; Ma, Z.; Zhao, Q.; Cao, X.; Cheng, C.; Liu, H.; Du, G. Mannoproteins interfering wine astringency by modulating the reaction between phenolic fractions and protein in a model wine system. *LWT* **2021**, *152*, 112217. [CrossRef]
50. Li, S.; Bindon, K.; Bastian, S.; Wilkinson, K. Impact of Commercial Oenotannin and Mannoprotein Products on the Chemical and Sensory Properties of Shiraz Wines Made from Sequentially Harvested Fruit. *Foods* **2018**, *7*, 204. [CrossRef] [PubMed]
51. Snyman, C.; Nguela, J.M.; Sieczkowski, N.; Marangon, M.; Divol, B. Optimised Extraction and Preliminary Characterisation of Mannoproteins from Non-Saccharomyces Wine Yeasts. *Foods* **2021**, *10*, 924. [CrossRef]
52. Perez-Jiménez, M.; Chaya, C.; Pozo-Bayón, M. Individual differences and effect of phenolic compounds in the immediate and prolonged in-mouth aroma release and retronasal aroma intensity during wine tasting. *Food Chem.* **2019**, *285*, 147–155. [CrossRef] [PubMed]
53. González-Muñoz, B.; Garrido-Vargas, F.; Pavez, C.; Osorio, F.; Chen, J.; Bordeu, E.; A O'Brien, J.; Brossard, N. Wine astringency: More than just tannin–protein interactions. *J. Sci. Food Agric.* **2021**, *102*, 1771–1781. [CrossRef] [PubMed]
54. Pérez-Jiménez, M.; Muñoz-González, C.; Pozo-Bayón, M.A. Oral Release Behavior of Wine Aroma Compounds by Using In-Mouth Headspace Sorptive Extraction (HSSE) Method. *Foods* **2021**, *10*, 415. [CrossRef] [PubMed]

Disclaimer/Publisher's Note: The statements, opinions and data contained in all publications are solely those of the individual author(s) and contributor(s) and not of MDPI and/or the editor(s). MDPI and/or the editor(s) disclaim responsibility for any injury to people or property resulting from any ideas, methods, instructions or products referred to in the content.

foods

MDPI

Article

Applying Different Vinification Techniques in Teran Red Wine Production: Impact on Bioactive Compounds and Sensory Attributes

Fumica Orbanić [1], Sara Rossi [1], Ena Bestulić [1], Irena Budić-Leto [2,*], Karin Kovačević Ganić [3], Ivana Horvat [1], Tomislav Plavša [1], Marijan Bubola [1], Igor Lukić [1], Ana Jeromel [4] and Sanja Radeka [1]

1 Institute of Agriculture and Tourism, Karla Huguesa 8, 52440 Poreč, Croatia; fumica@iptpo.hr (F.O.); sarar@iptpo.hr (S.R.); ena@iptpo.hr (E.B.); ihorvat@iptpo.hr (I.H.); tomislav@iptpo.hr (T.P.); marijan@iptpo.hr (M.B.); igor@iptpo.hr (I.L.); sanja@iptpo.hr (S.R.)
2 Institute for Adriatic Crops and Karst Reclamation, Put Duilova 11, 21000 Split, Croatia
3 Faculty of Food Technology and Biotechnology, University of Zagreb, Pierottijeva 6, 10000 Zagreb, Croatia; kkova@pbf.hr
4 Department of Viticulture and Enology, Faculty of Agriculture, University of Zagreb, Svetošimunska Cesta 25, 10000 Zagreb, Croatia; amajdak@agr.hr
* Correspondence: irena.budic-leto@krs.hr; Tel.: +385-21-434-420

Abstract: Six different vinification treatments, including a control treatment (7-day standard maceration) (K7), were performed to study the effects of non-standard techniques on bioactive compounds and sensory attributes of Teran red wine. Pre-fermentative mash cooling (8 °C; 48 h) and heating (50 °C; 48 h) followed by prolonged post-fermentative maceration of 13 days (C15;H15) or 28 days (C30;H30) were applied. In another treatment, after cooling, saignée was performed followed by 13-day prolonged maceration (CS15). Wine phenols and vitamins were analyzed by HPLC-DAD-FLD, minerals by ICP-OES, and sensory analysis was performed using the QDA and 100-point O.I.V./U.I.O.E. methods. Obtained results showed total phenolic concentration was the highest in the H30 treatment. The concentration of anthocyanins, flavan-3-ols and phenolic acids was significantly higher in wines of all vinification techniques compared to the control. Stilbene content was highly affected by pre-fermentative heating. Treatments CS15, H15, C30 and H30 resulted in the highest scores by both the QDA and 100-point sensory methods. The obtained results suggest that advanced non-standard vinification techniques have a significant impact on Teran wine by enhancing its composition of bioactive compounds and improving its sensory profile, which gives it an additional market value. Furthermore, a comprehensive comparison of such techniques applied simultaneously in one study is of substantial importance for additional research in wine production.

Keywords: pre-fermentative mash cooling; pre-fermentative mash heating; saignée; prolonged post-fermentative maceration; bioactive compounds; Teran red wine; wine sensory analysis

1. Introduction

Wine is characterized by its diverse chemical compounds that have piqued the interest of researchers seeking to unravel its complex chemical composition. This composition encompasses a range of compounds, such as phenols, macro- and microelements, and vitamins, which have been associated with potential health benefits and sensory attributes. The majority of the advantages can be attributed to phenolic compounds, which are generally categorized as either flavonoids or non-flavonoids [1]. These compounds are associated with potent antioxidant activity, including radical scavenging capacity, inhibition of lipid peroxidation, metal ion chelating ability and reduction capacity [2] (i.e., antioxidant, anti-inflammatory, antitumor, antithrombotic, antiatherogenic, antimicrobial, and antiviral activity). Nonetheless, their impact on human health depends on the quantity consumed and their bioavailability [3,4]. In the study conducted by [4], which involved moderate wine consumption, positive outcomes were observed including a decrease in systolic and diastolic blood pressure, as well as in total cholesterol and LDL (bad cholesterol) levels. Conversely, an increase in HDL (good cholesterol) and a rise in happiness hormones such as serotonin and dopamine were also noted.

The sensory structure of wine is affected by a complex interplay of various factors and compounds, including sugars, acids, phenolic and volatile compounds. The phenolic content of wine exerts a significant influence on red wine quality, affecting its organoleptic properties [3], such as appearance, color, astringency, bitterness, and flavor [5,6] as well as its stability during subsequent oxidative processes (oxidation in red wines) [3].

Flavonoids represent important constituents within grapes, and their presence is essential to wine quality. They have nutritional and pharmacological properties and may contribute to the health benefits associated with moderate wine consumption, potentially playing a role in these positive effects [4,7]. Flavonoids are classified into several groups of compounds (anthocyanins, flavanols, flavonols, among others).

The non-flavonoid group of phenolics consists of hydroxybenzoic acids, hydroxycinnamic acids, volatile phenols, stilbenes and miscellaneous compounds [8], which are recognized for their ability to enhance and stabilize the color of red wines through intra- and intermolecular reactions. Furthermore, they contribute to the flavor profile of wine, particularly the volatile phenolic acids and certain compounds among them, such as resveratrol, exhibit potent biological activities [8].

Aside from phenols, macro- and microelements contribute to enhancing the nutritional value of wine. Additionally, moderate daily wine consumption can make a substantial contribution to fulfilling the elemental requirements essential for the human organism [9,10] and to performing essential functions such as fortifying bone structure, facilitating nerve signal transmission, participating in the biosynthesis of various hormones, and regulating cardiac rhythm [10,11]. The concentration of these elements in wines relies upon numerous factors, including the specific production location, grape cultivation conditions, and the processes employed during winemaking [4,12].

Vitamins are a group of complex organic compounds found in food, and they are crucial for maintaining regular metabolic processes. Vitamins must be supplied by the diet because they cannot be produced in adequate amounts by the human body, except for vitamin D and vitamin K [13]. Grapes are rich in various vitamins, with a significant concentration found in the grape skin, this is why red wines typically have higher vitamin levels compared to white wines [14]. Although present in relatively low concentrations compared to other nutrients, they contribute to the overall nutritional content and health aspects of wine [15]. Over the past few decades, a range of advanced non-standard winemaking techniques have emerged with the aim of enhancing the production of high-quality red wines. These practices not only seek to improve the wine's physicochemical stability and sensory attributes but also to enhance its bioactive profile [1]. These practices include cold pre-fermentative mash cooling, known as cold maceration, cryomaceration or cold soaking, which is progressively employed to enhance some important quality attributes of wines, such as color and aroma [16]. Cryomaceration involves maceration without the presence of alcohol for a period that allows the selective diffusion of certain hydro-soluble compounds from the grape. The temperatures and durations of this practice can vary significantly, typically ranging from 3 to 10 °C for a period of three to seven days [17–19].

On the other hand, pre-fermentative heating, known as hot pre-fermentative maceration, can be defined as the heating of whole or crushed grape clusters before alcoholic fermentation to obtain musts or wines with a deeper color [18]. Due to heating, the cell walls of the grape skins undergo breakdown [20,21]; this process entails water-soluble phenolic compounds being released from the cells [3]. The temperatures to which the must is raised during the pre-fermentative stage typically range from 40 to 80 °C, with the duration of maceration being dependent on temperature (between 12 and 24 h or more) [18,22].

Pre-fermentation juice runoff, which is also referred to as saignée, is the procedure where the juice is removed before fermentation, leading to an increased ratio of skin to juice. Since anthocyanins and tannins, which contribute to color and structure, are primarily found in grape skins and seeds, this procedure theoretically enhances their concentration in the final wine. However, if extraction were solely determined by solubility, the opposite effect might occur, as there would be less liquid available to dissolve these phenolic compounds [17].

The contact between grape solids and the must is influenced by the duration of the pre-fermentation phase and the length of the alcoholic fermentation process [23,24]. The ideal maceration duration depends on the desired type of wine, i.e., on the character of the future wine [25]. The prolonged contact between the grape skins and seeds and the must allows a higher extraction of polyphenolic compounds, especially catechins and proanthocyanidins [26], in comparison to the short maceration times leading to lower levels of tannin extraction [27]. Higher extraction of these compounds plays a crucial role in providing stability of the color by the formation of anthocyanin–tannin complexes [27]. These phenomena also have different effects on the organoleptic impression, such as the impression of the body, bitterness and astringency of the wine [22].

Teran (*Vitis vinifera* L.) is the most widespread red autochthonous variety in Istria [28], traditionally grown in the north Adriatic area, including the Croatian Istria viticultural subregion [29]. In the present day, numerous producers are striving to elevate the phenolic concentration in Teran wine, aiming to attain high-quality wines with optimized levels of natural antioxidants.

The aim of this study was to determine how different vinification techniques such as pre-fermentative mash cooling and heating, saignée and prolonged post-fermentative macerations over two periods influence wine's bioactive composition, thus potentially increasing the health benefits and altering its sensory attributes. Although several papers reported the effect of pre-fermentative mash techniques or prolonged post-fermentative macerations on red wine, to our knowledge, there is a lack of a comprehensive comparison of such techniques applied simultaneously and their effect on bioactive compounds' concentration and the sensory profile of red wine.

2. Materials and Methods

2.1. Chemicals and Reagents

Methanol, formic acid, water, acetonitrile (all HPLC-grade purity), sodium dihydrogen phosphate, disodium hydrogen phosphate, *trans*-caftaric acid, caffeic acid, syringic acid, quercetin hydrate, quercetin-3-glucoside *trans*-piceid, and vitamin standards were purchased from Sigma-Aldrich (St. Louis, MO, USA). Gallic acid, protocatechuic acid, *p*-coumaric acid, ferulic acid, and taxifolin were purchased from Fluka (Buchs, Switzerland). Quercetin-3-glucuronide, procyanidins, (+)-catechin, (−)-epicatechin, piceatannol, and resveratrol were purchased from Extrasyn-these (Genay, France); *p*-hydroxybenzoic acid and myricetin were purchased from Acros Organics (Geel, Belgium). The *cis*-isomer of caftaric acid was obtained by UV illumination of a methanol solution containing the *trans*-isomer for four hours [30]. Anthocyanins (monoglucoside chlorides) were purchased from Biosynth Carbosynth (Bratislava, Slovakia). Ultrapure water and all the reagents (60% HNO_3) and standards for analysis using inductively coupled plasma optical emission spectroscopy (ICP-OES) were obtained from Merck (Darmstadt, Germany).

2.2. Plant Material

The experiment was performed on the red grape variety Teran (*Vitis vinifera* L.), grafted on SO4 rootstock and cultivated in the experimental vineyard of the Institute of Agriculture and Tourism, located in Poreč, in the Istria winegrowing region of Croatia. The vineyard was planted in 2006 in chromic luvisol (Terra rossa) soil, on a westerly exposed slope with a 5% inclination. Vines were planted with a spacing of 0.8 m within the row and 2.5 m between rows (plant density of 5000 vines/ha) and trained to a single Guyot training system, pruned to one spur containing two buds and one cane containing eight buds. The vineyard was not irrigated. Manual removal of approximately two leaves per shoot in the fruit zone was performed in all treatments at the berry set. Mechanical shoot trimming at a canopy height of 135 cm was performed at the berry set (the first decade of June) and three weeks thereafter. As Teran is a high-yielding variety that requires cluster thinning in order to regulate the yield [31], 25% of the clusters were removed from the vines at the onset of veraison. The average yield at harvest was 1.8 kg per vine. Fertilization of the vineyard was regularly performed each year in autumn with 800 kg of pelleted organic manure and 300 kg of P:K 20:30 fertilizer. Regular treatments against major fungal diseases were performed during the season. At harvest, grapes were phytosanitary, healthy, and in good condition, with no symptoms of any disease.

The manual harvest was held in 2020 on 30 September, when the sugar content was measured at 18.9 °Brix, total acidity expressed as tartaric acid at 8.0 g/L, and pH 3.20.

2.3. Minivinification

Grapes were subjected to destemming and crushing procedures employing standard equipment in the experimental wine cellar "Minivinification" situated at the Institute of Agriculture and Tourism. The grape mash was submitted to the addition of 5 g/hL of potassium metabisulfite (AEB SPA, Brescia, Italy) and an equivalent quantity of Aromax, a product composed of L-Ascorbic acid and potassium metabisulfite (AEB SPA, Brescia, Italy). Subsequent to these operations, the red grape mash was homogenized and evenly distributed within the 220 L-stainless steel tanks where vinification took place, in accordance with the predetermined plan of the experiment, which encompasses a total of six vinification treatments performed in three replications (Table 1).

Table 1. Overview of the experiment: applied vinification techniques in Teran wine treatments.

<table>
<tr><th rowspan="2">Treatment</th><th rowspan="2" colspan="2">Pre-Fermentative Procedure</th><th colspan="3">Fermentation and Maceration</th><th rowspan="2">Pre-Fermentative Procedure + Maceration Duration</th></tr>
<tr><th>Vinification Technique—Maceration</th><th>Fermentation/ Maceration Temperature</th><th>Maceration Duration</th></tr>
<tr><td>K7</td><td colspan="2">/</td><td>Standard maceration</td><td rowspan="4">24 °C</td><td>7 days</td><td>/</td></tr>
<tr><td>CS15</td><td rowspan="2">Cooling at 8 °C, 48 h</td><td>Saignée</td><td rowspan="3">Fermentation/maceration + prolonged post-fermentative maceration</td><td>13 days</td><td>15 days</td></tr>
<tr><td>C15
C30</td><td></td><td>13 days
28 days</td><td>15 days
30 days</td></tr>
<tr><td>H15
H30</td><td colspan="2">Heating at 50 °C, 48 h</td><td>13 days
28 days</td><td>15 days
30 days</td></tr>
</table>

K—control; C—cooling; H—heating; S—saignée; 7, 13, 15, 28, and 30 days of maceration duration.

Additionally, a pectolytic enzyme Endozym Rouge (AEB SPA, Brescia, Italy) was introduced to the mashes of all treatments, at a concentration of 4 g/hL, except those treatments that were submitted to pre-fermentative mash heating, where the enzyme was added subsequent to the heating process. Two of six vinification treatments were submitted to pre-fermentative mash heating at 50 °C for 48 h, followed bysimultaneous fermentation/maceration at 24 °C and prolonged post-fermentative maceration for two periods: 13 days, from the start of fermentation/maceration until the end of the prolonged post-fermentative maceration, accounting for 15 days in total including the pre-fermentative procedure (H15), and 28 days accounting for 30 days in total including the pre-fermentative procedure (H30). Conversely, three vinification treatments were submitted to pre-fermentative mash cooling at 8 °C (cryomaceration) for 48 h, followed by simultaneous fermentation/maceration at 24 °C and prolonged post-fermentative maceration over two periods: 13 days (C15) and 28 days (C30), summing up to durations of 15 and 30 days, respectively. In one of the treatments with pre-fermentative mash cooling, the saignée procedure was performed before fermentation (CS15). A proportion (33%) of the total juice quantity was racked and the concentrated mash was subjected to simultaneous fermentation/maceration (24 °C) and prolonged 13-day maceration, accounting for 15 days in total. This experiment also included a control treatment (K7), with a standard 7-day maceration and a maceration/fermentation at a temperature of 24 °C. Following a pre-fermentative procedure, all grape mashes underwent inoculation with a selected dry yeast, *Saccharomyces cerevisiae* Fermol Mediterranee (AEB SPA, Brescia, Italy), at a concentration of 30 g/hL, rehydrated with Fermol Plus Starter (AEB SPA, Brescia, Italy) at 10 g/hL and chaptalized with 3 kg/hL of saccharose, which was conducted according to prescribed provisions by the Croatian wine law (NN 32/2019), Regulation (EU) no. 1308/2013 and the ordinance on wine production (NN 2/2005) which was valid at the time when investigation was performed [32–34]. To further enhance the fermentation process, a yeast supplement Fermol Plus H_2S Free (AEB SPA, Brescia, Italy) was incorporated into the mash. This supplementation occurred on the fourth day of fermentation, each time at a dosage of 10 g/hL. Over the course of the maceration period, the cap was manually punched down three times daily, employing a wine cap punch-down tool that was integrated with the tank. To monitor the progression of fermentation and ascertain its completion, the analysis of reducing sugars was carried out. The determination of the end of fermentation occurred when the reducing sugar levels were around 2 g/L or below. Consequently, the grape mash underwent sulphitation with potassium metabisulfite and Aromax (AEB SPA, Brescia, Italy) to attain a dosage of 20 mg/L of free SO_2. Upon completion of the maceration, the wines underwent pressing, facilitated by a closed-type pneumatic press with a capacity of 500 L (Letina Inox d.o.o., Čakovec, Croatia). The pressing procedure was conducted under pressure conditions of 3 × 0.3 bar and 1 × 0.5 bar. Subsequent to this process, the wine was transferred and stored within stainless steel tanks, and following a period of 12 days, the wine was racked to separate it from the lees. Additionally, the sulfur content was monitored and regulated to 20 mg/L free SO_2. During the following period, the wine underwent two additional racking processes, and sulfur levels were controlled. Six months after the harvest, the wine was bottled into 0.75 L bottles and was stored in the cellar conditions until sensory analysis. Furthermore, for the purposes of laboratory analysis, 6 months after aging in the bottle, the wine was frozen at −18 °C.

2.4. Standard Physico-Chemical Analysis

In accordance with the methods prescribed by the International Organization of Vine and Wine (O.I.V./U.I.O.E.) [35], the following standard physico-chemical parameters were analyzed: alcoholic strength by volume (vol. %), reducing sugars (g/L), total dry extract (g/L), total dry extract excluding reducing sugars (g/L), total acidity (g/L), volatile acidity (g/L), free and total SO_2 (mg/L) and pH.

2.5. Analysis of Phenolic Compounds

The analysis of phenolic compounds (anthocyanins, stilbenes, flavan-3-ols, flavonols, phenolic acids) was carried out by high-performance liquid chromatography (HPLC), through an Agilent Infinity 1260 system (Agilent Technologies, Palo Alto, CA, USA), equipped with a G1311B quaternary pump, a G1329B auto sampler, a G1316A column oven, and G4212B DAD and G7121B FLD detectors. Prior to the analysis, wine samples were first filtered through 0.45 µm PTFE filters, following which, 10 µL of the filtered samples were injected into the system [4]. The column oven temperature was set at 26 °C for most compounds [36–38], except in the case of anthocyanins where the temperature was maintained at 40 °C during analysis [35].

The separation of hydroxycinnamic and hydroxybenzoic acids was conducted based on a modified method outlined by [36]. This separation was executed on a reversed-phase column, Poroshell 120 EC-C18 column (150 × 4.6 mm i.d., particle size 2.7 µm, Agilent Technologies, Palo Alto, CA, USA), which was equipped with a guard column of the same type (Poroshell 120 EC-C18, 5 × 4.6 mm i.d., particle size 2.7 µm, Agilent). The method's specific conditions are detailed in [4]. For detection, UV–VIS wavelengths were set at

280 nm (for hydroxybenzoic acids) and 330 nm (for hydroxycinnamic acids). The spectral range for detection spanned from 200 to 600 nm.

The anthocyanins were separated following a modified version of the OIV method [35]. This analysis was carried out using the Poroshell 120 EC-C18 column, accompanied by the previously mentioned guard column from Agilent Technologies, Palo Alto, CA, USA. A more detailed procedure can be found in [4]. The recording of chromatograms was undertaken at a wavelength of 518 nm to capture the relevant analytical information.

The separation of flavan-3-ols was performed following the procedure in [38]. This separation process was conducted on a reversed-phase Zorbax column (250 × 4 mm i.d., particle size 5 μm, Agilent, Technologies, Palo Alto, CA, USA). The methodology's comprehensive description can be found in [28]. For the purpose of detection, a fluorescence detector (FLD) was utilized, with excitation set at 280 nm and emission at 320 nm, maintaining a medium fluorescence intensity.

Using the same column, an analysis of stilbenes was conducted following the method outlined in [37]. The procedural steps were as described in [28]. The chromatographic separations were tracked at a wavelength of 306 nm.

When available, identification was performed by retention times and/or UV–Vis spectra with those of pure standards. Quantification was carried out using standard calibration curves. Identification was accomplished by comparing retention times and/or UV–Vis spectra with those of authentic standards, when accessible. Quantification was carried out using the standard calibration curves. For *cis*-caftaric acid and *cis*-piceid compounds, semi-quantitative analysis was performed with calibration curves for *trans*-caftaric and *trans*-piceid. Acetyl and *p*-coumaroyl derivates of anthocyanins were identified by comparing retention times reported in the characteristic chromatogram in the OIV method [35] and quantified using a standard curve of corresponding anthocyanins. In the Supplementary Materials and data, in Table S1, the validation parameters retention time, retention time CV, detection wavelengths, calibration curve equitation, and coefficient of determination are reported. Total HPLC phenolic concentration was presented as the sum of all identified phenolic compounds determined by HPLC.

2.6. Analysis of Macro- and Microelements

Analysis of the macroelements (K, Ca, Mg, Na), as well as microelements (Al, Cu, Fe, Mn), was performed within the Laboratory for Technology and Analysis of Wine at the Faculty of Food Technology and Biotechnology in Zagreb using the Optima DV 2000 inductively coupled plasma-optical emission spectrometer (ICP-OES) (Perkin Elmer, Shelton, CT, USA) equipped with a Meinhard spray chamber, nebulizer, and peristaltic sample delivery system. The samples were introduced into the plasma under operational conditions as outlined in [39] and following the procedure previously described in [28]. Both calibration solutions and wine samples with ethanol removed, in accordance with the method proposed by [40], were subjected to analysis using 2% HNO_3. The identification of elements was achieved using ICP-OES and PerkinElmer's WinLab 1.35 software, while quantification was carried out via a direct calibration approach.

2.7. Analysis of Vitamins

Analysis of the vitamins in wine was performed in Laboratory for Technology and Analysis of Wine at the Faculty of Food Technology and Biotechnology in Zagreb including chromatographic analyses performed on an Agilent 1100 Series liquid chromatography system (Agilent Technologies, Waldbronn, Germany) with a DAD and a single quadrupole mass detector equipped with an electrospray ionization interface (G1946D). Separation of vitamins was performed on a Luna Phenomenex C18 (5 μm, 150 × 4.6 mm) column at room temperature following the protocol by [41] described in [4]. Identification of four vitamins, B1 (thiamine), B2 (riboflavin), B3 (niacin), and B6 (pyridoxine), was carried out by comparison with the retention times of authentic standards and their spectral properties, respectively. Identified compounds were quantified via the direct calibration method.

2.8. Sensory Analysis

For the purpose of a comprehensive evaluation of wine quality, twelve months after the bottling process, a sensory analysis was conducted. The wine assessment took place at the Institute of Agriculture and Tourism according to all requirements prescribed by ISO standards [42,43] as described in [44]. The sensory analysis was performed on 18 wine samples (6 wines × 3 replicates) by both quantitative descriptive analysis (QDA) and hedonic 100-point O.I.V./U.I.O.E. (Organisation Internationale de la Vigne et du Vin/Union Internationale des Oenologues) methods [45]. Complete wine assessment results were elaborated; however, due to the intent of this paper, only results derived from the evaluation of color and taste (QDA and O.I.V. 100-point method) were employed. Both analyses were carried out by the accredited sensory panel of the Institute of Agriculture and Tourism, in accordance with the methodology previously outlined in [28,44]. The sensory panel consists of seven expert wine tasters who have experience in the evaluation of Teran wine. In the beginning, they attuned their sensory evaluation criteria through the examination of two samples of Teran red wine. These panel members are affiliated with the Croatian Viticultural and Enological Society and have been certified and authorized by the Croatian Ministry of Agriculture for official commercial wine sensory analysis in placing wines on the Croatian market. Additionally, the sensory panel is accredited according to the ISO standard "General requirements for the competence of testing and calibration laboratories" (ISO 17025:2017,) [46] for organoleptic (sensory) testing of wines using the method prescribed by the ordinance on wine and fruit wine sensory testing "Official Gazette" N. N. 106/04 with all amendments concluding with N.N. 1/15 [47], which was valid at the time when investigation was performed.

Quantitative descriptive analysis (QDA) was used to thoroughly evaluate wine color and taste using a 10-point structured scale (0 = attribute not perceptible, 10 = attribute strongly perceptible) including the following descriptors sorted into groups: color (ruby red, granite red, and dark red); color reflection (red reflection, purple reflection, and brick red reflection); taste attributes (freshness, acidity, body, sweetness, viscosity, bitterness, astringency, tannin presence, tannin quality, aftertaste quality, and aftertaste intensity); varietal typicity (taste typicity and overall varietal typicity); and wine overall impression. The O.I.V./U.I.O.E. 100-point evaluation method proposed by OIV, 2009 [45] was used to evaluate the visual and taste category with an adequate number of points. The total maximum score that these two categories can achieve is 59 points.

2.9. Statistical Data Analysis

The experiment was conducted in triplicate, and the subsequent data analysis was based on the mean values. Statistical assessment was undertaken through a one-way analysis of variance (ANOVA), followed by Fisher's least significance difference (LSD) test for comparison of mean values ($p \leq 0.05$). In the statistical analysis of the standard physico-chemical parameters, phenolic compounds, vitamins, macro- and microelements, and sensory attributes of the Teran wine, the factor was the vinification technique, i.e., treatment (control treatment and five treatments submitted to different vinification techniques). For enhanced data visualization, the dataset underwent principal component analysis (PCA). Statistics were performed using Statistica 10.0. software (Sta-Soft Inc., Tulsa, OK, USA). Pearson's correlation was used to investigate the relationship between several parameters. Results were presented using Pearson's correlation coefficient®.

3. Results and Discussion

3.1. Standard Physico-Chemical Parameters

All results of the standard physico-chemical analysis were within the prescribed limits for red wine by Croatian wine law, NN 32/2019, Regulation (EU) No 1308/2013, and Commission Regulation (EC) No 606/2009 [32,33,48] (Table 2).

Table 2. Parameters of standard physico-chemical analysis in Teran red wine.

Parameters	Treatments					
	K7	CS15	C15	H15	C30	H30
Alcohol (vol. %)	12.08 ± 0.03 [c]	11.62 ± 0.01 [e]	12.33 ± 0.03 [a]	12.03 ± 0.02 [c]	12.22 ± 0.04 [b]	11.94 ± 0.07 [d]
Total dry extract (g/L)	23.0 ± 0.15 [d]	24.6 ± 0.35 [c]	25.3 ± 0.4 [a]	25.3 ± 0.12 [ab]	25.6 ± 0.10 [a]	24.8 ± 0.30 [bc]
Reducing sugars (g/L)	1.3 ± 0.10 [b]	1.5 ± 0.06 [b]	1.5 ± 0.01 [b]	1.2 ± 0.15 [b]	2.6 ± 0.78 [a]	1.4 ± 0.10 [b]
Extract without reducing sugars (g/L)	20.7 ± 0.25 [d]	22.1 ± 0.36 [bc]	22.8 ± 0.40 [ab]	23.1 ± 0.10 [ab]	22.0 ± 0.78 [c]	22.4 ± 0.40 [abc]
Ash (g/L)	2.53 ± 0.03 [d]	2.71 ± 0.02 [c]	2.77 ± 0.05 [c]	2.94 ± 0.07 [a]	2.76 ± 0.01 [c]	2.85 ± 0.01 [b]
pH	3.17 ± 0.03 [e]	3.33 ± 0.01 [c]	3.31 ± 0.01 [d]	3.41 ± 0.01 [b]	3.34 ± 0.01 [c]	3.44 ± 0.01 [a]
Total acidity * (g/L)	8.6 ± 0.25 [a]	6.4 ± 0.06 [c]	6.2 ± 0.01 [c]	5.6 ± 0.01 [d]	6.6 ± 0.01 [b]	5.6 ± 0.01 [d]
Volatile acidity ** (g/L)	0.26 ± 0.02 [c]	0.38 ± 0.04 [b]	0.37 ± 0.05 [b]	0.30 ± 0.03 [c]	0.46 ± 0.01 [a]	0.40 ± 0.01 [b]
Free SO_2 (mg/L)	11 ±1 [a]	10 ± 1 [a]	10 ± 2 [a]	11 ± 2 [a]	9 ± 1 [b]	10 ± 2 [a]
Total SO_2 (mg/L)	80 ± 2 [a]	79 ± 3 [a]	78 ± 3 [a]	77 ± 3 [a]	78 ± 4 [a]	79 ± 3 [a]

* expressed as tartaric acid; ** expressed as acetic acid. Each value is the mean ± standard deviation, n = 3. Lower-case letters in superscript represent significant differences at $p \leq 0.05$ level (LSD test). Control treatment (K7), 48 h pre-fermentative mash cooling (8 °C) followed by prolonged post-fermentative maceration of 13 days (C15), 28 days (C30), saignée technique followed by prolonged post-fermentative maceration of 13 days (CS15), and 48 h heating (50 °C) followed by prolonged post-fermentative maceration of 13 (H15) and 28 days (H30).

The obtained wines differed significantly in alcohol content, with the C15 treatment displaying the statistically strongest alcoholic strength, similar to investigations by [49], and the lowest in the CS15 treatments, respectively. When observing treatments submitted to a particular pre-fermentative mash procedure, cooling or heating and two periods of maceration duration, it was noted that a 30-day prolonged post-fermentative maceration caused the decrease in alcoholic strength in comparison to 15-day prolonged post-fermentative maceration. It is possible that oxidation of ethanol occurred as reported in [50], which may partly explain the reduction in ethanol content observed in treatments subjected to prolonged post-fermentative maceration.

Total wine acidity is one of the most important factors in the determination of wine quality [51]. In this study, the significantly highest content of total acidity was found in the control wine (K7), 8.6 g/L, which negatively affected the sensory impression of acidity, which was conducted by sensory analysis of wine. On the other hand, all applied vinification techniques (pre-fermentative mash heating, cooling, saignée and prolonged post-fermentative macerations) strongly contribute to the decrease in total acidity in Teran red wine, especially pre-fermentative mash heating. This is of great importance because it is generally known that wines of the Teran variety are characterized by high to very high total acidity [28]. Therefore, such a result favorably reflected upon the sensory properties of those wines, thus enhancing the impression of acidity, being perceived as more harmonious, and subsequently increasing the overall impression of the wine. The decrease in the total acidity in treatments submitted to pre-fermentative cooling (CS15, C15 and C30) could be due to prolonged maceration, where the liberation of potassium from grape skins over time probably resulted in partial salification of tartaric acid [22,28]. Regarding results in pre-fermentative heating treatments (H15 and H30), several other authors noted that hot pre-fermentative maceration extracts higher levels of tartaric and malic acids, which decrease during fermentation to reach values similar to those in traditional winemaking due to increased cation extraction and acid reduction as outlined in [18,52]. Results of total acidity were partly in correspondence with pH; the control wine (K7) achieved the significantly lowest pH value, while other applied procedures affected pH value notably. Volatile acidity ranged from 0.26 g/L in control wine (K7) to 0.46 g/L in (C30). It is important to emphasize that volatile acid values in all treatments are notably below the upper limit for red wines (20 milliequivalents per liter; 1.2 g/L) prescribed by [32,53]. Dry extract is composed of the fixed constituents within wine, encompassing mineral compounds, organic acids, phenolic substances, and reducing sugars, that are responsible for defining the wine's body and mouthfeel [54]. The analysis results for total dry extract and extract without reducing sugars showed that control wine (K7) exhibited the significantly lowest content, indicating that every non-standard vinification technique applied significantly affected the extract in the wine. Similar results were obtained in terms of ash content in the wine; the control treatment (K7) had the significantly lowest ash content compared to the other treatments.

3.2. Phenolic Compounds

Numerous research efforts have been dedicated to increasing the concentration of phenolic compounds in red wine. While the primary objectives aimed to improve physico-chemical stability and sensory attributes, it is worth noting that they also have the potential to enhance the bioactive profile of the wine [1].

The concentration of individual phenolic compounds analyzed by HPLC is reported in Table 3. The concentration of total anthocyanins was the significantly lowest in the control treatment (K7), 33.76 mg/L, in comparison to concentrations in other treatments where impact of pre-fermentative mash heating, cooling, saignée and prolonged post-fermentative maceration over two periods was evident, ranging from 38.40, in treatment submitted to pre-fermentative mash cooling and 15-day (C15) to 41.50 mg/L pre-fermentative mash heating and 30-day maceration (H30). These results were in accordance with wine color evaluated by QDA sensory analysis, where it was found that the ruby red color showed decreased intensity in the control treatment (K7). Moreover, this could be correlated with the correlation coefficient, $r = 0.82$, which displayed a strong relationship. Among the treatments that underwent both pre-fermentative mash cooling and maceration over two periods (15 and 30 days), the duration of maceration had a more pronounced impact on the extraction of anthocyanins compared to the pre-fermentative procedure itself. These findings align with the research conducted by [27], which indicates that extended maceration generally results in a higher concentration of anthocyanins in wine. However, it is worth noting that certain researchers have not observed a direct correlation between maceration duration and anthocyanin content. This lack of correlation could potentially be attributed to the binding of anthocyanins to solid components and their conversion into colorless forms during the process [27]. Moreover, in treatments involving pre-fermentative heating and prolonged maceration (also over two periods, 15 and 30 days), the impact of maceration duration on anthocyanin content was not evident, H15 and H30 treatments were statistically equal. In [55], it was noted that monoglucosides (delphinidin, petunidin, peonidin and malvidin), which constitute the majority of the total anthocyanin content in the Babić variety, were not affected by the duration of maceration. According to these findings, it is possible that the extraction dynamic of anthocyanins during maceration and post-fermentative maceration in wines subjected to pre-fermentative cooling is slightly different than in wines subjected to pre-fermentative heating. The most represented anthocyanin compound detected in Teran wine was malvidin-3-*O*-glucoside, as was found by [56], with variations in concentration levels among treatments similar to those observed in total anthocyanins. In this study, malvidin-3-*O*-glucoside makes up 70% of the total anthocyanin concentration in Teran wine which, according to the literature, generally accounts for 60–80% of total anthocyanins [57].

Table 3. Concentration of anthocyanins, phenolic acids, flavonols, flavan-3-ols, stilbenes and total phenolic compounds in Teran red wines (mg/L).

Phenolic Compounds	Treatments					
	K7	CS15	C15	H15	C30	H30
Anthocyanins						
Delphinidin-3-*O*-glucoside	1.47 ± 0.06 [d]	1.65 ± 0.12 [c]	1.47 ± 0.10 [d]	1.95 ± 0.04 [b]	1.61 ± 0.06 [c]	2.08 ± 0.02 [a]
Cyanidin-3-*O*-glucoside	0.20 ± 0.01 [d]	0.26 ± 0.04 [bc]	0.30 ± 0.02 [a]	0.28 ± 0.01 [ab]	0.23 ± 0.01 [cd]	0.26 ± 0.01 [ab]
Petunidin-3-*O*-glucoside	1.62 ± 0.09 [d]	1.78 ± 0.05 [c]	1.76 ± 0.03 [c]	2.3 ± 0.05 [a]	1.77 ± 0.04 [c]	2.04 ± 0.01 [b]
Peonidin-3-*O*-glucoside	1.51 ± 0.07 [a]	1.92 ± 0.07 [a]	1.58 ± 0.05 [a]	1.64 ± 0.01 [a]	1.35 ± 0.03 [a]	1.57 ± 0.08 [a]
Malvidin-3-*O*-glucoside	23.34 ± 0.95 [d]	27.54 ± 0.67 [c]	27.09 ± 0.93 [c]	27.77 ± 0.74 [bc]	29.13 ± 0.54 [a]	29.04 ± 0.26 [ab]
Peonidin-3-*O*-acetylglucoside	1.64 ± 0.12 [c]	2.58 ± 0.10 [b]	2.65 ± 0.03 [b]	2.52 ± 0.05 [b]	2.88 ± 0.17 [a]	2.97 ± 0.03 [a]
Malvidin-3-*O*-acetylglucoside	0.50 ± 0.01 [b]	0.58 ± 0.02 [a]	0.47 ± 0.01 [cd]	0.49 ± 0.01 [bc]	0.46 ± 0.01 [d]	0.48 ± 0.01 [bcd]
Peonidin-3-*O*-cumarylglucoside	0.29 ± 0.03 [b]	0.35 ± 0.03 [a]	0.22 ± 0.01 [c]	0.26 ± 0.01 [b]	0.32 ± 0.01 [a]	0.26 ± 0.01 [b]
Malvidin-3-*O*-cumarylglucoside	3.18 ± 0.12 [a]	2.92 ± 0.06 [b]	2.86 ± 0.16 [b]	2.93 ± 0.23 [b]	2.84 ± 0.10 [b]	2.79 ± 0.02 [b]
Total detected anthocyanins	**33.76 ± 1.34 [d]**	**39.58 ± 0.73 [bc]**	**38.4 ± 1.17 [c]**	**40.14 ± 1.07 [ab]**	**40.6 ± 0.60 [ab]**	**41.50 ± 0.32 [a]**
Phenolic acids						
Gallic acid	12.41 ± 0.3 [e]	35.98 ± 0.5 [c]	35.54 ± 0.84 [c]	33.56 ± 1.25 [d]	44.96 ± 0.69 [a]	43.47 ± 0.66 [b]
Protocatechuic acid	3.10 ± 0.09 [d]	4.37 ± 0.32 [ab]	4.22 ± 0.20 [ab]	4.11 ± 0.33 [b]	4.57 ± 0.14 [a]	4.60 ± 0.08 [a]
p-Hydroxybenzoic acid	0.43 ± 0.01 [cd]	0.68 ± 0.07 [b]	0.89 ± 0.03 [a]	0.38 ± 0.04 [d]	0.63 ± 0.03 [b]	0.46 ± 0.02 [c]
Syringic acid	2.46 ± 0.04 [d]	3.75 ± 0.05 [c]	3.74 ± 0.19 [c]	4.30 ± 0.06 [ab]	3.96 ± 0.16 [bc]	4.32 ± 0.42 [a]
Total detected hydroxybenzoic acids	**18.39 ± 0.31 [d]**	**44.78 ± 0.69 [b]**	**44.39 ± 0.98 [b]**	**42.35 ± 1.66 [c]**	**54.12 ± 0.94 [a]**	**52.85 ± 1.07 [a]**
cis-Caftaric acid	0.46 ± 0.01 [ab]	0.47 ± 0.02 [a]	0.48 ± 0.04 [a]	0.43 ± 0.01 [bc]	0.40 ± 0.02 [c]	0.45 ± 0.02 [ab]
trans-Caftaric acid	39.94 ± 1.66 [d]	46.95 ± 0.22 [c]	46.86 ± 0.62 [c]	67.18 ± 1.78 [a]	39.33 ± 0.39 [d]	60.55 ± 0.53 [b]
Caffeic acid	1.70 ± 0.04 [c]	2.39 ± 0.04 [a]	2.18 ± 0.05 [b]	1.63 ± 0.10 [c]	2.42 ± 0.08 [a]	1.46 ± 0.08 [d]
p-Coumaric acid	1.11 ± 0.01 [c]	0.88 ± 0.03 [c]	1.49 ± 0.06 [b]	0.48 ± 0.08 [d]	2.13 ± 0.07 [a]	0.51 ± 0.03 [d]
Ferulic acid	1.40 ± 0.02 [e]	2.06 ± 0.02 [c]	1.80 ± 0.03 [d]	4.15 ± 0.18 [a]	1.14 ± 0.11 [f]	2.68 ± 0.08 [b]
Total detected hydroxycinnamic acids	**44.60 ± 1.69 [d]**	**52.75 ± 0.22 [c]**	**52.80 ± 0.71 [c]**	**73.88 ± 1.98 [a]**	**45.41 ± 0.49 [d]**	**65.65 ± 0.60 [b]**
Flavonols						
Quercetin 3-glucoside + Quercetin 3-glucuronide	3.78 ± 0.29 [e]	6.46 ± 0.21 [c]	4.99 ± 0.17 [d]	13.21 ± 0.64 [a]	2.65 ± 0.10 [f]	8.74 ± 0.15 [b]
Myricetin	1.54 ± 0.34 [ab]	2.14 ± 0.59 [a]	1.52 ± 0.19 [ab]	1.32 ± 0.30 [b]	1.20 ± 0.26 [b]	1.08 ± 0.31 [b]
Quercetin	9.19 ± 0.48 [cd]	11.48 ± 1.47 [a]	10.06 ± 0.47 [bc]	11.01 ± 0.61 [ab]	8.33 ± 0.84 [d]	9.60 ± 0.37 [bcd]
Total detected flavonols	**14.52 ± 0.67 [d]**	**20.08 ± 1.97 [b]**	**16.57 ± 0.81 [c]**	**25.54 ± 1.13 [a]**	**12.18 ± 1.12 [e]**	**19.41 ± 0.52 [b]**

Table 3. *Cont.*

Phenolic Compounds	Treatments					
	K7	CS15	C15	H15	C30	H30
Flavan-3-ols						
Procyanidin B1	8.97 ± 0.35^{f}	22.1 ± 0.64^{d}	20.95 ± 0.60^{e}	27.46 ± 0.72^{b}	25.45 ± 0.24^{c}	30.36 ± 0.45^{a}
Procyanidin B3	2.21 ± 0.03^{e}	7.26 ± 0.17^{b}	5.58 ± 0.28^{d}	8.99 ± 0.22^{c}	8.83 ± 0.43^{c}	10.3 ± 0.61^{a}
(+)-Catechin	11.84 ± 0.45^{e}	27.06 ± 0.41^{d}	27.98 ± 0.51^{d}	33.56 ± 1.12^{c}	37.98 ± 0.70^{b}	41.07 ± 0.91^{a}
Procyanidin B2	5.11 ± 0.16^{f}	15.91 ± 0.59^{e}	16.77 ± 0.31^{d}	18.06 ± 0.43^{c}	21.45 ± 0.26^{b}	25.27 ± 0.29^{a}
(−)-Epicatechin	3.71 ± 0.09^{f}	11.47 ± 0.28^{e}	12.79 ± 0.38^{d}	14.33 ± 0.45^{c}	20.34 ± 0.27^{b}	21.04 ± 0.39^{a}
Procyanidin C1	0.90 ± 0.03^{f}	3.45 ± 0.04^{d}	3.14 ± 0.02^{e}	3.92 ± 0.01^{c}	4.56 ± 0.13^{b}	5.47 ± 0.03^{a}
Total detected flavan-3-ols	$\mathbf{32.74 \pm 1.01}^{e}$	$\mathbf{87.24 \pm 1.89}^{d}$	$\mathbf{87.20 \pm 2.03}^{d}$	$\mathbf{106.32 \pm 2.85}^{c}$	$\mathbf{118.61 \pm 1.62}^{b}$	$\mathbf{133.51 \pm 2.42}^{a}$
Stilbenes						
trans-Piceid	10.48 ± 0.17^{e}	12.94 ± 0.55^{c}	12.65 ± 0.46^{d}	19.51 ± 0.18^{a}	9.31 ± 0.47^{f}	16.68 ± 0.06^{b}
Piceatannol	0.60 ± 0.08^{bc}	0.58 ± 0.07^{c}	0.81 ± 0.05^{a}	0.68 ± 0.04^{b}	0.69 ± 0.02^{b}	0.47 ± 0.02^{d}
trans-Resveratrol	1.12 ± 0.03^{c}	1.83 ± 0.36^{ab}	1.46 ± 0.18^{bc}	2.20 ± 0.16^{a}	1.54 ± 0.35^{b}	2.00 ± 0.1^{a}
cis-Piceid	5.74 ± 0.12^{bc}	5.74 ± 2.52^{bc}	6.38 ± 0.16^{b}	9.10 ± 0.06^{a}	4.27 ± 0.01^{c}	8.22 ± 0.13^{a}
Total detected stilbenes	$\mathbf{17.95 \pm 0.25}^{d}$	$\mathbf{21.09 \pm 2.19}^{c}$	$\mathbf{21.30 \pm 0.81}^{c}$	$\mathbf{31.49 \pm 0.33}^{a}$	$\mathbf{15.82 \pm 0.50}^{e}$	$\mathbf{27.38 \pm 0.09}^{b}$
Total detected phenolic compounds	$\mathbf{163.23 \pm 4.13}^{e}$	$\mathbf{266.54 \pm 0.76}^{d}$	$\mathbf{261.48 \pm 5.56}^{d}$	$\mathbf{321.17 \pm 6.32}^{b}$	$\mathbf{287.87 \pm 3.45}^{c}$	$\mathbf{341.20 \pm 4.70}^{a}$

Each value is the mean ± standard deviation, $n = 3$. Lower-case letters in superscript represent significant differences at $p \leq 0.05$ level (LSD test). Control treatment (K7), 48 h pre-fermentative mash cooling (8 °C) followed by prolonged post-fermentative maceration of 13 days (C15), 28 days (C30), saignée technique followed by prolonged post-fermentative maceration of 13 days (CS15), and 48 h heating (50 °C) followed by prolonged post-fermentative maceration of 13 (H15) and 28 days (H30).

Flavan-3-ols, mainly (+)-catechin and (−)-epicatechin, are also extracted from grape skin and seeds during winemaking. They can interact with anthocyanins through the copigmentation process [58]. Furthermore, they play a significant role in determining the sensory characteristics of red wine [3]. In the results of flavan-3-ols, the total content was the significantly highest in treatment submitted to pre-fermentative heating (H30), measuring 133.51 mg/L, which is 4.1-fold higher in comparison to the control wine (K7), measuring the lowest concentration of 32.74 mg/L. Equal results were obtained among all individual compounds detected, procyanidin B1, procyanidin B2, procyanidin B3, (+)-catechin, (−)-epicatechin and procyanidin C1, indicating that all procedures applied influenced the extraction of flavan-3-ols, but pre-fermentative heating and 30-day prolonged post-fermentative maceration (H30) the most. Additionally, it was observed that the concentration of flavan-3-ols increased along with the duration of maceration, among treatments submitted to pre-fermentative mash heating (H15 and H30), as well as pre-fermentative mash cooling (C15 and C30). Numerous studies have noted that enhancing the duration of skin contact and raising the ethanol concentration lead to improved extraction of flavan-3-ols. Furthermore, it is worth mentioning that these compounds continue to be extracted even after the maximum anthocyanin extraction has been achieved [58–60]. When comparing pre-fermentative mash cooling and heating, it was noted that concentrations of (+)-catechin, procyanidin B2, (−)-epicatechin, procyanidin C1, and their total concentration were higher within treatments that included pre-fermentative heating (H15 and H30). In [56], it was reported that concentrations of total flavan-3-ols were higher in wines obtained after pre-fermentative mash heating in comparison to cold pre-fermentative maceration. Several authors reported that maceration temperature correlates with the extractability of flavanols [56,61,62]. Procyanidin B1 and (+)-catechin were the most representative compounds, with concentrations ranging from 8.97 to 30.37 mg/L and 11.84 to 41.07 mg/L, respectively.

The total concentration of hydroxycinnamic acids (HC), which could act as copigments, varied from 44.60 mg/L in the control treatment (K7) to the significantly highest amount of 73.88 mg/L found in the H15 treatment. In general, the content of HC acids was higher in wine subjected to pre-fermentative treatment. In [56], it was reported that these compounds are located mostly in grape pulp and juice (caftaric and fertaric acid), as well as in skins (coutaric acid), and they tend to transfer more readily into the juice compared to other phenols [63]. Thus, intensive heating and stirring during thermal treatment may have been adequate to achieve elevated levels of these compounds. Additionally, regarding H15 treatment, submission to pre-fermentative mash heating exhibited a higher HC content compared to the treatment subjected to a 30-day prolonged maceration (H30). A similar observation was found among treatments employing pre-fermentative cooling, where the HC acid concentration was higher in C15 than in C30. Those results could be explained by the finding that HC acids, such as *p*-coumaric and caffeic acid, have also been reported as cofactors that enhance the color of red wine [64,65]. It might be possible that during 30-day prolonged maceration, copigmentation with anthocyanins occurred. Trans-caftaric acid emerged as the predominant HC acid, showing the significantly highest amount in H15 treatment. Similar results were observed in terms of ferulic acid content. On the other hand, concerning caffeic and *p*-coumaric acids, a statistically significant distinction was identified between treatments subjected to pre-fermentative mash cooling and heating, showing higher concentrations in cooling (CS15, C15, C30) compared to heating treatments (H15, H30).

Concerning total hydroxybenzoic acid (HB) content, which contributes to astringent perception, the control treatment showed the significantly lowest concentration, measuring 18.39 mg/L, in contrast to the range observed in other treatments, which extends up to 54.12 mg/L. Respectively, the highest content was significantly found in C30 and H30; both treatments were submitted to 30-day prolonged maceration, despite the pre-fermentative mash procedure. Such results were in agreement with the findings in the literature [28,66]. Gallic acid stood out as the most abundant HB acid, displaying an increase in concentration along with maceration duration when considering treatments submitted to pre-fermentative mash cooling or heating separately. Also, an increase in gallic acid concentration with length of maceration duration was reported by [67]. The increase in HB in prolonged post-fermentative macerations may be attributed to the presence of gallic acid, the most abundant compound, which is primarily located in grape seeds. Gallic acid exists in both free and esterified forms, as well as in complexed forms, which may necessitate a longer duration for extraction into juice and wine [56,61]. Concerning the other compounds, both protocatechuic and syringic acid concentrations were also significantly the lowest in the control treatment (K7), while *p*-hydroxybenzoic acid exhibited the most pronounced increase within the C15 treatment.

The major dietary sources of stilbenes for humans are grape berries, and wine [68]. Concerning the identified stilbenes in Teran wine, it was found that the pre-fermentative heating treatment exerted a remarkable impact on the extraction of the majority of individual stilbenes, as well as on the total stilbene content of the wine. On the other hand, in a similar investigation [56], no significant difference in concentrations of total stilbenes between treatments submitted to pre-fermentative cold maceration and pre-fermentative heating was found. In the present study, the significantly highest total concentration of stilbenes was found in the treatment submitted to pre-fermentative heating and 15-day maceration (H15), measuring 31.49 mg/L, while in the treatment subjected to 30-day maceration (H30), the total concentration was slightly lower. As reported by [69], this might be due to the rapid diffusion of stilbenes in wine reaching their maximum levels at 10–12 days of maceration. Regarding individual stilbenes, piceatannol, *trans*-resveratrol, *cis*-piceid, and *trans*-piceid were detected, with *trans*-piceid showing the most abundant content, ranging from 9.31 mg/L to the significantly highest intensities found in H15 treatment of 19.51 mg/L. *trans*- and *cis*-piceid are actually isomeric forms of resveratrol [70]. Additionally, concentrations determined in Teran wines were a few-fold higher than those obtained in macerated Merlot wine reported by [71]. In studies by [28,56], the *trans*-piceid compound was also detected as the most dominant in Teran red wine, with a high concentration. Furthermore, *trans*-resveratrol and *cis*-piceid concentrations were statistically the highest in treatments submitted to pre-fermentative mash heating (H15 and H30), despite the duration of maceration. Prolonged post-fermentative macerations had a notable impact on the extraction of those compounds in comparison to the control treatment (K7), as reported in [72]. Considering the results of the comprehensive analysis of stilbenes in the Teran wine, it can be concluded that this wine represents a rich source of these compounds, and therefore has beneficial effects.

In the context of flavonols, the total concentration was significantly highest in the H15 treatment, measuring 25.54 mg/L, followed by the concentrations found in the H30 and CS15 treatments where the statistical difference was not evident. According to the findings in [73], flavonols are typically extracted gradually during the initial 5 to 7 days of maceration under standard conditions. However, more significant extraction occurs after 8 or 9 days, or when the skin vacuoles are disrupted through specific treatments like flash-release or thermovinification. In [56], it was reported that significant amounts of flavonols were found in wines submitted to pre-fermentative cooling and heating, which included a non-alcoholic maceration phase, although these phenols are located in the grape skin cells and do not occur in pulp, and their glycosylated forms are more soluble in an alcoholic wine medium than in water. Among the individual flavonols detected, quercetin emerged as the most prevalent compound, with concentrations ranging from 8.33 mg/L to 11.48 mg/L. Remarkable results were found regarding both myricetin and quercetin content, where treatment submitted to the saignée procedure (CS15) displayed the highest level of these compounds. Contrarily, ref. [56] did not find that the saignée procedure notably affects concentrations of these individual compounds. From a gustative standpoint, quercetin derivatives have been generally linked with the perception of bitterness in red wines [73], which surely contributes to the moderate perception of bitterness in Teran wine.

The total concentration of phenols, i.e., the sum of individual compound concentrations, was significantly lowest in control wine (K7) in comparison to other treatments subjected to various above-mentioned vinification techniques, which notably affect the total extraction of phenols. Among them, the significantly highest content was found in the H30 treatment, where pre-fermentative mash heating and 30-day prolonged post-fermentative maceration were performed. It was reported that different maceration durations significantly influenced the chemical composition of the wine samples [74] and that heating the must above 40 °C increases the extraction of phenolic compounds from the grapes [18].

3.3. Macroelements and Microelements

The total content of microelements ranged from 3.09 to 6.37 mg/L, where it was found that the significantly highest were treatments submitted to pre-fermentative heating (H15 and H30), despite the duration of prolonged maceration (Table 4). Given that C15 and C30 obtained a lower microelement content, this suggests that pre-fermentative mash heating had a more significant impact on microelement extraction compared to pre-fermentative cooling.

Table 4. Concentrations of macro- and microelements in Teran red wine (mg/L).

Macroelements	Treatments					
	K7	CS15	C15	H15	C30	H30
K	822.50 ± 54.89 [a]	736.50 ± 43.02 [b]	785.33 ± 51.75 [a]	770.83 ± 50.51 [a]	794.00 ± 28.25 [a]	786 ± 28.51 [a]
Ca	129.50 ± 7.05 [b]	115.83 ± 6.53 [c]	127.33 ± 4.86 [b]	132.50 ± 6.56 [ab]	129.50 ± 4.00 [b]	140.33 ± 4.37 [a]
Mg	78.30 ± 1.00 [c]	78.75 ± 1.13 [c]	83.07 ± 1.35 [b]	83.42 ± 1.21 [ab]	85.22 ± 1.08 [a]	84.65 ± 1.08 [ab]
Na	8.27 ± 0.77 [a]	7.95 ± 0.71 [a]	7.51 ± 0.69 [a]	7.79 ± 0.74 [a]	7.62 ± 0.61 [a]	7.89 ± 0.53 [a]
Total macroelements	**1038.6 ± 63.59 [a]**	**939.03 ± 50.21 [b]**	**1003.3 ± 58.53 [ab]**	**994.54 ± 58.94 [ab]**	**1016.3 ± 32.17 [ab]**	**1018.9 ± 34.35 [ab]**
Microelements						
Al	0.38 ± 0.06 [b]	0.26 ± 0.03 [c]	0.32 ± 0.04 [bc]	0.65 ± 0.10 [a]	0.30 ± 0.02 [bc]	0.62 ± 0.06 [a]
Cu	0.03 ± 0.01 [a]	0.02 ± 0.01 [b]	0.01 ± 0.01 [e]	0.02 ± 0.01 [c]	0.02 ± 0.01 [d]	0.02 ± 0.01 [b]
Fe	2.49 ± 0.03 [e]	2.01 ± 0.03 [f]	2.88 ± 0.03 [d]	4.69 ± 0.02 [b]	2.98 ± 0.01 [c]	4.75 ± 0.02 [a]
Mn	0.97 ± 0.06 [a]	0.80 ± 0.07 [b]	0.90 ± 0.06 [ab]	0.98 ± 0.08 [a]	0.90 ± 0.07 [ab]	0.98 ± 0.08 [a]
Total microelements	**3.86 ± 0.12 [c]**	**3.09 ± 0.10 [d]**	**4.12 ± 0.08 [b]**	**6.33 ± 0.15 [a]**	**4.19 ± 0.08 [b]**	**6.37 ± 0.16 [a]**

Each value is the mean ± standard deviation, $n = 3$. Lower-case letters in superscript represent significant differences at $p \leq 0.05$ level (LSD test). Control treatment (K7), 48 h pre-fermentative mash cooling (8 °C) followed by prolonged post-fermentative maceration of 13 days (C15), 28 days (C30), saignée technique followed by prolonged post-fermentative maceration of 13 days (CS15), and 48 h heating (50 °C) followed by prolonged post-fermentative maceration of 13 (H15) and 28 days (H30).

In [28], the significantly highest total microelement content was also obtained in treatments submitted to pre-fermentative mash heating. It might be possible that the strong effect of heating on cell wall breakdown resulted in an increased extraction of microelements. The most abundant microelement in the investigated wine was iron (Fe), showing the highest concentration in the treatment which was equally affected by both pre-fermentative heating and prolonged post-fermentative maceration (H30). Moderate wine consumption contributes many essential metals, including iron, which is vital for almost all living organisms by participating in a wide variety of metabolic processes [75]. On the other hand, the obtained results showed that the total content of macroelements in the investigated wine ranged from 939.03 to 1038.57 mg/L. High levels of K, Ca, Cu, and Na can be associated with mineral levels in the soil, fertilization or fining, and clarifying substances added to wine [76]. It is important to note that all applied vinification techniques evenly influence the extraction of macroelements from grape berry cells. In treatment subjected to pre-fermentative cooling and the saignée procedure, a slight increase in macroelement content was evident in comparison to K7 treatment. This could be attributed to a higher proportion of solid parts in the must, on which elevated reabsorption of minerals could occur, or could be also due to assimilation by yeasts, precipitations, and their deposits [77].

3.4. Vitamins

The detected vitamins are part of the B-complex, which includes vitamin B1 (thiamine), vitamin B2 (riboflavin), vitamin B3 (niacin), and vitamin B6 (pyridoxine) (Figure 1).

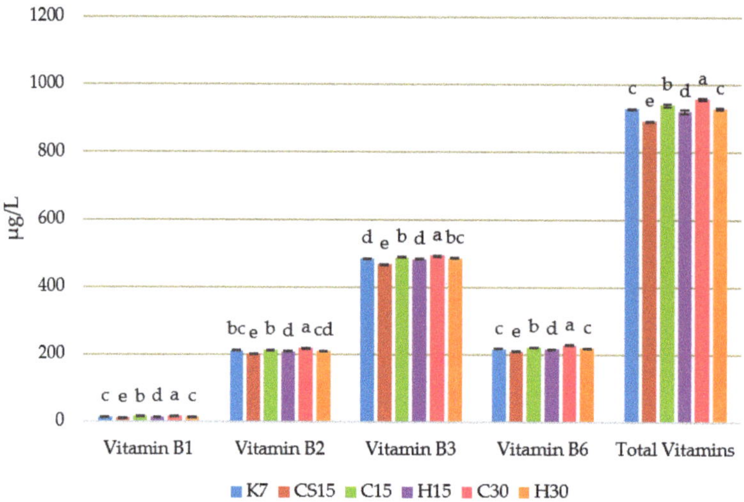

Figure 1. Concentration of B-complex vitamins in Teran red wine (μg/L). Each value is the mean ± standard deviation, $n = 3$. Lower-case letters represent significant differences at $p \leq 0.05$ level (LSD test). Control treatment (K7), 48 h pre-fermentative mash cooling (8 °C) followed by prolonged post-fermentative maceration of 13 days (C15), 28 days (C30), saignée technique followed by prolonged post-fermentative maceration of 13 days (CS15), and 48 h heating (50 °C) followed by prolonged post-fermentative maceration of 13 (H15) and 28 days (H30).

When considering the total vitamin content in the investigated Teran wine, the results revealed that the significantly highest amount was achieved in the treatment subjected to pre-fermentative mash cooling and a 30-day prolonged maceration (C30), followed by the concentration in C15 treatment, which also involved pre-fermentative cooling but had a shorter 15-day maceration period. It is noteworthy that treatments involving mash heating (H15 and H30) exhibited statistically lower vitamin content compared to treatments involving mash cooling (C15 and C30). This suggests that the extraction of vitamins was initially favored by pre-fermentative mash cooling and subsequently enhanced with a longer duration of maceration. It was noted that the stability of vitamins is often threatened due to various technological procedures during vinification, and especially due to temperature changes. The concentration of individual vitamins, vitamin B1, vitamin B2, vitamin B3, and vitamin B6 had an equal trend to the total sum of vitamins. Thus, C30 treatment had the most significantly elevated concentration of each vitamin, followed by concentrations found in C15 treatment. Vitamin contents of wine appear to be related to a long contact period between wines and lees after the completion of fermentation, thus allowing the exsorption of vitamin resources by yeasts into the liquid medium [78], as well as the transfer of vitamin contents from the solid part of the harvest to the wine [15,79]. Among these vitamins, the most abundant was vitamin B3, also known as niacin, with concentrations ranging from 467.80 μg/L to 493.43 μg/L.

3.5. Sensory Analysis

Wine is characterized by five major attributes that play a crucial role in creating consumer perception of wine quality, acceptability, and balance, including sweetness, acidity, tannin, alcohol content, and body. Additionally, factors such as color and mouthfeel also contribute significantly to the overall acceptability of wine [80]. Phenolic compounds impact wine body and mouthfeel [81], as they mostly elicit astringency and bitterness sensations [82]. In red winemaking, polyphenolic compounds, which contribute to wine's sensory properties, are extracted from the solid parts of grape barriers [21], and contact time between the skins and the must/wine is an important aspect during the transfer of polyphenols [1].

The sensory analysis, encompassing color and taste evaluation of Teran red wine obtained by six distinct vinification treatments, was conducted using quantitative descriptive analysis (QDA) and is presented in Figure 2.

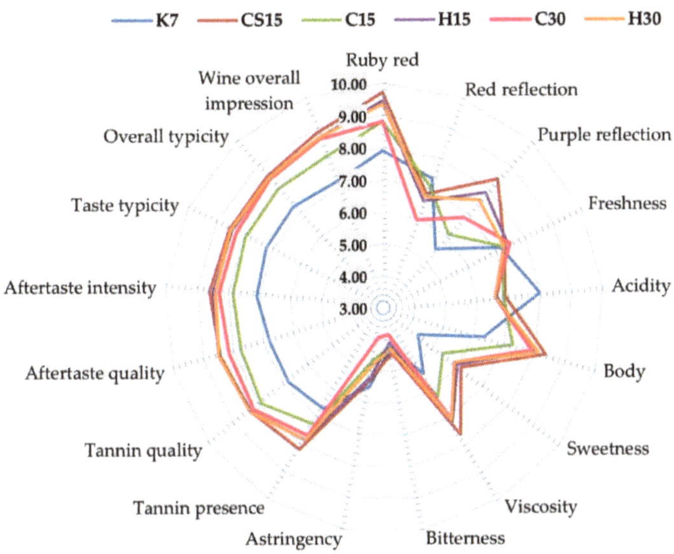

Figure 2. Perception of color and taste attributes intensity in Teran red wine obtained by QDA. Control treatment (K7), 48 h pre-fermentative mash cooling (8 °C) followed by prolonged post-fermentative maceration of 13 days (C15), 28 days (C30), saignée technique followed by prolonged post-fermentative maceration of 13 days (CS15), and 48 heating (50 °C) followed by prolonged post-fermentative maceration of 13 (H15) and 28 days (H30).

Anthocyanins are the main pigments responsible for the color of red grapes and red wines [22]. The color of the Teran red wine was characterized as ruby red, with intensity values spanning from 7.90 to 9.71 on a scale of ten points. Significantly, the strongest intensity values were achieved in wine, i.e., treatment where the saignée procedure (CS15) and pre-fermentative mash heating (H15 and H30), where the impact of maceration duration was not evident, were applied. This might be corroborated by the following findings in the literature; in [18], it was noted that the increased extraction of anthocyanins during hot pre-fermentative maceration resulted in wines with higher concentrations of these pigments and, consequently, greater color intensity. The wines produced with the saignée procedure were characterized by more polymeric pigments [17]. Furthermore, the remaining treatments, including pre-fermentative cooling (C15 and C30), were statistically the same regardless of maceration duration, and they displayed decreased color intensity compared with treatments submitted to mash heating (H15 and H30). These results could be attributed to the low temperature during the pre-fermentative stage, which results in lower solubilization of anthocyanins in the grape skins, causing a delay in their extraction, a phenomenon that has also been reported by other researchers [83,84]. The control wine (K7) demonstrated the significantly lowest intensity of ruby red color, which indicated that the strongest influence on ruby red wine color was a vinification procedure such as the saignée procedure or pre-fermentative heating, but also pre-fermentative cooling. The concentration of total anthocyanins is strongly correlated with the perception of ruby red wine color, displaying a correlation coefficient of r = 0.82. Moreover, the characterization of the wine color was enhanced by employing color reflection descriptors, and within the Teran red wine, the presence of red and purple reflections was observed. The strongest intensity of red reflection was obtained in the control wine (K7), but the lowest intensity of purple reflection. The highest intensity of the purple reflection was found in CS15, H15, C30 and H30, and it might be possible that those vinification procedures affected anthocyanin composition differently, thus expressing purple reflection more strongly. This could be corroborated by the findings of [85], who proposed that the higher color intensity observed in young wines produced after maceration may be attributed to improved condensation of anthocyanins and proanthocyanidins or catechins. This process can lead to the formation of new anthocyanin compounds that contribute to stabilizing the violet tinge of the wine.

The taste profile of Teran wines was evaluated using various taste descriptors, and the majority of these descriptors, including body, sweetness, viscosity, tannin presence, tannin quality, aftertaste quality, aftertaste intensity, taste typicity, overall varietal typicity, and overall impression of the wine exhibited the significantly lowest intensities in the control wine (K7). These results suggest that the majority of taste attributes were significantly influenced by the applied vinification procedures, pre-fermentative mash cooling, heating, saignée, and prolonged post-fermentative macerations, with each procedure almost equally contributing to the enhancement of the wine's taste. Regarding Teran wines, the significantly strongest intensities of the wine body, viscosity, and sweetness attributes were perceived in CS15, H15, C30 and H30 treatments, while correlation coefficient between wine body intensity and total concentration of phenols was r = 0.84, suggesting that wine structure was enriched with phenolic compounds. Additionally, a good indicator of wine structure, i.e., body, is total dry extract, whose correlation coefficient was r = 0.66.

The wine freshness descriptor showed no significant difference among treatments, but the perception of acidity (sourness) was the significantly strongest in the control wine (K7) which is in accordance with the total acidity parameter, which was also the highest in the control wine (K7). Hence, the analysis of overall acidity exhibited a strong correlation with the sensory evaluation results conducted by QDA, specifically with the perception of acidity, displaying a correlation coefficient of 0.94. Vinification techniques, pre-fermentative mash cooling and heating, saignée and the duration of maceration had a substantial impact on reducing the perception of acidity in wine, a factor of great importance in the case of Teran red wine, which is known for its pronounced sourness. In [28], a notable decrease in the perception of wine acidity was also reported in Teran wines subjected to prolonged 21-day maceration and pre-fermentative mash heating.

Tannins are widely recognized as the components in wines that contribute to the perception of bitterness and astringency, essentially comprising the textural elements of wine [86]. In this study of Teran wines, the perception of wine bitterness was statistically equal among all treatments, with scores ranging from 3.86 to 4.45 points, which fall within the range considered to represent a moderately strong level of bitterness. The perception of astringency in Teran wines was also noted to be at moderate intensities, with values from 3.95 to 5.50, contributing to pleasant astringency. In the case of certain beverages like red wine, bitterness is regarded as a necessary attribute, particularly when it is present with a moderate intensity [80].

In the evaluation of tannin perception in Teran wines, it was determined that in all treatments, a noticeable tannin presence was detected in comparison to the control wine. Intensity values ranged from 6.69 in the control wine (K7) to 8.17 in CS15. Furthermore, when tannin presence and concentration of total flavan-3-ols were correlated, a moderate to strong correlation was achieved, with a correlation coefficient of r = 0.73. Additionally, tannin quality, the desirable attribute of tannins, was also rated with high points, ranging from, 6.83, the significantly lowest in the control wine (K7) to 8.38 in CS15. Taste typicity and the overall varietal typicity showed higher values in all treatments, subjected to particular vinification procedures in comparison to control wine (K7), which showed that both pre-fermentative mash procedures, saignée and prolonged post-fermentative macerations, contribute positively to the varietal typicity of Teran wine.

The sensory evaluation conducted through QDA showed that wines derived from the treatments CS15, H15, C30, and H30 were notably ranked as superior in overall quality impression. These results are strongly correlated with the results of wine evaluation using the hedonic 100-point O.I.V./U.I.O.E. method (Figure 3), demonstrating a high correlation coefficient of r = 0.85. With the 100-point method, the sum of scores derived from both the visual and taste categories was considered; the cumulative maximal score that these two categories can achieve was 59 points. The control wine (K7) achieved the lowest score and was significantly different in comparison to other treatments. Conversely, the highest score was obtained in wines in which pre-fermentative mash cooling, heating, the saignée procedure and prolonged post-fermentative maceration were performed (CS15, H15 and H30). The following best rated wines were those statistically similar to H15 and H30, subjected to pre-fermentative mash cooling (C15 and C30). These results suggested that in comparison to the control standard vinification treatment (K), each applied vinification procedure positively affected to the overall score of the wines, implying that the use of the advanced non-standard vinification procedure has led to the production of high-quality Teran red wine.

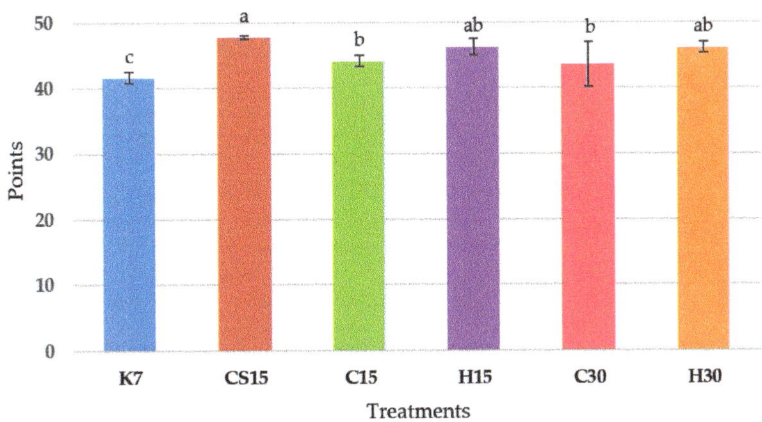

Figure 3. The total score including visual and taste category evaluated with hedonic 100-point O.I.V./U.I.O.E. method. The total maximal score that visual and taste categories can achieve is 59 points. Each value is the mean ± standard deviation, $n = 3$. Lower-case letters represent significant differences at $p \leq 0.05$ level (LSD test). Control treatment (K7), 48 h pre-fermentative mash cooling (8 °C) followed by prolonged post-fermentative maceration of 13 days (C15), 28 days (C30), saignée technique followed by prolonged post-fermentative maceration of 13 days (CS15), and 48 h heating (50 °C) followed by prolonged post-fermentative maceration of 13 (H15) and 28 days (H30).

3.6. Principal Component Analysis

The visualization of differences between treatments applied to Teran red wine and groups of determined bioactive compounds (total phenols, anthocyanins, flavan-3-ols, stilbenes, flavonols, hydroxycinnamic acids, hydroxybenzoic acids, macroelements, microelements and vitamins) was performed using unsupervised statistical analysis and PCA (Figure 4).

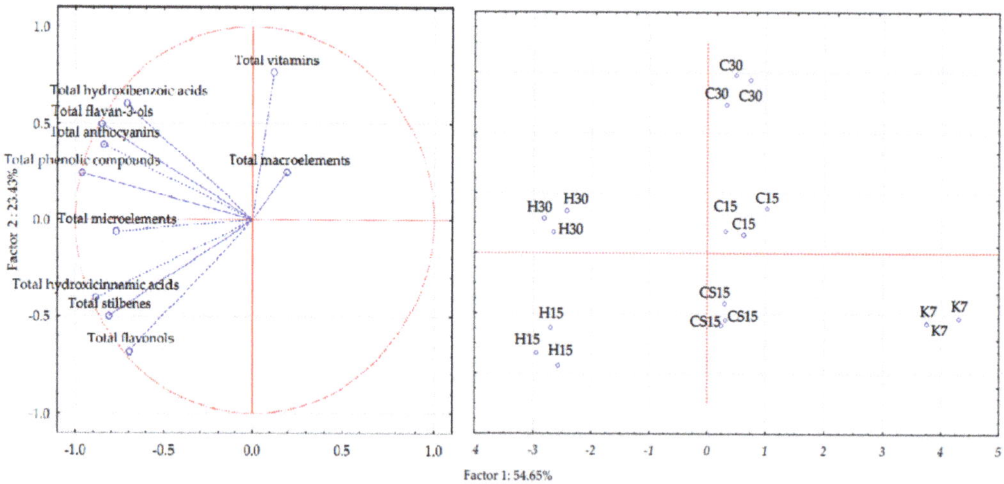

Figure 4. Separation of Teran red wines produced by different vinification techniques presented in three replications in two-dimensional space defined by the first two principal components (PC1 and PC2) and separation of bioactive compounds.

PC1 and PC2, the first two principal components, explained 78.08% of the total variance, thus enabling a good separation of wine treatments. The first principal component explained 54.65% of the variation, while the second principal component (PC2) explained 23.43% of the total variance. The control treatment (K7) was clearly separated from other treatments along the first and the second principal components, thus demonstrating the weakest correlation with the analyzed compound among all treatments. Such results are in accordance with the ones obtained by instrumental data analyses, given that K7 treatment wine showed the lowest concentration of all groups of bioactive compounds. According to the obtained plot, H30 treatment wine highly correlated with the majority of bioactive groups of compounds, such as total phenolic concentration, total anthocyanins, total flavan-3-ols, total hydroxycinnamic acids and total microelements. A similar trend was observed for H15 treatment, which was also placed on the left side of the Cartesian system, but correlated more with total stilbenes, total flavonols and total hydroxybenzoic acids. As for the other treatment wines, CS15 treatment wine and C15 gravitated towards the interception of the two axes, with the latter correlating highly with total macroelements. Among all treatments, C30 treatment correlated the most with total vitamins. The results are consistent with the data presented in Tables 3 and 4 and Figure 1.

4. Conclusions

The enhanced extraction and elevated content of bioactive compounds during winemaking processes offer opportunities to affect and improve wine quality and nutritional value, aligning with the growing interest of consumers in high-quality red wines and their potential health benefits. The obtained results showed that the concentrations of all the identified phenolic compounds in Teran red wine were significantly higher in wines of the treatments that included advanced non-standard techniques in relation to the control wine obtained by standard vinification techniques. Such results were in accordance with ones obtained from sensory analysis, since those treatments exhibited significantly higher results compared to the control wine and were ranked as superior using both QDA and O.I.V./U.I.O.E. methods. The obtained findings imply that the investigated advanced non-standard vinification techniques, such as prolonged post-fermentative maceration along with the application of mash cooling, heating, or saignée, could result in diverse wines with positively altered sensory profiles, and thus allow the production of superior quality Teran red wines. Moreover, the increased levels of bioactive compounds in these wines might have a pronounced impact on nutritional value and potential wine health benefits, when moderately consumed, which adds an additional value to such wines on the red wine market.

Supplementary Materials: The following supporting information can be downloaded at: https://www.mdpi.com/article/10.3390/foods12203838/s1, Table S1: Validation parameters: retention time, retention time CV, detection wavelengths, calibration curve equitation, coefficient of determination.

Author Contributions: Conceptualization, F.O. and S.R. (Sanja Radeka); Methodology, F.O. and S.R. (Sanja Radeka); Formal Analysis, I.L., K.K.G., F.O., I.H. and E.B.; Investigation, S.R. (Sanja Radeka), F.O. and T.P.; Data Curation, F.O., S.R. (Sanja Radeka) and I.B.-L.; Writing—Original Draft Preparation, S.R. (Sanja Radeka), F.O. and E.B.; Writing—Review and Editing, I.B.-L., S.R. (Sara Rossi), M.B., T.P., I.H., E.B., I.L. and A.J.; Visualization, S.R. (Sanja Radeka) and F.O.; Supervision S.R. (Sanja Radeka); Project Administration, S.R. (Sanja Radeka); Funding Acquisition, S.R. (Sanja Radeka). All authors have read and agreed to the published version of the manuscript.

Funding: Croatian Science Foundation, Research project "Influence of different vinification technologies on the qualitative characteristics of wines from Croatian autochthonous varieties: the role of wine in human diet"—VINUM SANUM (IP-2018-01-5049) and the project "Young Researchers' Career Development Project—Training New Doctoral Students" (DOK-2021-02-6937).

Data Availability Statement: The data used to support the findings of this study can be made available by the corresponding author upon request.

Acknowledgments: This work has been supported by the Croatian Science Foundation under the project "Influence of different vinification technologies on the qualitative characteristics of wines from Croatian autochthonous varieties: the role of wine in human diet"—VINUM SANUM (IP-2018-01-5049) and the project "Young Researchers' Career Development Project—Training New Doctoral Students" (DOK-2021-02-6937).

Conflicts of Interest: The authors declare no conflict of interest.

References

1. Vejarano, R.; Luján-Corro, M. Red Wine and Health: Approaches to Improve the Phenolic Content During Winemaking. *Front. Nutr.* **2022**, *9*, 890066. [CrossRef] [PubMed]
2. Ruskovska, T.; Budić-Leto, I.; Corral-Jara, K.F.; Ajdžanović, V.; Arola-Arnal, A.; Bravo, F.I.; Deligiannidou, G.-E.; Havlik, J.; Janeva, M.; Kistanova, E.; et al. Systematic Bioinformatic Analyses of Nutrigenomic Modifications by Polyphenols Associated with Cardiometabolic Health in Humans—Evidence from Targeted Nutrigenomic Studies. *Nutrients* **2021**, *13*, 2326. [CrossRef] [PubMed]
3. Gutiérrez-Escobar, R.; Aliaño-González, M.J.; Cantos-Villar, E. Wine Polyphenol Content and Its Influence on Wine Quality and Properties: A Review. *Molecules* **2021**, *26*, 718. [CrossRef]
4. Radeka, S.; Rossi, S.; Bestulić, E.; Budić-Leto, I.; Kovačević Ganić, K.; Horvat, I.; Lukić, I.; Orbanić, F.; Zaninović Jurjević, T.; Dvornik, Š. Bioactive Compounds and Antioxidant Activity of Red and White Wines Produced from Autochthonous Croatian Varieties: Effect of Moderate Consumption on Human Health. *Foods* **2022**, *11*, 1804. [CrossRef]
5. Alcalde-Eon, C.; García-Estévez, I.; Ferreras-Charro, R.; Rivas-Gonzalo, J.C.; Ferrer-Gallego, R.; Escribano-Bailón, M.T. Adding Oenological Tannin vs. Overripe Grapes: Effect on the Phenolic Composition of Red Wines. *J. Food Compos. Anal.* **2014**, *34*, 99–113. [CrossRef]
6. Avizcuri, J.-M.; Sáenz-Navajas, M.-P.; Echávarri, J.-F.; Ferreira, V.; Fernández-Zurbano, P. Evaluation of the Impact of Initial Red Wine Composition on Changes in Color and Anthocyanin Content during Bottle Storage. *Food Chem.* **2016**, *213*, 123–134. [CrossRef] [PubMed]
7. Cheynier, V. Flavonoids in Wine. In *Flavonoids*; Andersen, Ø., Markham, K., Eds.; CRC Press: Boca Raton, FL, USA, 2005; pp. 263–318, ISBN 978-0-8493-2021-7.
8. Rentzsch, M.; Wilkens, A.; Winterhalter, P. Non-Flavonoid Phenolic Compounds. In *Wine Chemistry and Biochemistry*; Moreno-Arribas, M.V., Polo, M.C., Eds.; Springer: New York, NY, USA, 2009; pp. 509–527, ISBN 978-0-387-74118-5.
9. Galgano, F.; Favati, F.; Caruso, M.; Scarpa, T.; Palma, A. Analysis of Trace Elements in Southern Italian Wines and Their Classification According to Provenance. *LWT Food Sci. Technol.* **2008**, *41*, 1808–1815. [CrossRef]
10. Vrček, I.V.; Bojić, M.; Žuntar, I.; Mendaš, G.; Medić-Šarić, M. Phenol Content, Antioxidant Activity and Metal Composition of Croatian Wines Deriving from Organically and Conventionally Grown Grapes. *Food Chem.* **2011**, *124*, 354–361. [CrossRef]
11. Gharibzahedi, S.M.T.; Jafari, S.M. The Importance of Minerals in Human Nutrition: Bioavailability, Food Fortification, Processing Effects and Nanoencapsulation. *Trends Food Sci. Technol.* **2017**, *62*, 119–132. [CrossRef]

12. Núñez, M.; Peña, R.M.; Herrero, C.; García-Martín, S. Analysis of Some Metals in Wine by Means of Capillary Electrophoresis. Application to the Differentiation of Ribeira Sacra Spanish Red Wines. *Analusis* **2000**, *28*, 432–437. [CrossRef]
13. Ball, G.F.M. *Vitamins: Their Role in the Human Body*; John Wiley & Sons: Hoboken, NJ, USA, 2008; ISBN 978-1-4051-4810-8.
14. Jackson, R.S. Wine, Food, and Health. In *Wine Science*; Elsevier: Amsterdam, The Netherlands, 2020; pp. 947–978, ISBN 978-0-12-816118-0.
15. Evers, M.S.; Roullier-Gall, C.; Morge, C.; Sparrow, C.; Gobert, A.; Alexandre, H. Vitamins in Wine: Which, What for, and How Much? *Compr. Rev. Food Sci. Food Saf.* **2021**, *20*, 2991–3035. [CrossRef] [PubMed]
16. Heredia, F.J.; Escudero-Gilete, M.L.; Hernanz, D.; Gordillo, B.; Meléndez-Martínez, A.J.; Vicario, I.M.; González-Miret, M.L. Influence of the Refrigeration Technique on the Colour and Phenolic Composition of Syrah Red Wines Obtained by Pre-Fermentative Cold Maceration. *Food Chem.* **2010**, *118*, 377–383. [CrossRef]
17. Sacchi, K.L.; Bisson, L.F.; Adams, D.O. A Review of the Effect of Winemaking Techniques on Phenolic Extraction in Red Wines. *Am. J. Enol. Vitic.* **2005**, *56*, 197–206. [CrossRef]
18. Piccardo, D.; González-Neves, G. Extracción de polifenoles y composición de vinos tintos Tannat elaborados por técnicas de maceración prefermentativa. *Agrociencia* **2013**, *17*, 36–44. [CrossRef]
19. de Beer, D.; Joubert, E.; Marais, J.; Manley, M. Maceration Before and During Fermentation: Effect on Pinotage Wine Phenolic Composition, Total Antioxidant Capacity and Objective Colour Parameters. *SAJEV* **2017**, *27*, 137–150. [CrossRef]
20. Girard, B.; Yuksel, D.; Cliff, M.A.; Delaquis, P.; Reynolds, A.G. Vinification Effects on the Sensory, Colour and GC Profiles of Pinot Noir Wines from British Columbia. *Food Res. Int.* **2001**, *34*, 483–499. [CrossRef]
21. Maza, M.; Álvarez, I.; Raso, J. Thermal and Non-Thermal Physical Methods for Improving Polyphenol Extraction in Red Winemaking. *Beverages* **2019**, *5*, 47. [CrossRef]
22. Ribéreau-Gayon, P.; Glories, Y.; Maujean, A.; Dubourdieu, D. (Eds.) *Handbook of Enology, Volume 2: The Chemistry of Wine—Stabilization and Treatments*, 2nd ed.; Wiley: Chichester, UK; Hoboken, NJ, USA, 2006; ISBN 978-0-470-01037-2.
23. Casassa, L.F.; Huff, R.; Steele, N.B. Chemical Consequences of Extended Maceration and Post-Fermentation Additions of Grape Pomace in Pinot Noir and Zinfandel Wines from the Central Coast of California (USA). *Food Chem.* **2019**, *300*, 125147. [CrossRef]
24. Jagatić Korenika, A.-M.; Kozina, B.; Preiner, D.; Tomaz, I.; Volarević, J.; Jeromel, A. The Effect of Seed Removal and Extraction Time on the Phenolic Profile of Plavac Mali Wine. *Appl. Sci.* **2023**, *13*, 5411. [CrossRef]
25. Scudamore-Smith, P.D.; Hooper, R.L.; McLaran, E.D. Color and Phenolic Changes of Cabernet Sauvignon Wine Made by Simultaneous Yeast/Bacterial Fermentation and Extended Pomace Contact. *Am. J. Enol. Vitic.* **1990**, *41*, 57–67. [CrossRef]
26. Ivanova, V.; Dörnyei, Á.; Márk, L.; Vojnoski, B.; Stafilov, T.; Stefova, M.; Kilár, F. Polyphenolic Content of Vranec Wines Produced by Different Vinification Conditions. *Food Chem.* **2011**, *124*, 316–325. [CrossRef]
27. Gómez-Plaza, E.; Gil-Muñoz, R.; López-Roca, J.M.; Martínez-Cutillas, A.; Fernández-Fernández, J.I. Maintenance of Colour Composition of a Red Wine During Storage. Influence of Prefermentative Practices, Maceration Time and Storage. *LWT Food Sci. Technol.* **2002**, *35*, 46–53. [CrossRef]
28. Rossi, S.; Bestulić, E.; Horvat, I.; Plavša, T.; Lukić, I.; Bubola, M.; Ganić, K.K.; Ćurko, N.; Jagatić Korenika, A.-M.; Radeka, S. Comparison of Different Winemaking Processes for Improvement of Phenolic Composition, Macro- and Microelemental Content, and Taste Sensory Attributes of Teran (*Vitis vinifera* L.) Red Wines. *LWT* **2022**, *154*, 112619. [CrossRef]
29. Rusjan, D.; Bubola, M.; Janjanin, D.; Užila, Z.; Radeka, S.; Poljuha, D.; Pelengić, R.; Javornik, D.; Štajner, N. Ampelographic Characterisation of Grapevine Accessions Denominated "Refošk", "Refosco", "Teran" and "Terrano" (*Vitis vinifera* L.) from Slovenia, Croatia and Italy. *VITIS J. Grapevine Res.* **2015**, *54*, 77–80. [CrossRef]
30. Vrhovšek, U. Extraction of Hydroxycinnamoyltartaric Acids from Berries of Different Grape Varieties. *J. Agric. Food Chem.* **1998**, *46*, 4203–4208. [CrossRef]
31. Bubola, M.; Sivilotti, P.; Janjanin, D.; Poni, S. Early Leaf Removal Has a Larger Effect than Cluster Thinning on Grape Phenolic Composition in Cv. Teran. *Am. J. Enol. Vitic.* **2017**, *68*, 234–242. [CrossRef]
32. Croatian Wine Law (NN 32/2019). Available online: https://narodne-novine.nn.hr/clanci/sluzbeni/2019_03_32_641.html (accessed on 16 August 2023).
33. Regulation (EU) No 1308/2013 of the European Parliament and of the Council of 17 December 2013 Establishing a Common Organisation of the Markets in Agricultural Products and Repealing Council Regulations (EEC) No 922/72, (EEC) No 234/79, (EC) No 1037/2001 and (EC) No 1234/2007. 2013. Available online: https://eur-lex.europa.eu/legal-content/EN/TXT/HTML/?uri=CELEX:32013R1308 (accessed on 16 July 2023).
34. Ordinance on Wine Production (NN 2/2005). Available online: https://narodne-novine.nn.hr/clanci/sluzbeni/2005_01_2_17.html (accessed on 8 October 2023).
35. International Organisation of Vine and Wine. *OIV Compendium of International Methods of Wine and Must Analysis*; International Organisation of Vine and Wine (OIV): Paris, France, 2018; Volume 1.
36. Pati, S.; Crupi, P.; Benucci, I.; Antonacci, D.; Di Luccia, A.; Esti, M. HPLC-DAD–MS/MS Characterization of Phenolic Compounds in White Wine Stored without Added Sulfite. *Food Res. Int.* **2014**, *66*, 207–215. [CrossRef]
37. Mark, L.; Nikfardjam, M.S.P.; Avar, P.; Ohmacht, R. A Validated HPLC Method for the Quantitative Analysis of Trans-Resveratrol and Trans-Piceid in Hungarian Wines. *J. Chromatogr. Sci.* **2005**, *43*, 445–449. [CrossRef]

38. Ćurko, N.; Kovačević Ganić, K.; Gracin, L.; Đapić, M.; Jourdes, M.; Teissedre, P.L. Characterization of Seed and Skin Polyphenolic Extracts of Two Red Grape Cultivars Grown in Croatia and Their Sensory Perception in a Wine Model Medium. *Food Chem.* **2014**, *145*, 15–22. [CrossRef]
39. Larcher, R.; Nicolini, G. Survey of 22 Mineral Elements in Wines from Trentino (Italy) Using ICP-OES. *Ital. J. Food Sci.* **2001**, *13*, 233–241.
40. Leder, R.; Kubanović, V.; Petric, I.V.; Vahčić, N.; Banović, M. Chemometric Prediction of the Geographical Origin of Croatian Wines through Their Elemental Profiles. *J. Food Nutr. Res.* **2015**, *54*, 229–238.
41. Trang, H. Development of HPLC Methods for the Determination of Water-Soluble Vitamins in Pharmaceuticals and Fortified Food Products. Master's Thesis, Clemson University, Clemson, SC, USA, 2013.
42. *ISO 3591:1977*; Sensory Analysis. Apparatus Wine-Tasting Glass. Group B; International Organization for Standardization: Geneva, Switzerland, 1997. Available online: https://www.iso.org/standard/9002.html (accessed on 18 August 2023).
43. *ISO 8589:2007*; Sensory Analysis—General Guidance for the Design of Test Rooms. International Organization for Standardization: Geneva, Switzerland, 2007. Available online: https://www.iso.org/cms/render/live/en/sites/isoorg/contents/data/standard/03/63/36385.html (accessed on 18 August 2023).
44. Radeka, S.; Bestulić, E.; Rossi, S.; Orbanić, F.; Bubola, M.; Plavša, T.; Lukić, I.; Jeromel, A. Effect of Different Vinification Techniques on the Concentration of Volatile Aroma Compounds and Sensory Profile of Malvazija Istarska Wines. *Fermentation* **2023**, *9*, 676. [CrossRef]
45. *OIV Resolution 332a/2009*; OIV Standard for International Wine and Spirituous Beverages of Vitivinicultural Origin Competitions. International Organisation of Vine and Wine: Paris, France, 2009.
46. *ISO 17025:2017*; ISO/IEC 17025:2017—General Requirements for the Competence of Testing and Calibration Laboratories. International Organization for Standardization: Geneva, Switzerland, 2017. Available online: https://www.iso.org/standard/66912.html (accessed on 18 August 2023).
47. Ordinance on Wine and Fruit Wine Sensory Testing (NN 106/04). Available online: https://narodne-novine.nn.hr/clanci/sluzbeni/2004_07_106_2061.html (accessed on 17 July 2023).
48. Commission Regulation (EC) No 606/2009 of 10 July 2009 Laying down Certain Detailed Rules for Implementing Council Regulation (EC) No 479/2008 as Regards the Categories of Grapevine Products, Oenological Practices and the Applicable Restrictions. 2009. Available online: https://eur-lex.europa.eu/legal-content/EN/TXT/HTML/?uri=CELEX:32009R0606 (accessed on 17 July 2023).
49. Parenti, A.; Spugnoli, P.; Calamai, L.; Ferrari, S.; Gori, C. Effects of Cold Maceration on Red Wine Quality from Tuscan Sangiovese Grape. *Eur. Food Res. Technol.* **2004**, *218*, 360–366. [CrossRef]
50. Giriboni, P.P.; Xavier, A.; Roque, V.; Vargas, G.; Souza, F.; Costa, V.B. Physical and Chemical Analyzis of Chardonnay Wine with Different Periods of Skin Contact. *BIO Web Conf.* **2016**, *7*, 02021. [CrossRef]
51. Luque de Castro, M.D.; Gonzalez-Rodriguez, J.; Perez Juan, P. Analytical Methods in Wineries: Is It Time to Change? *Food Rev. Int.* **2005**, *21*, 231–265. [CrossRef]
52. Hidalgo Togores, J. *Tratado de Enología. Volumen I y II*; Mundi-Prensa: Madrid, Spain, 2003; ISBN 978-84-8476-752-7.
53. Commission Delegated Regulation (EU) 2019/934 of 12 March 2019 Supplementing Regulation (EU) No 1308/2013 of the European Parliament and of the Council as Regards Wine-Growing Areas Where the Alcoholic Strength May Be Increased, Authorised Oenological Practices and Restrictions Applicable to the Production and Conservation of Grapevine Products, the Minimum Percentage of Alcohol for By-Products and Their Disposal, and Publication of OIV Files. 2019. Available online: https://eur-lex.europa.eu/legal-content/EN/TXT/HTML/?uri=CELEX:32019R0934 (accessed on 8 October 2023).
54. Neto, F.S.; de Castilhos, M.B.; Telis, V.R.; Telis-Romero, J. Effect of Ethanol, Dry Extract and Reducing Sugars on Density and Viscosity of Brazilian Red Wines. *J. Sci. Food Agric.* **2015**, *95*, 1421–1427. [CrossRef]
55. Budić-Leto, I.; Lovrić, T.; Kljusurić, J.G.; Pezo, I.; Vrhovšek, U. Anthocyanin Composition of the Red Wine Babić Affected by Maceration Treatment. *Eur. Food Res. Technol.* **2006**, *222*, 397–402. [CrossRef]
56. Lukić, I.; Budić-Leto, I.; Bubola, M.; Damijanić, K.; Staver, M. Pre-Fermentative Cold Maceration, Saignée, and Various Thermal Treatments as Options for Modulating Volatile Aroma and Phenol Profiles of Red Wine. *Food Chem.* **2017**, *224*, 251–261. [CrossRef]
57. González-Neves, G.; Gil, G.; Barreiro, L.; Favre, G. Pigment Profile of Red Wines Cv. Tannat Made with Alternative Winemaking Techniques. *J. Food Compos. Anal.* **2010**, *23*, 447–454. [CrossRef]
58. Gonzalez-Manzano, S.; Duenas, M.; Rivasgonzalo, J.; Escribanobailon, M.; Santosbuelga, C. Studies on the Copigmentation between Anthocyanins and Flavan-3-Ols and Their Influence in the Colour Expression of Red Wine. *Food Chem.* **2009**, *114*, 649–656. [CrossRef]
59. Sims, C.A.; Bates, R.P. Effects of Skin Fermentation Time on the Phenols, Anthocyanins, Ellagic Acid Sediment, and Sensory Characteristics of a Red *Vitis Rotundifolia* Wine. *Am. J. Enol. Vitic.* **1994**, *45*, 56–62. [CrossRef]
60. Şener, H. Effect of Temperature and Duration of Maceration on Colour and Sensory Properties of Red Wine: A Review. *S. Afr. J. Enol. Vitic.* **2018**, *39*, 1–8. [CrossRef]
61. Koyama, K.; Goto-Yamamoto, N.; Hashizume, K. Influence of Maceration Temperature in Red Wine Vinification on Extraction of Phenolics from Berry Skins and Seeds of Grape (*Vitis vinifera*). *Biosci. Biotechnol. Biochem.* **2007**, *71*, 958–965. [CrossRef] [PubMed]
62. Vrhovsek, U.; Vanzo, A.; Nemanic, J. Effect of Red Wine Maceration Techniques on Oligomeric and Polymeric Proanthocyanidins in Wine, Cv. Blaufränkisch. *Vitis* **2002**, *41*, 47–51.

63. Garrido, J.; Borges, F. Wine and Grape Polyphenols—A Chemical Perspective. *Food Res. Int.* **2011**, *44*, 3134–3148. [CrossRef]
64. Bloomfield, D.G.; Heatherbell, D.A.; Nikfardjam, M.S.P. Effect of P-Coumaric Acid on the Color in Red Wine. *Mitteilungen Klosterneubg.* **2003**, *53*, 195–198.
65. Bimpilas, A.; Panagopoulou, M.; Tsimogiannis, D.; Oreopoulou, V. Anthocyanin Copigmentation and Color of Wine: The Effect of Naturally Obtained Hydroxycinnamic Acids as Cofactors. *Food Chem.* **2016**, *197*, 39–46. [CrossRef]
66. Darias-Martín, J.; Rodr'iguez, O.; D'Iaz, E.; Lamuela-Raventós, R.M. Effect of Skin Contact on the Antioxidant Phenolics in White Wine. *Food Chem.* **2000**, *71*, 483–487. [CrossRef]
67. Lisov, N.; Čakar, U.; Milenković, D.; Čebela, M.; Vuković, G.; Despotović, S.; Petrović, A. The Influence of Cabernet Sauvignon Ripeness, Healthy State and Maceration Time on Wine and Fermented Pomace Phenolic Profile. *Fermentation* **2023**, *9*, 695. [CrossRef]
68. Benbouguerra, N.; Hornedo-Ortega, R.; Garcia, F.; El Khawand, T.; Saucier, C.; Richard, T. Stilbenes in Grape Berries and Wine and Their Potential Role as Anti-Obesity Agents: A Review. *Trends Food Sci. Technol.* **2021**, *112*, 362–381. [CrossRef]
69. Sun, X.; Chen, X.; Li, L.; Ma, T.; Zhao, F.; Huang, W.; Zhan, J. Effect of Ultra-High Pressure Treatment on the Chemical Properties, Colour and Sensory Quality of Young Red Wine. *S. Afr. J. Enol. Vitic.* **2015**, *36*, 393–401. [CrossRef]
70. Carrizzo, A.; Izzo, C.; Vecchione, C. Protective Activity of Resveratrol in Cardio- and Cerebrovascular Diseases. In *Resveratrol—Adding Life to Years, Not Adding Years to Life*; IntechOpen: London, UK, 2018; ISBN 978-1-78984-995-0.
71. Kostadinović, S.; Wilkens, A.; Stefova, M.; Ivanova, V.; Vojnoski, B.; Mirhosseini, H.; Winterhalter, P. Stilbene Levels and Antioxidant Activity of Vranec and Merlot Wines from Macedonia: Effect of Variety and Enological Practices. *Food Chem.* **2012**, *135*, 3003–3009. [CrossRef] [PubMed]
72. Petrović, A.; Lisov, N.; Čakar, U.; Marković, N.; Matijašević, S.; Cvejić, J.; Atanacković, M.; Gojković-Bukarica, L. The Effects of Prokupac Variety Clones and Vinification Method on the Quantity of Resveratrol in Wine. *Food Feed. Res.* **2019**, *46*, 189–198. [CrossRef]
73. Casassa, L.F.; Harbertson, J.F. Extraction, Evolution, and Sensory Impact of Phenolic Compounds During Red Wine Maceration. *Annu. Rev. Food Sci. Technol.* **2014**, *5*, 83–109. [CrossRef]
74. Kocabey, N.; Yilmaztekin, M.; Hayaloglu, A.A. Effect of Maceration Duration on Physicochemical Characteristics, Organic Acid, Phenolic Compounds and Antioxidant Activity of Red Wine from *Vitis vinifera* L. Karaoglan. *J. Food Sci. Technol.* **2016**, *53*, 3557–3565. [CrossRef]
75. Lieu, P.T.; Heiskala, M.; Peterson, P.A.; Yang, Y. The Roles of Iron in Health and Disease. *Mol. Asp. Med.* **2001**, *22*, 1–87. [CrossRef]
76. Skendi, A.; Papageorgiou, M.; Stefanou, S. Preliminary Study of Microelements, Phenolics as Well as Antioxidant Activity in Local, Homemade Wines from North-East Greece. *Foods* **2020**, *9*, 1607. [CrossRef]
77. Bora, F.-D.; Bunea, C.-I.; Rusu, T.; Pop, N. Vertical Distribution and Analysis of Micro-, Macroelements and Heavy Metals in the System Soil-Grapevine-Wine in Vineyard from North-West Romania. *Chem. Cent. J.* **2015**, *9*, 19. [CrossRef]
78. Ournac, A.; Flanzy, M. The Increase in Vitamin B1 in Wine Stored with Its Lees. *Ann. Technol. Agric.* **1967**, *16*, 41–54.
79. Lafourcade, S.; Peynaud, E.; Ribereau-Gayon, J. Different forms of nicotinamide in wine. *Bull. Soc. Chim. Biol.* **1956**, *38*, 923–930.
80. Soares, S.; Brandão, E.; Mateus, N.; de Freitas, V. Sensorial Properties of Red Wine Polyphenols: Astringency and Bitterness. *Crit. Rev. Food Sci. Nutr.* **2017**, *57*, 937–948. [CrossRef] [PubMed]
81. Olejar, K.J.; Fedrizzi, B.; Kilmartin, P.A. Enhancement of Chardonnay Antioxidant Activity and Sensory Perception through Maceration Technique. *LWT Food Sci. Technol.* **2016**, *65*, 152–157. [CrossRef]
82. Ferrer-Gallego, R.; Hernández-Hierro, J.M.; Rivas-Gonzalo, J.C.; Escribano-Bailón, M.T. Sensory Evaluation of Bitterness and Astringency Sub-Qualities of Wine Phenolic Compounds: Synergistic Effect and Modulation by Aromas. *Food Res. Int.* **2014**, *62*, 1100–1107. [CrossRef]
83. Casassa, F.; Sari, S.; Avagnina, S.; Díaz, M.; Jofré, V.; Fanzone, M. Influencia de dos técnicas de maceración sobre la composición polifenólica, aromática y las características organolépticas de vinos cv. Merlot. *Vitic. Enol. Prof.* **2007**, 5–20.
84. González-Neves, G.; Ferrer, M. Efectos del sistema de conducción y del raleo de racimos en la composición de uvas Merlot. *Agrociencia* **2008**, *12*, 10–18. [CrossRef]
85. Budić-Leto, I.; Gracin, L.; Lovric, T.; Vrhovsek, U. Effects of Maceration Conditions on the Polyphenolic Composition of Red Wine "Plavac Mali". *Vitis J. Grapevine Res.* **2008**, *47*, 245–250.
86. Ofoedu, C.E.; Ofoedu, E.O.; Chacha, J.S.; Owuamanam, C.I.; Efekalam, I.S.; Awuchi, C.G. Comparative Evaluation of Physicochemical, Antioxidant, and Sensory Properties of Red Wine as Markers of Its Quality and Authenticity. *Int. J. Food Sci.* **2022**, *2022*, e8368992. [CrossRef]

Disclaimer/Publisher's Note: The statements, opinions and data contained in all publications are solely those of the individual author(s) and contributor(s) and not of MDPI and/or the editor(s). MDPI and/or the editor(s) disclaim responsibility for any injury to people or property resulting from any ideas, methods, instructions or products referred to in the content.

foods

MDPI

Article

Winemaking Technologies for the Production of Cabernet Sauvignon and Feteasca Neagra Wines Enriched with Antioxidant Active Principles Due to the Addition of Melatonin

Sandra A. V. Eremia [1], Camelia Albu [1,*], Gabriel-Lucian Radu [1], Andreia Alecu [1], Alice G. Stoica [1] and Elena Brinduse [2]

[1] Centre of Bioanalysis, National Institute of Research and Development for Biological Sciences–Bucharest, 296 Splaiul Independentei, 060031 Bucharest, Romania; sandra.eremia@incdsb.ro (S.A.V.E.); lucian.radu@incdsb.ro (G.-L.R.); andreia.alecu@incdsb.ro (A.A.); alice.stoica@incdsb.ro (A.G.S.)

[2] Institute for Research and Development for Viticulture and Wine Making, 2 Valea Mantei, Valea Calugareasca, 107620 Valea Calugareasca, Romania; elabrinduse@gmail.com

* Correspondence: camelia.albu@incdsb.ro; Tel.: +40-212-200-900

Abstract: In recent years, various studies have been carried out to increase the concentration of antioxidant active principles in red wines as a consequence of the effects of winemaking techniques on the polyphenols content. In this study, in order to obtain the most optimal wine in terms of content and efficiency of antioxidant activity, various winemaking technologies (punching-down and pumping-over maceration) were tried with diverse gradations (Feteasca Neagra and Cabernet Sauvignon wines) and the addition of different concentrations of melatonin in must. Suitable HPLC and spectrophotometric methods were used to follow the evolution of the antioxidant compounds from wines during aging (for 12 months). After comparing the acquired results, an increase was observed in the antioxidant compound concentrations, particularly in resveratrol (85%), peonidin-3-glucoside (100%) or cyanidin-3-glucoside (100%), and antioxidant activity (10–40%). The most enriched wine was obtained in the case of Feteasca Neagra by the addition of 0.5 mg of melatonin per 1 kg of must using the punch-down technology and, in the case of Cabernet Sauvignon, by the addition of 0.05 mg of melatonin per 1 kg of must using the pumping-over technique. This study can provide winemakers with an approach to enhance red wines with antioxidant compounds.

Keywords: winemaking process; Feteasca Neagra; polyphenols; antioxidant capacity; Cabernet Sauvignon; melatonin treatment

1. Introduction

Wine is an economically relevant beverage that contains a variety of natural antioxidants and makes an important contribution to a healthy diet. The starting point for studies on red wine and health was the introduction of the term "French paradox", which linked the low incidence of coronary heart disease to the population that has a moderate consumption of red wine [1]. Phenolic compounds are known to act as exogenous natural antioxidants [2], and research on wines has mainly focused on polyphenols [3], mainly flavonoids [4,5] and non-flavonoids [6,7]. An important antioxidant active principle contained, especially in red wines, is resveratrol, which reduces the risk of cardiovascular diseases and acts as a powerful antioxidant, both through a novel glutathione-sparing mechanism and through the classical uptake of hydroxyl radicals [8]. Red wines are the main source of resveratrol in the human diet.

Winemaking techniques are important for the concentration of phenolic compounds in red wines and can increase their extraction, particularly from the solid parts of the grapes (seeds and skins). The maceration conditions influence the phenolic composition and the biological activities of grape musts. For example, higher temperatures and longer maceration times result in a higher concentration of total anthocyanins [9]. Two classic methods of maceration–fermentation techniques are punching down and pumping over. During fermentation, a cap is formed from the solid components of the grapes, which is raised to the top of the fermentation tank via carbon dioxide production. To ensure contact between the juice and the skins, or seeds, they are usually mixed several times a day, either by pushing the cap under the juice (punching down) or by pumping the fermenting must out from the bottom of a tank and spraying it over the top of the tank (pumping over). Studies have been published comparing punching-down vinification with pumping-over. It was found that the extraction of polyphenolic compounds mainly depends on the grape varieties [10].

Feteasca Neagra (FN) is an old grape variety that mainly grows in Romania and the Republic of Moldova and from which a red wine with ruby-red reflections and an aroma of black raisins is produced, which becomes richer and softer with age. The FN wine is considered one of the best red wines from Romania and has a higher content of phenolic compounds than Pinot Noir [11]. Cabernet Sauvignon (CS) is the most widely cultivated grape variety in the world, producing a dark red wine with vanilla and dark fruit flavors. CS is considered the king of wines with a high tannin concentration and higher antioxidant activity compared to other red wines [12].

Some studies have been carried out to observe the effects of different winemaking techniques on the polyphenol content of CS. It was found that the vinification technique involving prolonged maceration (21 or 150 days) and implicitly the solid parts during the post-fermentative maceration phase significantly modulated the phenolic content and provided a higher content of polyphenolic compounds in steel tanks [13,14]. In wines with 16% saignée, a winemaking technique that involves the removal of a proportion of juice from a tank of crushed red grapes [15] increased the tannin content, and the results were long-lasting. Saignée also increased anthocyanin content in the first few days, but these results diminished by day 120 [16]. In young wines, the total anthocyanin content can be increased by the Ganimede fermenter, but the content decreases significantly after two years of aging, making the Ganimede fermenter suitable for the production of CS wines for early consumption [17]. It has been observed that the application of continuous treatment with pulsed electric fields resulted in obtaining a CS wine with a higher content of polyphenolic compounds and anthocyanins at the end of alcoholic fermentation [18]. In FN wines, it has been shown that the use of oak staves during vinification can lead to an increase in phenolic content, total antioxidant activity, the number of volatile compounds, and color intensification over time [19].

Many studies have demonstrated significant biological effects and evidenced that resveratrol from wines modulates various targets related to different biochemical pathways and can be considered as a therapeutic agent and as a protective agent against numerous diseases [20,21]. However, it is not possible to absorb the recommended therapeutic doses of resveratrol by drinking wine [22], and various studies have been focused on enhancing its concentration in wines. One factor that affects the presence of resveratrol in wine is the selected yeast strain used in alcoholic fermentation. To increase resveratrol content, yeast can be used with a higher ethanol-producing capacity, a pectolytic enzyme can be added, or transgenic microorganisms can also be added during maceration. Also, the extended maceration and malolactic fermentation could contribute to increased resveratrol levels in wine [23].

However, these approaches have various extraction yields and efficiency, and they may be feasible to increase the concentration of antioxidant active principle in red wine by increasing the maceration time and using the expensive tanks/oak in the winemaking process.

Melatonin, an important indoleamine compound with strong antioxidant activity [24], is present naturally in wines because it is extracted from grapes [25] and because it is biosynthesized from its precursors in some stages of the winemaking process [26,27]. Melatonin may have a synergistic effect with other antioxidants naturally present in wines [28] and may enhance the production of antioxidant compounds in the winemaking process. The exogenous administration of melatonin in must at the initial stage of the vinification process allows for the biosynthesis of polyphenolic compounds via phenylpropanoid metabolism and stimulates the activity of the enzymes phenylalanine ammonia lyase and cinnamic acid 4-hydroxylase to considerably modify the expression of specific genes for anthocyanin biosynthesis [29]. However, the research is still in an early stage, and it has not established what level of melatonin should be used and which winemaking process should be applied for the different grape varieties to improve red wines with antioxidant compounds. Currently, the addition of melatonin is not a legal practice foreseen on the list and description of the OIV Code of Oenological Practices referred to in the Commission Delegated Regulation [30]. However, the presence of melatonin in foodstuffs is allowed in certain European countries, and the European Food Safety Authority has issued a favorable opinion for two claims relating to the presence of melatonin in foodstuffs [31].

Through the recent detection of melatonin in beverages such as wine [32,33], an exciting new field of research has emerged, including the need for supplementary studies on melatonin-positive effects on polyphenolic profiles through its application in the winemaking process for quality improvements of wines with potential health benefits.

This study aims to investigate different winemaking technologies with diverse gradations, starting with the addition of melatonin to the must to obtain a wine that is the most suitable in terms of content and efficiency of antioxidant activity. The aim of this study is to establish the most optimal winemaking technologies to produce red wines enhanced with antioxidant-active principles starting from the application of melatonin. Therefore, different melatonin concentrations, punching-down and pumping-over maceration techniques, and FN and CS grape varieties were evaluated to obtain the most efficient approach to improve wines with antioxidant activities. Appropriate chromatographic methods (HPLC) and spectrophotometry were used to monitor the evolution of the antioxidant compounds of the control and treated wines at the key aging stages of the wines (3, 6, 9, and 12 months).

2. Materials and Methods

2.1. Reagents

Melatonin (Sigma, Steinheim, Germany, M5250), chlorogenic acid (C3878), caffeic acid (C0625), (+)-catechin (43412), (−)-catechin (C0567), quercetin 3-glucoside (17793), quercitrin (Q3001), quercetin (Q4951), rutin (78095), resveratrol (R5010), cyanidin (79457), pelargonidin (PHL80084), Trolox (6-hidroxi-2,5,7,8-tetrametillchroman-2-carboxilic acid), chloroform (366919), ABTS (2,2-azinobis(3-ethylbenzothiaziline-6-sulfonate)), NaCl (S7653), sodium carbonate (Na_2CO_3), DPPH (2,2-diphenyl-1-picrylhydrazyl), potassium persulfate (K2S2O8), Folin & Ciocalteu's phenol reagent (FCR), acetic acid (A6283), and formic acid (F0507) were bought from Sigma-Aldrich (Steinheim, Germany). Ellagic acid (45140), (−)-epicatechin (68097), gallic acid (48630), and myricetin (70050) were acquired from Fluka. Roth (Karlsruhe, Germany) has provided malvidin (6140.1), delphinidin (4537.1), petunidin-3-glucoside (89755), and peonidin-3-glucoside (1619.1) from PhytoLab (Dutendorfer, Germany). Delphinidin-3-glucoside, malvidin-3-glucoside and cyanidin-3-glucoside were procured from Polyphenols AS (Sandnes, Norway). All other used reagents, MeCN, methanol, and ethanol were chromatographic purity from Riedel-de Haen (Berlin, Germany). In total, 1 mg mL^{-1} of stock solutions were prepared in ethanol, taking into account the specific solubility of each standard and the chemical composition of the wines.

2.2. Wine Sampling

To accomplish the goal of this work, a $2 \times 2 \times 4$ polyfactorial experiment was carried out with 16 variants in 4 replicates, in which the must was treated with melatonin in the preliminary stage. Three experimental factors were included with the following gradations:

- The variety of vines with gradations: FN and CS;
- The maceration–fermentation technique, with gradations: punching down (I) and pumping over (II);

- Melatonin concentration with gradations: 0 (M), 0.05 (V1), 0.1 (V2), and 0.5 (V3) mg of melatonin were applied per 1 kg of must.

FN and CS grape varieties from the geographical region of Valea Calugareasca, Romania, were used. Duplicate replications for each treatment/variant were carried out.

The grapes from FN and CS varieties were aleatorily harvested at technological ripeness (soluble sugar content according to the OIV-MA-AS2-02 Method was 210.1 ± 0.05 g L^{-1} for FN and 212.4 ± 0.14 g L^{-1} for CS; the total acidity according to the OIV-MA-AS313-01 Method [34] was 4.48 ± 0.12 g L^{-1} for FN and 4.10 ± 0.06 g L^{-1} for CS), destemmed, crushed and sulphurised (50 mg L^{-1} SO_2). The wines were obtained for both grape varieties by the traditional method of maceration (duration of maceration in every treatment was 7 days), with half of the variants obtained by punching down, i.e., breaking the cap and immersing it in the must twice a day for 7 days, and the remaining samples by pumping over, i.e., extracting the juice from the bottom of the fermentation tank and pumping it over the cap once a day for 7 days. In the pre-vinification phase, 0.05, 0.1, and 0.5 mg of melatonin were applied per 1 kg of must. After decanting of experimental must, Actiflore F 33, *Saccharomyces cerevisiae* (20 g/hL) was added to start fermentation, and the wines were stirred in a stainless steel red wine tank to complete the fermentation and maturation process (12 months). Malolactic fermentation occurred spontaneously within 25–30 days. When the malolactic fermentation was complete and before bottling, the young wines were sulfited up to a level of 30 mg L^{-1} of free SO_2. In this stage, the wines obtained from the FN variety were dry wines, with a content of residual sugar ranging between 2.7 and 4.0 g L^{-1}, an alcoholic content of 12.5–13.1% vol., and a volatile acidity with values between 0.24 and 0.37 g L^{-1} (acetic acid). The wines obtained from the CS variety were dry wines, with a content of residual sugar ranging between 1.8 and 4.0 g L^{-1}, an alcoholic content of 12.0–13.8% vol., and a volatile acidity with values between 0.20 and 0.42 g L^{-1} (acetic acid). The red wines produced were bottled after 12 months of aging. The storage conditions during the 12 months of the experiment were constant, and the wines were kept at a controlled cellar temperature between 10 and 15 °C.

For the monitoring of melatonin, a dispersive liquid–liquid microextraction with MeCN as the dispersing solvent and chloroform as the extraction solvent was used. A total of 8.5 mL of the centrifuged wine sample was transferred to a centrifuge tube and 1.5 mL of the chloroform/MeCN mixture at a ratio of 1:1, was added. After vortexing the samples for 10 min at 1200 rpm, 1 g of NaCl salt was added and then mixed again for 5 min. The next step was centrifugation at 5000 rpm for 5 min, after which the analyte of interest was found in the lower phase [35]. All wine samples were filtered into HPLC vials with a 0.2 µm Syringe Filter Unit (PTFE, Agilent, Beijing, China) and injected into the HPLC system.

To achieve the aim of this work and select the winemaking protocol of the most improved wine in terms of content and efficiency of antioxidant activity, monitoring of the active ingredients was carried out at the main aging stages of the wines: three, six, nine, and 12 months.

2.3. HPLC Analysis

The HPLC analyses were accomplished using a Shimadzu complete system with two detectors, an RF-20A XS fluorescence detector and a mass spectrometer detector, LCMS-2010. For melatonin analysis, the fluorescence detector parameters were as follows: λ_{ex} = 285 nm and λ_{em} = 340 nm; cell temperature control at 250 °C; 1.5 s response; Gain × 4; medium sensitivity. For polyphenol compounds analysis, the MS parameters were as follows: ESI interface, CDL temperature, 200 °C; heat block temperature, 200 °C; nebulization gas (N_2) flow rate, 1.5 L min^{-1}; detector voltage, 1.8 kV, interface voltage 4 kV and interface temperature, 250 °C.

The identification and quantification of melatonin by HPLC-FL were performed using a previously developed method [35], a Kromasil 100-5-C18 2.1 × 100 mm column, an elution gradient of mobile phase (methanol, solvent A, and 1% acetic acid, solvent B) and flow rate of 1 mL min^{-1}. The analysis time was 30 min and the column temperature was 20 °C.

The analysis of polyphenolic compounds was achieved on a C18 column, Kromasil, 100-3.5, 2.1 × 100 mm, with a mobile phase, pH = 3, composed from water with formic acid, and MeCN with formic acid. An elution and a flow rate gradient of the mobile phase was used [11].

The MS detection was performed using negative ionization mode and the selected ion monitoring (SIM) mode was used to obtain the corresponding peaks of the polyphenolic compounds fragment ions ([M − H]: 169, 179, 227, 289, 301, 317 353, 447, 463 and 609).

The anthocyanidins and anthocyanins quantification was achieved through a C18 Kromasil column, 100-3.5 4.6 × 50 mm, with a gradient of mobile phase (5% formic acid in water and 5% formic acid in methanol), a flow rate of 0.2 mL min^{-1}, and the temperature column at 40 °C [36]. The positive ionization mode was used to obtain the following corresponding peaks of the compound fragment ions ($[M - H]^+$: 493, 479, 465, 463, 449, 331, and 303) in SIM mode.

2.4. *Trolox Equivalent Antioxidant Capacity (TEAC) Assay Using ABTS·+*

ABTS·+ is a radical cation formed by the oxidation of ABTS (7 × 10^{-3} M) with $K_2S_2O_8$ (2.5 × 10^{-3} M). For the stabilization of ABTS·+, the solution was kept for at least 20 h at room temperature in the dark. The solution was obtained by diluting (1:40 *v/v)* the stock solution and combining it with water and sampled at a ratio of 25:4:1, ABTS·+/solvent:/sample, *v/v/v*. The absorbance was measured at 735 nm with a Thermo Spectrophotometer (Thermo Fischer Scientific, Waltham, MA, USA). The obtained values were presented in relation to a compound considered a reference antioxidant, Trolox equivalent (8.5 × 10^{-6} M) [37].

2.5. *TEAC Assay Using DPPH•*

The stable radical, DPPH• (2.5 × 10^{-4} M), was formed by dissolving DPPH in ethanol. The solution was mixed with ethanol and sampled at a ratio of 1:1.9:0.1, DPPH•/solvent/sample, *v/v/v*. The decrease in DPPH was measured at 515 nm. The results were expressed as Trolox (8.5 × 10^{-6} M) equivalent [38].

2.6. *Folin & Ciocalteu Assay*

The FCR was diluted in ultrapure water (1:10) and mixed at a ratio of 5:1, *v/v*, with the phenol compound. After 10 min, to obtain a blue complex between the polyphenol compounds and the Folin and Ciocalteu, 7.8% Na_2CO_3 was pipetted in the alkaline medium [39]. After 1 h, the complex absorbance was measured at 766 nm. The results attained were presented in relation to a reference polyphenol in wines, namely gallic acid.

Each sample of standards and wines was analyzed in triplicate. The results were presented using descriptive statistics, which summarizes data using indexes such as mean, median, and standard deviation (SD), as mean ± SD. Also, the data were analyzed by a one-way analysis of variance (ANOVA) using Excel software, 2021. ANOVA was used to determine the significant differences, the probability ($p < 0.05$), and the variability (F and F-critical values) between the means.

3. Results and Discussion

For the monitoring of active principles from red wines, the performance characteristics of the HPLC methods required for the analytical analyses were first determined. From chromatograms obtained, a clear difference was observed between the specific peaks, and the retention times were between 3.27 and 35.83 min for polyphenolic compounds, between 23.98 and 35.25 min for anthocyanidins/anthocyanins, and 23.58 min for melatonin. The peaks were properly resolved with a suitable resolution in the case of catechin compounds ([M − H] = 289) and quercetin and ellagic acid ([M − H] = 301). The values obtained for the range of response (0.5–50 µg mL^{-1} for polyphenolic compounds and 1–30 ng mL^{-1} for melatonin analysis), the correlation coefficients (R) between 0.9991 and 0.9999, the limit of detection (LoD), which was less than 0.44 μg mL^{-1} for polyphenolic compounds and 0.01 ng mL^{-1} for melatonin analysis, and the limit of quantification (LoQ) (0.05 ng mL^{-1} for melatonin analysis and less than 0.48 μg mL^{-1}) prove that these three methods are appropriate for the analysis of polyphenols compounds, anthocyanidins/anthocyanins, and melatonin from the 16 wine samples.

3.1. The Evolution of Melatonin Contents in Profiles in the Aging Process of FN and CS Wines

The HPLC-FL analysis of the wine samples provides important information about the melatonin reference values in the grape variety and, in our case, in the 16 variants of the wine samples. Based on this information, we can determine the most promising melatonin concentration to be applied in the must and identify which grape varieties lead to the highest melatonin content obtained in winemaking. These results also help to understand the evolution of the polyphenol compounds or antioxidant activity during the aging process of red wines. The results from HPLC-FL analyses are presented in Table 1.

Table 1. The melatonin values obtained by the HPLC-FL method from 16 variant wine samples (mean ± standard deviation and ANOVA results).

			Melatonin ng mL^{-1}				
FNIM	FNIV1	FNIV2	FNIV3	FNIIM	FNIIV1	FNIIV2	FNIIV3
			3 months				
0.74 ± 0.01	12.99 ± 0.09	30.31 ± 0.12	193.65 ± 1.11	1.09 ± 0.02	20.50 ± 0.14	30.22 ± 0.17	158.17 ± 1.58
			6 months				
1.92 ± 0.02	14.90 ± 0.14	31.41 ± 0.21	192.33 ± 1.57	2.10 ± 0.01	24.49 ± 0.15	35.43 ± 0.24	154.33 ± 1.85
			9 months				
1.80 ± 0.03	9. 67 ± 0.10	20.18 ± 0.15	124.69 ± 1.91	3.13 ± 0.03	19.28 ± 0.17	27.24 ± 0.23	107.55 ± 1.62
			12 months				
1.63 ± 0.02	9. 18 ± 0.11	20.08 ± 0.17	111.1 ± 1.31	1.14 ± 0.02	18.59 ± 0.13	26.88 ± 0.31	102.76 ± 1.28
F	*p*-value		F crit	F	*p*-value		F crit
42.31	1.17×10^{-6}		3.49	59.373	1.8×10^{-7}		3.49
CSIM	CSIV1	CSIV2	CSIV3	CSIIM	CSIIV1	CSIIV2	CSIIV3
			3 months				
0.84 ± 0.02	16.19 ± 0.09	39.28 ± 0.11	158.71 ± 1.26	1.36 ± 0.04	24.41 ± 0.22	40.16 ± 0.31	165.34 ± 1.72
			6 months				
2.53 ± 0.02	27.89 ± 0.41	43.63 ± 0.38	154.99 ± 1.57	1.43 ± 0.01	25.67 ± 0.24	47.05 ± 0.38	177.47 ± 1.57
			9 months				
2.06 ± 0.5	11.28 ± 0.28	30.67 ± 0.27	114.64 ± 1.37	2.96 ± 0.02	17.12 ± 0.23	26.87 ± 0.15	102.39 ± 0.95
			12 months				
0.31 ± 0.01	6.65 ± 0.22	25.46 ± 0.23	89.66 ± 0.95	1.40 ± 0.02	16.93 ± 0.13	25.99 ± 0.22	92.31 ± 0.82
F	*p*-value		F crit	F	*p*-value		F crit
42.53	1.14×10^{-6}		3.49	28.00	1.06×10^{-5}		3.49

FN—Feteasca Neagra, CS—Cabernet Sauvignon, I—punching down, II—pumping over, M—control samples; V1—0.05, V2—0.1, and V3—0.5 mg of melatonin were applied to 1 kg of must.

Based on the results obtained after 3 and 6 months, the biosynthesis of melatonin was observed (except for variant 3, FNIV3, FNIIV3, and CSIV3, where the addition of melatonin was 0.5 mg per 1 kg of must and biosynthesis at the level of ng mL^{-1} was not observed), confirming the fact that the low temperatures made the fermentation extremely slow, requiring a longer period in the winemaking process. During fermentation, melatonin is synthesized by the yeasts from L-tryptophan and serotonin [40]. At this stage, L-tryptophan is released into the wine, and serotonin is obtained from the decarboxylation of L-tryptophan by the action of yeast, lactic acid bacteria, or other contaminating microorganisms [41]. The final melatonin concentration can be influenced by the fermentation conditions, the L-tryptophan concentration, and the sugar content [42]. A total of 9 months after the start of vinification, a decrease in melatonin concentration of about 30% was observed in the wines, which correlates with the positive trend observed between 6 and 9 months for the concentrations of different polyphenolic compounds. After 12 months, the melatonin concentrations observed show a slight downward trend, with values comparable to those observed after 9 months, which indicates the end of the wine's aging process and the slowing down of biochemical and chemical reactions.

3.2. The Evolution of Phenolic Acids, Flavanols, Flavonols, and Stilbenes Profiles in the Aging Process of FN and CS Wines

Three months after the start of the vinification, monitoring of the polyphenolic profiles of the FN wines revealed that the concentrations of polyphenolic compounds in the melatonin-treated wines were similar to those in the control wines. Higher concentrations were obtained for the wine treated with 0.5 mg of melatonin per 1 kg of must, with the best results obtained using the punching down technique (FNIV3).

In the 6-month phase, the previous results were maintained with FNIV3 as the most improved wine. It was found that in a young wine, when the concentration of catechins, caffeic acid, or chlorogenic acid from FN is sought, the pumping-over technique can be used, and when the concentration of quercetin compounds is desired, the punching down technique is recommended. Nine months after the start of vinification, the results showed that the concentration values for 11 polyphenolic compounds, except gallic acid, were higher in all treated FN wines than in the control wines, indicating that the punching-down technique is the most suitable to obtain better wines. Quantitatively, the monitored compounds were 50% higher than those of the control sample when 0.5 mg of melatonin as applied per 1 kg of must. After 12 months, the concentrations of polyphenolic compounds began to decrease in all wines and the FNIV3 was the most improved wine.

In the case of CS wines, slight improvements in the concentrations of polyphenolic compounds were observed after 3 months in the treated wines in comparison to the control wines. Even if the differences are small between the extraction yield of polyphenolic compounds when punching down is used and the one obtained using pumping over, which depends on different variables, it was shown that the pumping-over technique gave all varieties of wines significantly higher quercetin levels [43]. CS compared to FN has a higher content of quercetin compounds, which makes pumping over more suitable for obtaining wines with increased antioxidant compounds for both the control wine and the variants. The highest concentrations for these wines were obtained for the wine treated with 0.05 mg of melatonin per 1 Kg of must when the pumping-over technique was used (CSIIV1). The results obtained after 6 months remained constant, confirming that pumping over is the most optimal technique in the case of CS wine.

Based on 9 months of data, it was found that the maturation process in all CS wines continued up to 12 months, the reactions were still implicated in the biosynthesis of the monitored polyphenols compounds, and the CSIIV1 wine improved by 15% through melatonin treatment. After 12 months, the concentrations of polyphenolic compounds were generally highest in all wines, except for quercetin compounds where a decrease was observed and rutin was not identified, and better results were obtained in the treated wines than in the control wines. In the CSIIV1 wine, the quantitative values of the monitored compounds increased by about 25%.

The results obtained after 3 months show that the most improved wines were FNIV3 and CSIIV1 and that the most optimal techniques were punching down for FN and pumping over for CS. Figures 1 and 2 show the polyphenolic profile evolution of these red wines over time (3, 6, 9, and 12 months) compared to the control wines.

In these wines, the analytes of interest were quantified at higher concentrations than in the control samples, except for gallic acid in both FNIV3 and CSIIV1 and catechin, especially in FNIV3. An increase was observed in gallic acid during the aging of the wines, and a lower amount was observed in the treated wines, but this is in agreement with the results obtained for the monitoring of anthocyanidins and anthocyanins, considering the shikimate/phenylpropanoid metabolism. It is possible that, in the presence of melatonin, the conversion into ellagic acid of gallic acid increased.

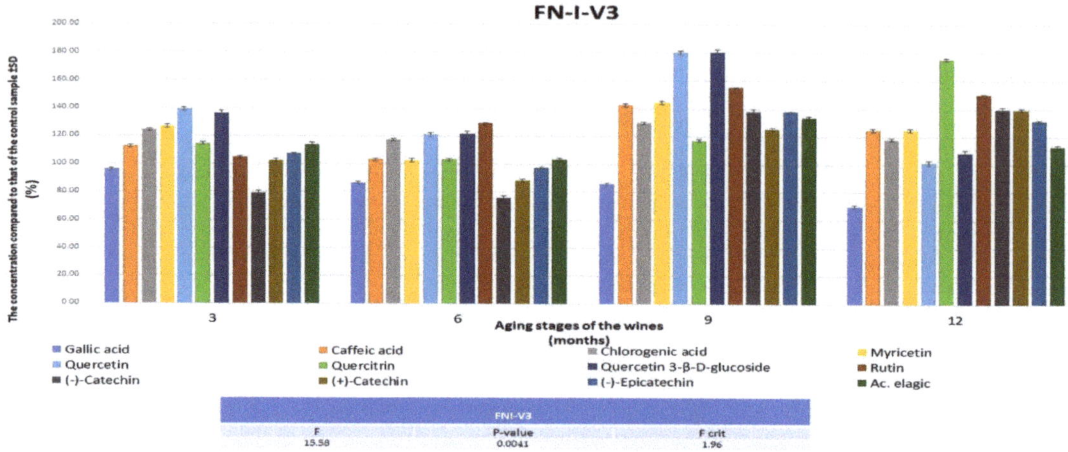

Figure 1. The evolution of the concentrations of polyphenols compounds in FNIV3 compared to the data obtained for the control wine during 12 months. FN—Feteasca Neagra; I—punching down; V3—0.5 mg of melatonin was applied to 1 kg of must.

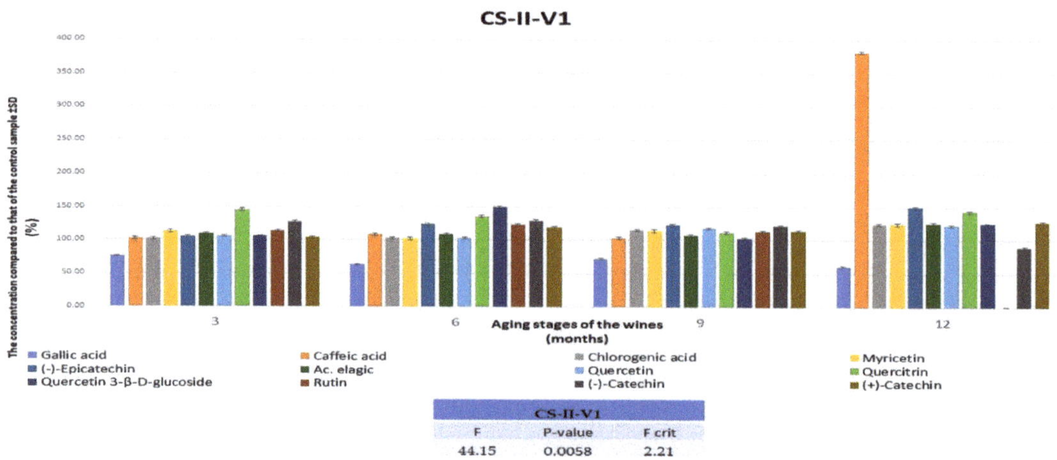

Figure 2. The evolution of the concentrations of polyphenols compounds in CSIIV1 compared to the data obtained for the control wine during 12 months. CS—Cabernet Sauvignon; II—pumping over; V1—0.05 mg of melatonin were applied to 1 kg of must.

Regarding the evolution of resveratrol concentration from the 16 varieties of the wines monitored, the results confirmed that the addition of melatonin leads to an increase in the concentration of this principle active in red wines. In FN wines, the highest concentration of resveratrol was obtained at 3 months after the start of the vinification process (9.57 ± 0.02–19.28 ± 0.21 μg mL^{-1}), followed by a linear decrease in the next 6 months and an improvement after 12 months of wine aging (7.98 ± 0.08–14.81 ± 0.12 μg mL^{-1}). In CS wines, the content of resveratrol increased during the aging of the wines, both in the control and in the treated wines, and at 12 months, the best results were obtained with values of 11.83 ± 0.14–21.40 ± 0.21 μg mL^{-1}. The best improvements in terms of resveratrol content in wines were obtained for FNIV3 and CSIIV1, respectively (Table 2). In the case of FNIV3, a 50% increase (19.28 ± 0.21 vs. 12.85 ± 0.18 μg mL^{-1}) was achieved in young wines (3 months), and after 12 months, an 85% intensification in the resveratrol content in the treated wine (14.81 ± 0.24 μg mL^{-1}) was observed compared to the control wine (7.98 ± 0.03 μg mL^{-1}). In the case of CSIIV1, approximately 45% increase was observed in all the monitored significant steps of the vinification process, and after 12 months, a 54% intensification was obtained in the resveratrol content in the treated wine (21.4 ± 0.28 μg mL^{-1}) compared to the control wine (13.9 ± 0.21 μg mL^{-1}). The treatment with melatonin improved the content of resveratrol both in FN wines and in CS wines, proving that a better winemaking technology for CS resulted in increased resveratrol compared to the thermovinification and separation of must from pomace [44]. The results attained are comparable to those previously published after studying the effect of 70 different strains of Saccharomyces cerevisiae on the content of resveratrol in CS wine [45].

Table 2. The evolution of the resveratrol concentrations in FNIV3, respectively, and CSIIV1, in comparison with the data obtained for the control wines during 12 months (mean ± standard deviation and ANOVA results).

Resveratrol μg mL^{-1}			
FNIM	**FNIV3**	**CSIIM**	**CSIIV1**
3 months			
12.85 ± 0.18	19.28 ± 0.21	9.29 ± 0.06	13.2 ± 0.12
6 months			
9.33 ± 0.05	11.5 ± 0.08	10.52 ± 0.07	15.73 ± 0.15
9 months			
7.58 ± 0.08	10.33 ± 0.09	13.68 ± 0.11	19.13 ± 0.14
12 months			
7.98 ± 0.03	14.81 ± 0.24	13.9 ± 0.21	21.4 ± 0.28
F	*p*-value		F crit
4.49	0.24		3.49

FN—Feteasca Neagra, CS—Cabernet Sauvignon, I—punching down, II—pumping over, and M—control samples; V1—0.05 and V3—0.5 mg of melatonin were applied to 1 kg of must.

3.3. *The Evolution of Anthocyanidins/Anthocyanins Profiles in the Aging Process of FN and CS Wines*

From the results obtained following the monitoring of seven anthocyanidins/anthocyanins in the key points of the wine maturation process, it appears that in the case of the treated FN wines, to obtain wines with higher concentrations of anthocyanins, the recommended technique is punching down (Figure 3). The wines obtained by pumping over generally have lower concentrations of the analytes of interest than the control wines. An explanation of this would be the fact that anthocyanin biosynthesis appears to be in competition with kaempferol 3-*O*-glycosides production usually glycosylated by the catalysis of UDP-D-glucose flavonoid 3-*O*-glycosyltransferase [46]. The treatment with melatonin in red wines leads to an increase in the enzyme activity of UDP-D-glucose flavonoid 3-*O*-glycosyltransferase, and it is possible that kaempferol 3-*O*-glycosides synthesis may be intensified in treated wines to the detriment of anthocyanins. The only promising wine is the one treated with 0.5 mg of melatonin per 1 kg of must, but only if it is in the first months of aging (3 or 6 months).

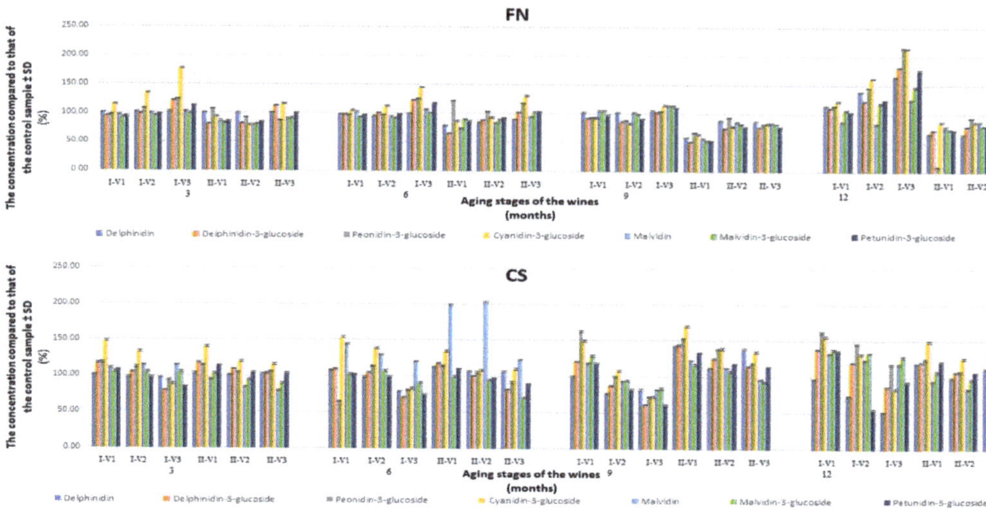

Figure 3. The concentration values of the 7 anthocyanidins/anthocyanins obtained for the 12 varieties of treated wines compared to the data obtained for the control wines, during the aging process (3, 6, 9, and 12 months, *p*-value = 0.0098 for FN, respectively, *p*-value = 0.0038 for CS). FN—Feteasca Neagra, CS—Cabernet Sauvignon, I—punching down, II—pumping over, V1—0.05, V2—0.1, and V3—0.5 mg of melatonin were applied to 1 kg of must.

The V3 variant differs from the wines obtained using the punching down technique, especially after 12 months, when a quantitative improvement in anthocyanins of over 60% was achieved compared to the concentrations in the control wine (Figure 4). These results are in line with those observed in the monitoring of polyphenolic compounds.

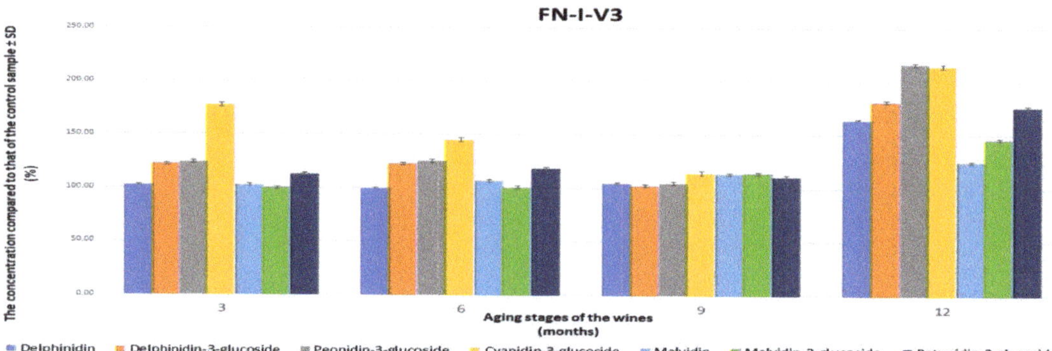

FN-I-V3		
F	P-value	F crit
8.77	0.045	2.04

Figure 4. The evolution of the concentrations of anthocyanidins/anthocyanins in FNIV3, in comparison with the data obtained for the control wine during 12 months. FN—Feteasca Neagra; I—punching down; V3—0.5 mg of melatonin was applied to 1 kg of must.

Regarding the CS wines, following HPLC-MS analyses for the quantification of anthocyanidins/anthocyanins for the 6 wine varieties, it was found that both maceration–fermentation techniques led to improved wines, and the lowest values were observed after the addition of 0.5 mg of melatonin per 1 kg of must. The best results were consistently achieved with the addition of only 0.05 mg of melatonin per 1 kg of must, with wines being enhanced by at least 12% after 3 months, reaching 40% after 9 months, and then reducing to 25% at 12 months. Comparing the two V1variants, and considering that at 3, 6, and 9 months, respectively, better results were obtained for the wine produced with the pumping-over technique, the CS-II-V1 variant was selected as the optimal improved wine. The evolution of anthocyanidins/anthocyanins for this wine variant over 12 months is presented in Figure 5. The results obtained were in accordance with those obtained from the HPLC-MS analyses of the polyphenol compounds (Table S1). The results prove that the addition of melatonin has effects comparable to those obtained by the addition of caffeic and rosmarinic acids before alcoholic fermentation in terms of the anthocyanin content of CS wines [47].

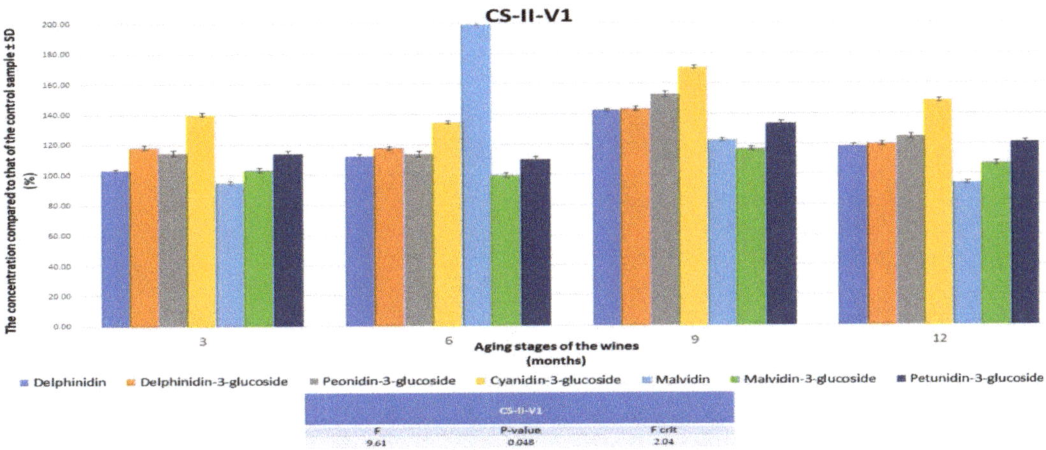

Figure 5. The evolution of the concentrations of anthocyanidins/anthocyanins in CSIIV1, in comparison with the data obtained for the control wine during 12 months. CS—Cabernet Sauvignon; II—pumping over; V1—0.05 mg of melatonin was applied to 1 kg of must.

3.4. The Evolution of Antioxidant Activity in the Aging Process of FN and CS Wines

When studying the antioxidant active principles, it is advisable to conduct qualitative/quantitative HPLC-MS analyses with the total phenolic content (TPC) assay and to use more than one method for antioxidant activity. In this study, the DPPH and ABTS methods, which hinge on electron transfer and imply the reduction in an oxidizing agent, were selected to estimate the antioxidant activity of red wines and to correlate these assays.

To evaluate the antioxidant activity of FN and CS wines, operational parameters such as the dilution factor used depending on the sample content, the specific measurement wavelength, and the maximum absorbance value were first determined. To determine the total content of polyphenols in the sample, the calibration curve of the reference polyphenol gallic acid was also established for seven different concentrations in the range of 0.01–0.1 mg mL^{-1}.

Based on the data values obtained from the TPC analysis over 12 months (Figure 6), the results were found to be consistent with those of the HPLC analyses for the polyphenolic compounds. The gallic acid equivalent (GAE) values found were in the range of 1687 ± 22–2574 ± 31 GAE mg L^{-1} for the FN wines and 2353 ± 29–3378 ± 35 GAE mg L^{-1} in the case of the CS wines. Similar to the HPLC analysis, the most improved wines with melatonin treatment were FNIV3 and CSIIV1. After 12 months from the start of vinification, the values in these wines were 14% higher for FNIV3 and 9% higher for CSIIV1 compared to the control wines.

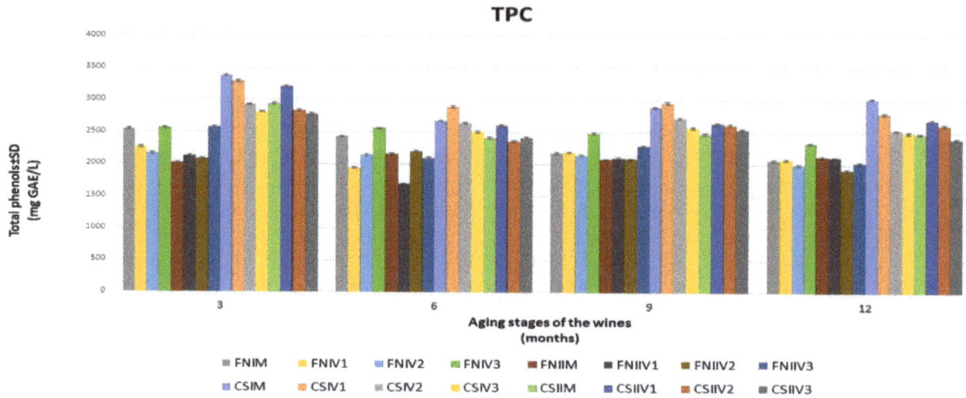

TPC- total phenolic content, FN- Feteasca Neagra, CS- Cabernet Sauvignon, I-punch down, II-pumping over, M-control samples, V1- 50, V2-100, V3-500 µg of melatonin were applied to 1 kg of must.

FNI			FNII			CSI			CSII		
F	P-value	F crit	F	P-value	F crit	F	P-value	F crit	F	P-value	F crit
6.03	0.0042	3.05	3.28	0.040	3.05	3.32	0.038	3.05	4.33	0.0157	3.05

Figure 6. Comparison of the total content of polyphenols in the 12 varieties of treated wines compared to the data obtained for the control wines during the aging process (3, 6, 9, and 12 months and ANOVA results).

Figure 7 shows that one of the effects of treating the wines with melatonin is to increase the capability to scavenge DPPH free radicals, with the highest increases in FNIV3 at 24% and CSIIV1 at 11%. Furthermore, it is observed that the data values obtained after 12 months for CS wines (CSIIM: 2.02 $\pm$ 0.05 TEAC mg L^{-1} and CSIIV1: 2.25 $\pm$ 0.05 TEAC mg L^{-1}, respectively) are higher than those for FN wines (FNIM: 1.6 $\pm$ 0.01 TEAC mg L^{-1} and FNIV3: 1.99 $\pm$ 0.02 TEAC mg L^{-1}, respectively), and the results are consistent with those obtained for TPC.

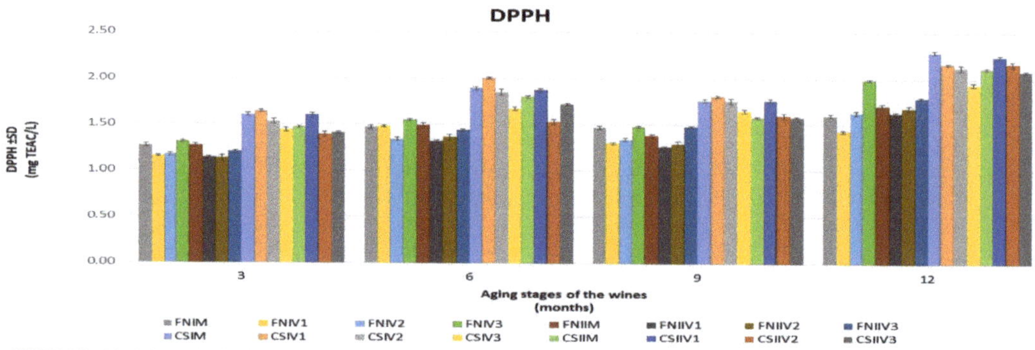

DPPH-2,2-diphenyl-1-picrylhydrazyl, FN- Feteasca Neagra, CS- Cabernet Sauvignon, I-punch down, II-pumping over, M-control samples, V1- 50, V2-100, V3-500 µg of melatonin were applied to 1 kg of must.

FNI			FNII			CSI			CSII		
F	P-value	F crit	F	P-value	F crit	F	P-value	F crit	F	P-value	F crit
5.83	0.0048	3.05	6.05	0.004	3.058	5.007	0.0091	3.05	3.12	0.046	3.05

Figure 7. Comparison of the results obtained from the DPPH assay for the 12 varieties of treated wines compared to the data obtained for the control wines during the aging process (3, 6, 9, and 12 months and ANOVA results).

The results obtained for the ABTS·+ radical scavenging capacity in the FN and CS wine samples did not differ significantly, and an intensification of the antioxidant activity by treating the wines with melatonin was obtained for the FNIV3 wine after 12 months and for the CSIIV1 wine after 9 months (Figure 8). Also, in this case, higher values were found for the CS wines (4.72 ± 0.05–6.52 ± 0.04 TEAC mg L^{-1}) compared to the FN wines (3.48 ± 0.02–5.15 ± 0.04 TEAC mg L^{-1}). The differences correlate with the data obtained for TPC and DPPH.

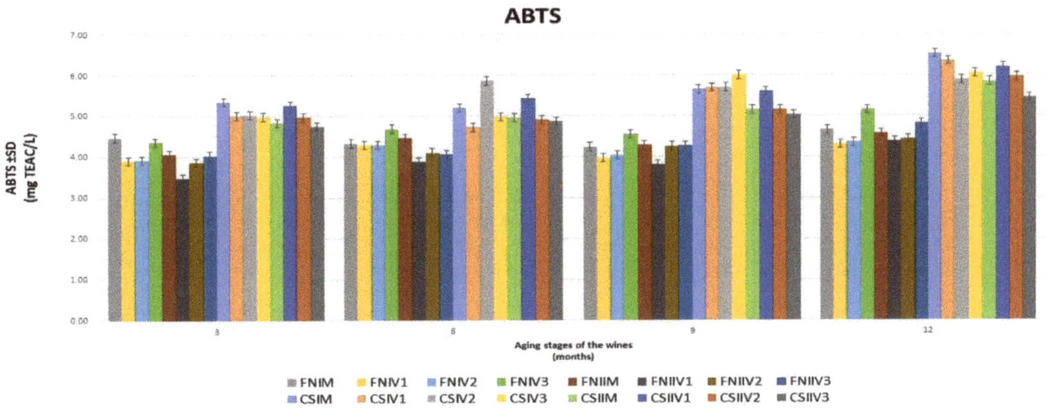

ABTS-2,2-azinobis(3-ethylbenzothiazaline-6-sulfonate), FN- Feteasca Neagra, CS- Cabernet Sauvignon, I-punch down, II-pumping over, M-control samples, V1-0.05, V2-0.1, V3-0.5 mg of melatonin were applied to 1 kg of must.

FNI			FNII			CSI			CSII		
F	P-value	F crit	F	P-value	F crit	F	P-value	F crit	F	P-value	F crit
4.17	0.018	3.05	6.64	0.0027	3.05	3.95	0.021	3.05	5.59	0.0058	3.05

Figure 8. Comparison of the results obtained from ABTS assay for the 12 varieties of treated wines compared to the data obtained for the control wines during the aging process (3, 6, 9, and 12 months and ANOVA results).

The intensification in the antioxidant activity obtained by treatment with melatonin (~10%), reflected in the DPPH and ABTS results, is not significantly different from that obtained in CS wines that were produced by using a mixed fermentation of *Pichia* kudriavzevii M759 and *Saccharomyces cerevisiae* 7VA [48].

After the quantitative monitoring of polyphenolic compounds, based on anthocyanidins/anthocyanins, total polyphenol content, and antioxidant activity of the 12 red wines produced by the addition of melatonin in the preliminary phase of vinification, it can be concluded that in the case of FN wines, antioxidant activity can be significantly increased by the addition of 0.5 mg of melatonin per 1 Kg of must and the application of the punch-down maceration method. In the case of CS wines, remarkable increases are achieved by adding 0.05 mg of melatonin per 1 Kg of must using both maceration and fermentation techniques, with the antioxidant-active ingredients being retained for 12 months by using the pumping-over technique.

4. Conclusions

In this work, after studying diverse technologies with different gradations, it can be concluded that the most suitable wine in terms of content and efficiency of antioxidant activity can be obtained, in the case of FN, by the addition of 0.5 mg of melatonin per 1 kg of must and the application of the punch-down maceration technology, and in the case of CS, by the addition of 0.05 mg of melatonin per 1 kg of must using the pumping-over technique. The addition of melatonin in the initial period of vinification process significantly increases the concentrations of resveratrol (85%), peonidin-3-glucoside (over 100%), or cyanidin-3-glucoside (over 100%) in FN (12 months), and resveratrol (54%) or cyanidin-3-glucoside (50%) in CS (12 months) compared with those from control wines. Also, regarding antioxidant activity, reflected in the DPPH and ABTS results, FN wine improved by at least 40% after 9 months and CS wine improved by ~10% compared to control wines.

The data obtained for Cabernet Sauvignon is comparable to data previously reported in the literature and support that the studied technology is an optimal tool to obtain a wine improved in terms of content and efficiency of antioxidant activity. However, further studies are necessary to monitor the repeatability and the reproducibility of antioxidant activity and the evolution of the profile of polyphenol compounds during the aging process of these two red wines produced using the proposed optimal technologies.

Supplementary Materials: The following supporting information can be downloaded at: https://www.mdpi.com/article/10.3390/foods13060884/s1, Table S1: The values obtained by the HPLC-MS methods from red wines I add citation in main.

Author Contributions: Conceptualization, S.A.V.E. and C.A.; methodology, S.A.V.E., A.A., E.B., A.G.S. and C.A.; validation, C.A. and G.-L.R.; writing—original draft preparation, C.A.; writing—review and editing, S.A.V.E. and E.B.; visualization, A.A. and A.G.S.; supervision, G.-L.R.; project administration, C.A. and G.-L.R. All authors have read and agreed to the published version of the manuscript.

Funding: This work was supported by a grant of the Ministry of Research, Innovation and Digitization, CNCS—UEFISCDI, project number PN-III-P1-1.1-TE-2021-0321, within PNCDI III and by a grant of the Core-Program, developed with the support of Ministry of Research, Innovation and Digitization, project PN 23-02-0101-Contract No. 7N/2023.

Institutional Review Board Statement: Not applicable.

Informed Consent Statement: Not applicable.

Data Availability Statement: The original contributions presented in the study are included in the article/Supplementary Materials, further inquiries can be directed to the corresponding author.

Conflicts of Interest: The authors declare no conflicts of interest.

References

1. Renaud, S.; de Lorgeril, M. Wine, alcohol, platelets, and the French paradox for coronary heart disease. *Lancet* **1992**, *339*, 1523–1526. [CrossRef] [PubMed]
2. Giovinazzo, G.; Carluccio, M.A.; Grieco, F. Wine Polyphenols and Health. In *Bioactive Molecules in Food. Reference Series in Phytochemistry*; Mérillon, J., Ramawat, K., Eds.; Springer Nature: Cham, Switzerland, 2019; pp. 1135–1155, ISBN 9783319780306.
3. Gutiérrez-Escobar, R.; Aliaño-González, M.J.; Cantos-Villar, E. Wine Polyphenol Content and Its Influence on Wine Quality and Properties: A Review. *Molecules* **2021**, *26*, 718. [CrossRef] [PubMed]
4. Cheynier, V. Flavonoids. In *Chemistry, Biochemistry and Applications*; Andersen, O., Markham, K., Eds.; Taylor&Francis: Oxford, UK, 2005; pp. 263–318.
5. Jordão, A.M.; Cosme, F. *Handbook of Anthocyanins: Food Sources, Chemical Applications and Health Benefits*; Warner, L., Ed.; Nova Science Pub Inc.: Hauppauge, NY, USA, 2014; pp. 89–117.
6. Guerrero, R.F.; García-Parrilla, M.; Puertas, B.; Cantos-Villar, E. Wine, Resveratrol and Health: A Review. *Nat. Prod. Commun.* **2009**, *4*, 635–638. [CrossRef] [PubMed]
7. Bimpilas, A.; Panagopoulou, M.; Tsimogiannis, D.; Oreopoulou, V. Anthocyanin copigmentation and color of wine: The effect of naturally obtained hydroxycinnamic acids as cofactors. *Food Chem.* **2016**, *197*, 39–46. [CrossRef]
8. Burkitt, M.J.; Duncan, J. Effects of trans-resveratrol on copper-dependent hydroxyl-radical formation and DNA damage: Evidence for hydroxyl-radical scavenging and a novel, glutathione-sparing mechanism of action. *Arch. Biochem. Biophys.* **2000**, *381*, 253–263. [CrossRef]
9. Ghanem, C.; Bouajila, J.; Rizk, Z.; El Beyrouthy, M.; Sadaka, C.; Gürer, E.S.; Souchard, J.P.; Taillandier, P.; Sharifi-Rad, J.; Nehme, N.; et al. Comparative analysis of pre-fermentation treatments on phenolic compounds and bioactivity in *Vitis vinifera* var. Syrah and var. Cabernet Sauvignon grapes. *Nutrire* **2023**, *48*, 25. [CrossRef]
10. Sacchi, K.L.; Bisson, L.F.; Adams, D.O. A Review of the Effect of Winemaking Techniques on Phenolic Extraction in Red Wines. *Am. J. Enol. Vitic.* **2005**, *56*, 197–206. [CrossRef]
11. Alecu, A.; Albu, C.; Litescu, S.C.; Eremia, S.A.V.; Radu, G.L. Phenolic and Anthocyanin Profile of Valea Calugareasca Red Wines by HPLC-PDA-MS and MALDI-TOF Analysis. *Food Anal. Methods* **2016**, *9*, 300–310. [CrossRef]
12. Pajović-Šćepanović, R.; Wendelin, S.; Eder, R. Phenolic composition and varietal discrimination of Montenegrin red wines (*Vitis vinifera* var. Vranac, Kratošija, and Cabernet Sauvignon). *Eur. Food Res. Technol.* **2018**, *244*, 2243–2254. [CrossRef]
13. Lisov, N.; Čakar, U.; Milenković, D.; Čebela, M.; Vuković, G.; Despotović, S.; Petrović, A. The Influence of Cabernet Sauvignon Ripeness, Healthy State and Maceration Time on Wine and Fermented Pomace Phenolic Profile. *Fermentation* **2023**, *9*, 695. [CrossRef]

14. Bambina, P.; Pollon, M.; Squadrito, M.; Barone, S.; Cinquanta, L.; Corona, O. The effect of prolonged (150 days) post-fermentative maceration in steel tanks and oak barrels on Cabernet Sauvignon wine quality: Mathematical modelization of the phenolic compounds' behaviour. *J. Wine Res.* **2023**, *34*, 247–271. [CrossRef]
15. Singleton, V.L. Effects on red wine quality of removing juice before fermentation to simulate variation in berry size. *Am. J. Enol. Vitic.* **1972**, *23*, 106–113. [CrossRef]
16. Casassa, L.F.; Larsen, R.C.; Harbertson, J.F. Effects of Vineyard and Winemaking Practices Impacting Berry Size on Evolution of Phenolics during Winemaking. *Am. J. Enol. Vitic.* **2016**, *67*, 257–268. [CrossRef]
17. Bai, B.; He, F.; Yang, L.; Chen, F.; Reeves, M.J.; Li, J. Comparative study of phenolic compounds in Cabernet Sauvignon wines made in traditional and *Ganimede fermenters. Food Chem.* **2013**, *15*, 3984–3992. [CrossRef]
18. Puértolas, E.; Hernandez-Orte, P.; Sladana, G.; Alvarez, I.; Raso, J. Improvement of wine making process using pulsed electric fields at pilot-plant scale. Evolution of chromatic parameters and phenolic content of Cabernet Sauvignon red wines. *Food Res. Int.* **2010**, *43*, 761–766. [CrossRef]
19. Dumitriu, G.D.; Teodosiu, C.; Gabur, I.; Cotea, V.V.; Peinado, R.A.; López de Lerma, N. Alternative Winemaking Techniques to Improve the Content of Phenolic and Aromatic Compounds in Wines. *Agriculture* **2021**, *11*, 233. [CrossRef]
20. Hrelia, S.; Di Renzo, L.; Bavaresco, L.; Bernardi, E.; Malaguti, M.; Giacosa, A. Moderate Wine Consumption and Health: A Narrative Review. *Nutrients* **2023**, *15*, 175. [CrossRef] [PubMed]
21. Haunschild, R.; Marx, W. On Health Effects of Resveratrol in Wine. *Int. J. Environ. Res. Public Health* **2022**, *19*, 3110. [CrossRef]
22. Weiskirchen, S.; Weiskirchen, R. Resveratrol: How Much Wine Do You Have to Drink to Stay Healthy? *Adv. Nutr.* **2016**, *7*, 706–718. [CrossRef]
23. Tiras, Z.S.E.; Okur, H.H.; Günay, Z.; Yildirim, H.K. Different approaches to enhance resveratrol content in wine. *Cienc. Tec. Vitivinic.* **2022**, *37*, 13–28.
24. Galano, A.; Tan, D.X.; Reiter, R.J. Melatonin as a natural ally against oxidative stress: A physicochemical examination. *J. Pineal. Res.* **2011**, *51*, 1–16. [CrossRef] [PubMed]
25. Iriti, M.; Rossoni, M.; Faoro, F. Melatonin content in grape: Myth or panacea? *J. Sci. Food Agric.* **2006**, *86*, 1432–1438. [CrossRef]
26. James, A.; Yao, T.; Ke, H.; Wang, Y. Microbiota for production of wine with enhanced functional components. *Food Sci. Hum. Wellness* **2023**, *12*, 1481–1492. [CrossRef]
27. Xiang, G.; Jia, R.; Wang, F.; Wang, S.; Li, Y.; Yao, Y. Exogenous L-tryptophan treatment increases the concentrations of melatonin and aroma compounds in grape berries and wine. *Food Qual. Saf.* **2024**, *8*, fyad042. [CrossRef]
28. Marhuenda, J.; Villaño, D.; Arcusa, R.; Zafrilla, P. Melatonin in wine and beer: Beneficial effects. *Molecules* **2021**, *26*, 343. [CrossRef] [PubMed]
29. Eremia, S.A.V.; Albu, C.; Radu, G.-L.; Alecu, A.; Brinduse, E. The Influence of Melatonin Treatment in the Vinification of Feteasca Neagra and Cabernet Sauvignon Wines on the Profile of Polyphenolic Compounds and Antioxidant Activity. *Antioxidants* **2023**, *12*, 1214. [CrossRef] [PubMed]
30. EUR-Lex. Available online: https://eur-lex.europa.eu/legal-content/EN/TXT/PDF/?uri=CELEX:52022XC0506(04)&from=EN (accessed on 10 January 2024).
31. EUR-Lex. Available online: https://eur-lex.europa.eu/LexUriServ/LexUriServ.do?uri=OJ:L:2012:136:0001:0040:en:PDF (accessed on 10 January 2024).
32. Viegas, O.; Esteves, C.; Rocha, J.; Melo, A.; Ferreira, I.M.P.L.V.O. Simultaneous determination of melatonin and trans-resveratrol in wine by dispersive liquid–liquid microextraction followed by HPLC-FLD. *Food Chem.* **2021**, *339*, 128091. [CrossRef]
33. Carneiro, C.N.; Gomez, F.J.V.; Spisso, A.; Silva, M.F.; Azcarate, S.M.; Dias, F.D.S. Geographical characterization of South America wines based on their phenolic and melatonin composition: An exploratory analysis. *Microchem* **2020**, *158*, 105240. [CrossRef]
34. OIV. *Compendium of International Methods of Wine and Must Analysis*; International Organisation of Vine and Wine: Paris, France, 2009; Volumes 61–82, pp. 433–434.
35. Eremia, S.A.V.; Albu, C.; Radu, G.L.; Ion, M. Different Extraction Approaches for the Analysis of Melatonin from Cabernet Sauvignon and Feteasca Neagra Wines Using a Validated HPLC-FL Method. *Molecules* **2023**, *28*, 2768. [CrossRef]
36. Albu, C.; Eremia, S.A.V.; Penu, R.; Vasilescu, I.; Litescu, S.C.; Radu, G.L. Characterization of the Phenolics and Free Radical Scavenging of Romanian Red Wine. *Anal. Lett.* **2017**, *50*, 591–606. [CrossRef]
37. Litescu, S.C.; Eremia, S.A.V.; Tache, A.; Vasilescu, I.; Radu, G.L. The Use of Oxygen Radical Absorbance Capacity (ORAC) and Trolox Equivalent Antioxidant Capacity (TEAC) Assays in the Assessment of Beverages' Antioxidant Properties. In *Processing and Impact on Antioxidants in Beverages*; Preedy, V., Ed.; Academic Press: London, UK, 2014; pp. 245–251.
38. Wu, X.; Beecher, G.R.; Holden, J.M.; Haytowitz, D.B.; Gebhardt, S.E.; Prior, R.L. Lipophilic and hydrophilic antioxidant capacities of common foods in the United States. *J. Agric. Food Chem.* **2004**, *52*, 4026–4037. [CrossRef]
39. Singleton, V.L.; Orthofer, R.; Lamuela-Raventós, R.M. Analysis of total phenols and other oxidation substrates and antioxidants by means of folin-ciocalteu reagent. *Methods Enzymol.* **1999**, *299*, 152–178.
40. Rodriguez-Naranjo, M.I.; Torija, M.J.; Mas, A.; Cantos-Villar, E.; Garcia-Parrilla, C. Production of melatonin by *Saccharomyces* strains under growth and fermentation conditions. *J. Pineal Res.* **2012**, *53*, 219–224. [CrossRef]
41. Vilela, A. The Importance of Yeasts on Fermentation Quality and Human Health-Promoting Compounds. *Fermentation* **2019**, *5*, 46. [CrossRef]

42. Rodriguez-Naranjo, M.I.; Gil-Izquierdo, A.; Troncoso, A.M.; Cantos-Villar, E.; Garcia-Parrilla, M.C. Melatonin is synthesised by yeast during alcoholic fermentation in wines. *Food Chem.* **2011**, *126*, 1608–1613. [CrossRef] [PubMed]
43. Fischer, U.; Strasser, M.; Gutzler, K. Impact of fermentation technology on the phenolic and volatile composition of German red wines. *Int. J. Food Sci. Technol.* **2000**, *35*, 81–94. [CrossRef]
44. Atanacković, M.; Petrović, A.; Jović, S.; Gojković-Bukarica, L.; Bursać, M.; Cvejić, J. Influence of winemaking techniques on the resveratrol content, total phenolic content and antioxidant potential of red wines. *Food Chem.* **2012**, *131*, 513–518. [CrossRef]
45. Caridi, A.; Sidari, R.; Giuffrè, A.M.; Pellicanò, T.M.; Maria, T.; Sicari, V.; Zappia, C.; Poiana, M. Test of four generations of *Saccharomyces cerevisiae* concerning their effect on antioxidant phenolic compounds in wine. *Eur. Food Res. Technol.* **2017**, *243*, 1287–1294. [CrossRef]
46. Chen, W.; Xiao, Z.; Wang, Y.; Wang, J.; Zhai, R.; Lin-Wang, K.; Espley, R.; Ma, F.; Li, P. Competition between anthocyanin and kaempferol glycosides biosynthesis affects pollen tube growth and seed set of Malus. *Hortic. Res.* **2021**, *8*, 173. [CrossRef] [PubMed]
47. Zhang, B.; Wang, X.Q.; Yang, B.; Li, N.N.; Niu, J.M.; Shi, X.; Han, S.Y. Copigmentation evidence of phenolic compound: The effect of caffeic and rosmarinic acids addition on the chromatic quality and phenolic composition of Cabernet Sauvignon red wine from the Hexi Corridor region (China). *J. Food Compost. Anal.* **2021**, *102*, 104037. [CrossRef]
48. Liu, W.; Ji, R.; Aimaier, A.; Sun, J.; Pu, X.; Shi, X.; Cheng, W.; Wang, B. Adjustment of impact phenolic compounds, antioxidant activity and aroma profile in Cabernet Sauvignon wine by mixed fermentation of *Pichia kudriavzevii* and *Saccharomyces cerevisiae*. *Food Chem. X* **2023**, *18*, 100685. [CrossRef] [PubMed]

Disclaimer/Publisher's Note: The statements, opinions and data contained in all publications are solely those of the individual author(s) and contributor(s) and not of MDPI and/or the editor(s). MDPI and/or the editor(s) disclaim responsibility for any injury to people or property resulting from any ideas, methods, instructions or products referred to in the content.

foods

MDPI

Article

Yeast Starter Culture Identification to Produce of Red Wines with Enhanced Antioxidant Content

Giuseppe Romano [1,†], Marco Taurino [1,†], Carmela Gerardi [1], Maria Tufariello [1], Marcello Lenucci [2] and Francesco Grieco [1,*]

1 National Research Council, Institute of Sciences of Food Production (ISPA), Via Prov. Lecce-Monteroni, 73100 Lecce, Italy; giuseppe.romano@ispa.cnr.it (G.R.); marco.taurino@ispa.cnr.it (M.T.); carmela.gerardi@ispa.cnr.it (C.G.); maria.tufariello@ispa.cnr.it (M.T.)

2 Department of Biological and Environmental Sciences and Technologies, University of Salento, Via Prov. Lecce-Monteroni, 73100 Lecce, Italy; marcello.lenucci@unisalento.it

* Correspondence: francesco.grieco@ispa.cnr.it; Tel.: +39-08-3242-2612

† These authors contributed equally to this work.

Abstract: Grape variety, quality, geographic origins and phytopathology can influence the amount of polyphenols that accumulate in grape tissues. Polyphenols in wine not only shape their organoleptic characteristics but also significantly contribute to the positive impact that this beverage has on human health. However, during the winemaking process, the total polyphenol content is substantially reduced due to the adsorption onto yeast wall polymers and subsequent lees separation. Despite this, limited information is available regarding the influence of the yeast starter strain on the polyphenolic profile of wine. To address this issue, a population consisting of 136 Saccharomyces cerevisiae strains was analyzed to identify those with a diminished ability to adsorb polyphenols. Firstly, the reduction in concentration of polyphenolic compounds associated to each strain was studied by assaying Total Phenolic Content (TPC) and Trolox Equivalent Antioxidant Capacity (TEAC) in the wines produced by micro-scale must fermentation. A total of 29 strains exhibiting a TPC and TEAC reduction $\leq$ 50%, when compared to that detected in the utilized grape must were identified and the nine most-promising strains were further validated by larger-scale vinification. Physico-chemical analyses of the resulting wines led to the identification of four strains, namely ITEM6920, ITEM9500, ITEM9507 and ITEM9508 which showed, compared to the control wine, a TPC and TEAC reduction $\leq$ 20 in the produced wines. They were denoted by a significant ($p < 0.05$) increased amount of anthocyanin, quercetin and trans-coutaric acid, minimal volatile acidity (<0.2 g/L), absence of undesirable metabolites and a well-balanced volatile profile. As far as we know, this investigation represents the first clonal selection of yeast strains aimed at the identifying "functional" fermentation starters, thereby enabling the production of regional wines with enriched polyphenolic content.

Keywords: autochthonous yeast; *Saccharomyces cerevisiae*; wine polyphenols

1. Introduction

Wine shows a complex chemical profile marked by different classes of metabolites that influence its nutritional, physiological, and aromatic value. Some of these compounds, known as varietal molecules, originate from grapes, while others are produced during the fermentation and wine aging processes. Phenolic compounds emerge as a key factor in the quality of wines, especially red ones. They play a dual role, not only enhancing wine quality properties like colour, clarity, taste structure, and resistance to oxidation but also contributing to the prevention of chronic diseases and promoting healthy aging. These physiological effects are particularly associated with flavonoid and stilbene contents, including molecules such as quercetin, catechins, resveratrol and trans-resveratrol [1,2]. The amount of the different polyphenolic classes is an important index of the nutritional quality of the product, since wine stands out as one of the main sources of antioxidants in the Mediterranean diet [3]. The polyphenolic extracts of red wine are a complex mixture of structurally different compounds, some of which exhibit important biological activities, such as the prevention of cardiovascular diseases [4,5], inhibition of inflammatory processes, and protection against certain cancers [6,7]. However, the specific roles of individual polyphenols, whether flavonoids or non-flavonoids, remain not entirely clear, and are often attributed to their synergistic action. Several factors influence the phenolic content of wine, with some of the most important ones being the physical or enzymatic interventions during maceration and the yeast strain used in the fermentation process. Currently, ongoing research delves into the mechanisms through which yeast influences the colour and polyphenolic compound content of wine. Three modes of interaction between yeast and the polyphenolic component have already been described. The first mechanism involves the adsorption of polyphenols on the cell wall of yeasts. However, although yeast has been shown as a factor capable of inducing the loss of part of polyphenols in wines, it remains unclear whether anthocyanin adsorption on the cell wall is the only mechanism at play.

The cell wall of *S. cerevisiae* is characterised by outward-facing mannoproteins bound to oligosaccharides, glucans, and chitin [8]. The different polarities of these cell wall components affect the yeast's capacity to absorb and retain certain classes of molecules, such as polyphenols, volatile compounds, and fatty acids [9]. The adsorption of molecules onto the yeast cell wall is further influenced by its porosity, with greater interstitial spaces providing an increased surface area that favours adsorption [10]. During alcoholic fermentation, the substantial biomass generated leads to a significant proportion of polyphenols being adsorbed onto the cell walls and subsequently removed from the wine along with the lees. It is plausible that different yeast strains have distinct cell wall composition, influencing the varying degrees of adsorption of phenolic compounds. Moreover, the potential for certain strains to demonstrate distinct adsorption patterns, selectively interacting with specific classes of polyphenols, cannot be ruled out.

Another form of interaction is associated with the enzymatic activity of β-glucosidase, which is released by the yeasts themselves [11]. Most of the anthocyanins in wine exist in a glycosylated form, i.e., bound to a sugar. In this state, they are much less susceptible to chemical or enzymatic oxidation. Therefore, the action of β-glucosidase, which generates the respective aglycones (anthocyanidins) in the wine, can promote their removal during winemaking [12].

Finally, some yeast strains release polysaccharides capable of binding with polyphenols, forming stable complexes over time. The presence of these complexes is directly related to the sensations of volume and roundness in the mouth, as well as the stability of wine over time. Research has demonstrated that several yeast metabolites, including pyruvic acid, can react with anthocyanins in grapes to form stable pigmentation and contribute to the aging of red wines.

Unfortunately, the absorption of these molecules onto yeast cell walls, and consequently, their decrease in the produced wine, poses a significant challenge in the fermentation process. This issue has garnered attention from researchers and winemakers. In this light, to prevent or reduce the loss of bioactive molecules through absorption, we studied the absorption capacity of some indigenous yeast strains selected in the Apulia region (Southern Italy).

The use of autochthonous selected yeasts initially gained popularity in white wine production and later extended to the crafting of red wines [13]. In recent years, several studies have underlined the pivotal role played by the microbiota associated with the "terroir" where a particular grape cultivar is cultivated. This microbiota imparts unique sensory properties to the resulting wine [14,15].

The employment of selected autochthonous yeast strains emerges as a powerful tool to enhance the organoleptic and sensory attributes of distinctive regional wines, establishing a stronger connection between these wines and their terroir [16].

The natural biodiversity of Apulian autochthonous yeast strains has been widely investigated [17–25]. It therefore appeared important to exploit our knowledge to identify yeast strains able to enhance the phenolic compound content and, consequently, the functional properties of produced wines.

As a first step, we proceeded with the characterization of a population of selected starter cultures according to their ability to minimize the reduction in the concentration of polyphenolic compounds during a micro-scale must fermentation. Then, the performances of the most effective strains were further validated through larger-scale vinification, and the resulting wines were analysed for their polyphenolic profiles. To the best of our knowledge, this study represents the first clonal selection of yeast strains directed towards the production of regional wines enriched in their polyphenolic content, to be used in the near future for the development of "functional" fermentation starter.

2. Materials and Methods

2.1. Yeast Strains

Yeast strains used in the present study were deposited in Agro-Food Microbial Culture Collection of ISPA (http://www.ispacnr.it/collezioni-microbiche, accessed on 18 December 2023). Yeasts were cultured in YPD broth (10 g/L yeast extract, 20 g/L peptone, 20 g/L glucose, 20 g/L agar) at 28 °C for 24 h, and maintained at −80 °C in glycerol 50%. Yeast populations were sampled at the end of the alcoholic fermentation process. Yeast total genomic DNA was extracted according to De Benedictis et al. [17] and isolates were genetically distinguished at strain level by inter-delta typing [26].

2.2. Vinifications

Primitivo (*Vitis vinifera*) grapes were sampled in a vineyard, with deep, clay-limestone soil, located in Cutrofiano (Lecce, Apulia, southern Italy), an area with a temperate climate. The vineyard was organically managed, fertilized with organic manure and without the use of fungicides.

The strains were firstly tested by a microfermentation assay in Primitivo grapes must (sugars 190 g/L, 20° Brix, pH 3.31) Then, the selected yeast strains were further assayed by inoculating two liters of Primitivo grape must (sugars 206 g/L, 21° Brix, pH 3.25).

Both grape musts were previously added with 100 mg/L potassium metabisulphite and the alcoholic fermentation was carried out in triplicate as described by Grieco et al. [27]. The samples of fermented must were stored at −20 °C until required for analysis. Each fermentation experiment was carried out by performing three simultaneous independent repetitions. The commercial starter CM was used as control since it was the most used strain by the winemakers in the sampled area. WineScanTM Flex (FOSS Italia S.r.l., Padova, Italy) was used to determine the total acidity and volatile compounds as well as the concentration of ethanol, reducing sugars, malic and lactic acids, and glycerol. Samples were centrifuged at 8000× *g* for 10 min and then analysed. The analyses were performed in triplicate.

2.3. Total Polyphenols Content

The total amount of polyphenols was measured by the optimized Folin–Ciocalteu method [28]. The total phenolic content in wine extracts was determined by measuring the absorbance at 765 nm according to the Folin–Ciocalteu colorimetric method. Results were expressed as milligram gallic acid equivalents per liter (mg GAEs/L).

2.4. Total Anthocyanin Content

The total content of anthocyanins was determined on wine samples using the pH differential method [29]. The wine samples were mixed using the appropriate dilution factor, with two different solutions to obtain different pH values, prepared as previously described [30]: pH 1.0 potassium chloride buffer (0.025 M KCl) and pH 4.5 sodium acetate buffer (0.4 M $CH_3CO_2Na{\cdot}3H_2O$). After 15 min incubation at room temperature, the absorbance of the samples was measured at 520 nm and 700 nm (Shimadzu UV-1800, spectrophotometer, Kyoto, Japan). The total content of anthocyanins, expressed as oenin equivalents, was calculated according to the formula described in Lee et al. [29] using a MW (molecular weight) = 493.5 g/mol for malvidin-3-glucoside and a molar extinction coefficient $\varepsilon = 2690\ L\ mol^{-1} \cdot cm^{-1}$.

2.5. TEAC Antioxidant Capacity Determination

The Trolox equivalent antioxidant capacity (TEAC) assay is based on the scavenging ability of antioxidants to quench the radical cationic activity of 2,20-azinobis (3-ethylbenzoithiazolone 6-sulphonate) ($ABTS^+$). The assay was performed as previously described [31] with some modifications. To generate the $ABTS^+$ radical cation, ABTS was dissolved in water (7 mM) and incubated with 2.45 mM potassium persulfate (final concentration) in the dark at room temperature for 12–16 h before use. For the calibration curve, the $ABTS^+$ solution was diluted with water to an absorbance value of 0.70 (±0.02) at 734 nm and mixed with 20 µL of Trolox standard solutions (from 0 to 25 µM). The assay was performed with extracts from wine, and absorbance was determined at 734 nm. Values were expressed as µmol Trolox equivalents (TE)/L.

2.6. Determination of Polyphenolic Profile of Wines

A reversed-phase HPLC analytical method was used for the analysis of polyphenolic compounds. The apparatus was an Agilent-1100 liquid chromatograph (Agilent Technologies, Santa Clara, CA, USA) equipped with a DAD detector (Agilent 1260 Infinity) and the separation was performed on a C18 column (5 µm UltraSphere 80 Å, 4.6 i.d. × 250 mm length) following the conditions described by Gerardi et al. [32]. Chromatograms were acquired at 520, 280, 320, 370 and 306 nm. The following reference compounds (purchased from Sigma-Aldrich, Saint Louis, MO, USA) were used, each with its retention time indicated in parentheses: quercetin (37.17 min), gallic acid (5.57 min), catechin (12.09 min), oenin (27.60 min), trans-resveratrol (35.74 min), caftaric acid (10.42 min), coutaric acid (14.43 min).

Identification of compounds was based on the comparison of peak retention time with the retention time and UV–vis spectra of pure standards while quantification was performed by adopting the external standard method.

2.7. Volatile Profile

The extraction of volatile compounds was performed by a solid phase microextraction in combination with gas chromatography coupled to mass spectrometry (SPME-GC/MS). According to Palombi et al. [33], 100 µL of internal standard solution (IS, 4-methyl-2-pentanol, 300 mg L^{-1}) was added to a volume of 5 mL of wine in a 20 mL headspace vial (Alltech Corp., Deerfield, IL, USA). A 50/30 DVB-CAR-PDMS solid phase microextraction (SPME) fiber (Supelco, Bellofonte, PA, USA) was inserted into the vial and let to adsorb volatiles for 30 min at 40 °C and then transferred to the injector port (250 °C) where desorption occurred in 2 min. Splitless mode was selected as injection mode. GC-MS analyses were performed on a GC 6890 (Agilent Technologies, Santa Clara, CA, USA) coupled to an Agilent MSD 5973 Network detector using a HP-INNOWAX capillary column (60 m × 0.25 mm, 0.25 µm, J & W Scientific Inc., Folsom, CA, USA) as reported by Tufariello et al. [34]. The annotation of the volatile compounds was achieved by comparing mass spectra with those of the data system library (NIST 98, $p > 90\%$), with the retention data of commercially available standards and MS data reported in the literature. Concentration of each volatile compound was assessed by the internal standard method [35,36].

2.8. Statistical Analysis

Values reported, represent the mean ± standard deviation of three independent replicates. The Analysis of Variance (ANOVA) and Tukey's post hoc method were applied to highlight significant differences between each yeast strain versus control for chemical parameters (p value < 0.05). Principal Component Analysis (PCA) was applied to separate yeast strains tested according to the results of the chemical analyses. Tukey's post hoc method was carried out for assessing significant differences between means ($p < 0.05$) by using SigmaStat software Version 3.1 (Jandel Corp., Erkrath, Germany). The principal component analysis of volatile compounds was carried out using the Statistica 6.0 software package.

3. Results

A laboratory-scale test was set up in liquid, using natural must, and the capacity of 136 strains from the ITEM Collection to adsorb polyphenols onto their cell wall was subsequently evaluated by chemical assays on the resulting wines. A starter yeast population was then characterized with a total number of 136 strains (Table S1). The adopted procedure included the following steps: establishing of a population of starter yeasts extracted from the ITEM Collection, seedling on a YPD agar plate, preparing liquid cultures in YPD, inoculating selected must, initiating fermentation in triplicate, and assessing the total polyphenolic content (TPC). For inoculum preparation, 5 mL of Primitivo must was placed in sterile tubes with a 13 mL volume cap. Each thesis was individually inoculated with a yeast concentration of 0.4×10^6 CFU/mL. Fermentations were carried out for 15 days, following which the determination of TPC was carried out for each sample using the Folin-Ciocalteu colorimetric assay and quantified as micrograms of gallic acid equivalent per mL (μgGAE/mL).

Upon analyzing the concentrations of polyphenolic compounds in the wines produced by each yeast, we identified 29 strains exhibiting a TPC reduction of ≤50%, when compared to that detected in the utilized grape must. Additionally, the determination of the antioxidant capacity (TEAC), quantified as μmol of Trolox Equivalents per mL (μmolTE/mL), performed by $ABTS^+$ colorimetric assay, was also reserved for the wines produced with the above 16 strains (Table 1).

In view of the lower reduction of both TPC and TEAC, the strains 83, 84, 86, 99, 105, 106, 112, 113 and 135 were selected for the subsequent experiments.

One liter of pasteurized Primitivo must was separately inoculated with 0.4×10^6 CFU/mL of each of the above nine strains. Fermentations were conducted in triplicate and they took a regular course completing the alcoholic fermentation process in above eight days. The wines were analyzed by Fourier transform infrared spectrophotometry (FT-IR) for the main chemical parameters. Table 2 shows the results of the main physical-oenological and color parameters of the finished product, respectively.

The ethanol amounts (g/100 mL) ranged from 11.97 (135) to 11.43 (86). The residual sugar contents determined for all the obtained wines was found in all wines below 2 g/L, a concentration consistent with a completed fermentation. The total acidity (TA) showed an amount varying from 7.50 g/L to 6.76 g/L, whereas the levels of volatile acidity (VA) ranged from 0.10 g/L to 0.14 g/L. As expected, all produced wine showed an acetic acid (VA) concentration < 0.2 g/L value. The nine selected yeast strain also produced a satisfactory quantity of glycerol, whose concentration ranged ranging from 7.75 to 6.38 g/L.

The nine resulting wines were then analyzed by assessing the concentration of total polyphenols (μgGAE/mL) by the Folin–Ciocalteu colorimetric assay and the antioxidant power (μmol TE/mL) through the TEAC assay (Table 3). As expected, the comparison of TPC and antioxidant activity in wines obtained by different yeast strain revealed a positive correlation.

When inoculated into a larger volume of must, the nine yeast strains demonstrated a decreased capacity to adsorb wine polyphenol compounds on their external surfaces compared to the results reported in Table 1. Specifically, strains 6920, 9500, 9507, and 9508 exhibited a reduction rate of TPC and TEAC of less than 20%.

Table 1. Analysis of Total Phenolic Content (TPC) and Trolox Equivalent Antioxidant Capacity (TEAC) of wines obtained from selected yeasts based on the least reduction of the polyphenolic component. The results of the nine best-performing strains were reported in bold.

ID	Strain	TPC (µgGAE/mL)	%TPC Reduction	TEAC (nmolTE/mL)	%TEAC Reduction
20	6978	607.60 ± 5.26	−48%	47.63 ± 0.97	−46%
63	8754	634.99 ± 5.43	−46%	63.70 ± 2.43	−27%
64	8744	623.57 ± 7.46	−47%	54.24 ± 3.01	−38%
69	8766	601.61 ± 5.76	−48%	63.50 ± 2.09	−27%
70	8767	612.87 ± 8.09	−48%	59.98 ± 1.54	−31%
71	8769	620.08 ± 5.73	−47%	72.75 ± 2.22	−17%
74	8772	627.44 ± 5.90	−46%	74.66 ± 2.00	−15%
77	8775	609.70 ± 8.78	−48%	72.22 ± 1.76	−17%
78	8776	651.59 ± 4.94	−44%	73.20 ± 1.05	−16%
80	8778	605.66 ± 3.76	−48%	61.80 ± 0.97	−29%
81	8779	616.54 ± 3.87	−47%	55.76 ± 0.90	−36%
82	8780	639.91 ± 5.85	−45%	63.57 ± 1.02	−27%
83	**6993**	**785.14 ± 8.47**	**−33%**	**64.04 ± 1.00**	**−27%**
84	**6920**	**789.55 ± 9.76**	**−32%**	**64.81 ± 0.99**	**−26%**
86	**8766**	**699.95 ± 6.73**	**−40%**	**61.60 ± 2.37**	**−30%**
87	6977	607.68 ± 2.38	−48%	63.23 ± 1.76	−28%
89	8795	612.56 ± 5.77	−48%	59.79 ± 1.43	−32%
90	17,292	626.68 ± 4.38	−46%	56.06 ± 0.98	−36%
91	17,293	636.65 ± 5.65	−45%	66.38 ± 1.76	−24%
92	9502	608.41 ± 2.90	−48%	64.86 ± 2.02	−26%
96	9531	643.67 ± 3.35	−45%	64.81 ± 1.74	−26%
98	1407	629.36 ± 4.84	−46%	62.40 ± 1.64	−29%
99	**14,093**	**746.44 ± 2.37**	**−36%**	**64.79 ± 1.23**	**−26%**
105	**9500**	**786.23 ± 5.74**	**−33%**	**62.95 ± 2.09**	**−28%**
106	**9501**	**755.92 ± 6.68**	**−35%**	**57.90 ± 0.98**	**−34%**
107	9502	607.47 ± 2.64	−48%	59.17 ± 0.78	−32%
112	**9507**	**766.47 ± 5.89**	**−34%**	**68.38 ± 2.20**	**−22%**
113	**9508**	**779.70 ± 3.65**	**−33%**	**65.78 ± 1.43**	**−25%**
135	**9530**	**729.32 ± 5.74**	**−38%**	**58.66 ± 0.98**	**−33%**
Must		1167.45 ± 8.46		87.49 ± 3.01	

Table 2. Analysis by FT-IR assay of the main chemical parameters of wines produced from the nine selected *S. cerevisiae* strains and the commercial yeast strain (CM).

Strain	Ethanol	Sugars	TA	VA	Malic Acid	Lactic Acic	Glycerol
6993	11.45 ± 0.023	1.6 ± 0.034	7.11 ± 0.008	0.14 ± 0.013	2.58 ± 0.005	0.15 ± 0.004	6.38 ± 0.055
6920	11.58 ± 0.004	1.34 ± 0.059	7.14 ± 0.056	0.07 ± 0.15	2.64 ± 0.002	0.08 ± 0.001	6.54 ± 0.088
8766	11.43 ± 0.061	1.28 ± 0.334	7.41 ± 0.022	0.15 ± 0.008	2.61 ± 0.025	0.2 ± 0.057	7.03 ± 0.028
14093	11.47 ± 0.004	1.39 ± 0.105	7.11 ± 0.03	0.11 ± 0.003	2.56 ± 0.035	0.07 ± 0.062	6.68 ± 0.055
9500	11.75 ± 0.033	1.35 ± 0.136	7.24 ± 0.006	0.15 ± 0.024	2.49 ± 0.039	0.18 ± 0.017	7.21 ± 0.116
9501	11.58 ± 0.014	1.40 ± 0.130	6.76 ± 0.013	0.11 ± 0.004	2.05 ± 0.012	0.39 ± 0.007	7.48 ± 0.128
9507	11.74 ± 0.019	1.48 ± 0.160	7.25 ± 0.017	0.16 ± 0.011	2.54 ± 0.016	0.18 ± 0.059	6.97 ± 0.09
9508	11.96 ± 0.028	1.67 ± 0.073	7.5 ± 0.039	0.13 ± 0.002	2.67 ± 0.016	0.12 ± 0.057	7.19 ± 0.046
9530	11.97 ± 0.004	1.24 ± 0.055	7.97 ± 0.022	0.10 ± 0.018	2.83 ± 0.055	0.33 ± 0.022	7.75 ± 0.033
CM	8.14 ± 0.012	39.51 ± 0.014	7.48 ± 0.023	0.14 ± 0.002	2.47 ± 0.047	0	5.88 ± 0.043

TA, total acidity. VA, volatile acidity. The ethanol concentration is expressed as g/100 mL. The other values are expressed as g/L. CM, commercial starter control. Initial sugar content in the must was 21° Brix.

Table 3. Analysis of Total Phenolic Content (TPC) and Trolox Equivalent Antioxidant Capacity (TEAC) of wines obtained from the nine selected yeasts and the commercial strain (CM).

ITEM	TPC (μgGAE/mL)	TPC % Reduction	TEAC (μmolTE/mL)	TEAC % Reduction
6993	600.81 ± 1.10 ^{dc}	−32%	4.49 ± 0.06 ^{cd}	−21%
6920	**701.70** ± 0.60 ^{b}	**−20%**	**4.56** ± 0.16 ^{cd}	**−19%**
8766	611.59 ± 37.60 ^{cde}	−30%	4.38 ± 0.02 ^{cd}	−22%
14093	630.71 ± 0.76 ^{bcde}	−28%	4.44 ± 0.07 ^{cd}	−21%
9500	**699.74** ± 1.41 ^{bc}	**−20%**	**4.62** ± 0.02 ^{bc}	**−18%**
9501	551.04 ± 20.00 ^{ef}	−37%	4.22 ± 0.14 ^{d}	−25%
9507	**712.13** ± 2.49 ^{b}	**−19%**	**4.51** ± 0.02 ^{cd}	**−20%**
9508	**801.64** ± 53.60 ^{a}	**−9%**	**4.97** ± 0.20 ^{b}	**−12%**
9530	639.39 ± 26.60 ^{bcd}	−27%	4.43 ± 0.04 ^{cd}	−22%
CM	502.04 ± 0.13 ^{f}	−43%	3.12 ± 0.02 ^{f}	−44%
Must	876.57 ± 5.15 ^{a}	//	5.66 ± 0.04 ^{a}	//

The reported values are means (n = 3) ± standard deviation. Mean that do not share a letter are significantly different. The results of the four best-performing strains were reported in bold. Data were submitted to one-way analysis of variance (ANOVA), Tukey's post hoc method was applied to establish differences between each yeast strain versus control. Different letters indicate statistically significant differences ($p < 0.05$).

Table 4 reports the mean values and standard deviations of phenolic molecules identified and quantified in wines obtained by the nine selected yeast strains. Among the hydroxybenzoic acids, gallic acid was detected, while in the hydroxycinnamic acid derivatives, caffeoyl tartaric (caftaric) and *p*-coumaroyl tartaric (coutaric) acids were detected in the wine samples. Gallic acid showed higher amounts in wine produced by the 6993 strain. Caftaric acid was more concentrated in wine obtained from the 6993, 6920, 8766, 9507, and 9508 strains compared to the control strain.

Table 4. Polyphenols content (mg/L) of wines obtained by the nine selected yeasts and the commercial strain (CM).

Strain	Quercetin	Gallic Acid	Catechin	Oenin	Trans-Resveratrol	Trans-Coutaric Acid	Caftaric Acid
				mg/L			
6993	2.857 ± 0.091 ^{b}	32.872 ± 0.486 ^{b}	58.207 ± 2.593 ^{b}	115.251 ± 1.096 ^{b}	1.412 ± 0.118 ^{a}	14.097 ± 0.113 ^{b}	52.723 ± 0.900 ^{b}
6920	2.433 ± 0.152 ^{b}	30.381 ± 0.549 ^{a}	49.160 ± 4.002 ^{a}	120.966 ± 0.643 ^{a}	2.766 ± 0.097 ^{b}	16.039 ± 0.214 ^{b}	56.158 ± 0.751 ^{b}
8766	2.369 ± 0.391 ^{b}	25.187 ± 0.590 ^{b}	44.896 ± 0.375 ^{a}	134.179 ± 1.729 ^{b}	1.710 ± 0.069 ^{a}	9.447 ± 0.157 ^{b}	43.842 ± 0.614 ^{b}
14093	2.456 ± 0.295 ^{b}	30.530 ± 0.344 ^{a}	50.545 ± 3.924 ^{a}	121.750 ± 0.924 ^{a}	1.721 ± 0.016 ^{a}	8.806 ± 0.059 ^{b}	42.246 ± 0.198 ^{a}
9500	3.375 ± 0.441 ^{b}	27.015 ± 0.427 ^{b}	48.366 ± 0.335 ^{a}	131.042 ± 2.551 ^{b}	3.085 ± 0.067 ^{b}	8.341 ± 0.052 ^{b}	42.143 ± 0.332 ^{a}
9501	2.132 ± 0.391 ^{b}	29.516 ± 0.673 ^{a}	42.259 ± 3.506 ^{a}	121.123 ± 3.725 ^{a}	3.528 ± 0.088 ^{b}	0.563 ± 0.000 ^{b}	3.130 ± 0.113 ^{b}
9507	2.184 ± 0.115 ^{b}	24.349 ± 0.179 ^{b}	44.707 ± 0.086 ^{a}	127.983 ± 1.007 ^{a}	2.139 ± 0.446 ^{b}	14.093 ± 0.046 ^{b}	60.821 ± 1.713 ^{b}
9508	2.902 ± 0.189 ^{b}	26.484 ± 0.875 ^{b}	47.336 ± 0.815 ^{a}	125.542 ± 3.056 ^{a}	1.442 ± 0.028 ^{a}	8.730 ± 0.138 ^{b}	54.666 ± 1.254 ^{b}
9530	0.671 ± 0.101 ^{a}	26.277 ± 0.834 ^{b}	46.072 ± 0.704 ^{a}	92.511 ± 0.632 ^{b}	0.515 ± 0.007 ^{b}	7.157 ± 0.154 ^{a}	37.062 ± 0.796 ^{b}
CM	0.677 ± 0.059 ^{a}	29.559 ± 0.009 ^{a}	47.319 ± 3.573 ^{a}	124.157 ± 0.709 ^{a}	1.565 ± 0.079 ^{a}	6.832 ± 0.017 ^{a}	40.754 ± 0.010 ^{a}

CM, commercial yeast strain. Data are mean ± S.D. and are representative of three different assays performed. Data were submitted to one-way analysis of variance (ANOVA), Tukey's post hoc method was applied to establish differences between each yeast strain versus control. Different letters indicate statistically significant differences ($p < 0.05$).

Coutaric acid showed a higher content in wines fermented by all strains, except 9501 and 9530, when compared to the control. The identified flavonol compound quercetin showed higher content in all samples in comparison to the control strain, except for the wine fermented by the 9530 strain. Catechin, belonging to the flavanol group of flavonoids, had a significantly higher amount only in wine fermented by 6993 strain.

Oenin was the non-acylated anthocyanin more abundant in all samples; its content was higher than control in wines fermented by the 8766 and 9500 strains. Trans-resveratrol, belonging to the stilbene class of polyphenols, showed a higher amount in wines fermented by the 6920, 9500, 9501, and 9507 strains compared to the control.

The obtained data (Table 4) indicated that wines produced using the 9530 yeast strain starter contains significantly lower or comparable amounts of all identified polyphenol molecules compared to the control. The total anthocyanin content detected in each of the produced wines is shown in Figure 1. Total anthocyanin content was significantly higher than the control for all samples, except for the wine fermented by the 8766 strain, while the sample fermented by the 9530 strain exhibited a significantly lower anthocyanin concentration.

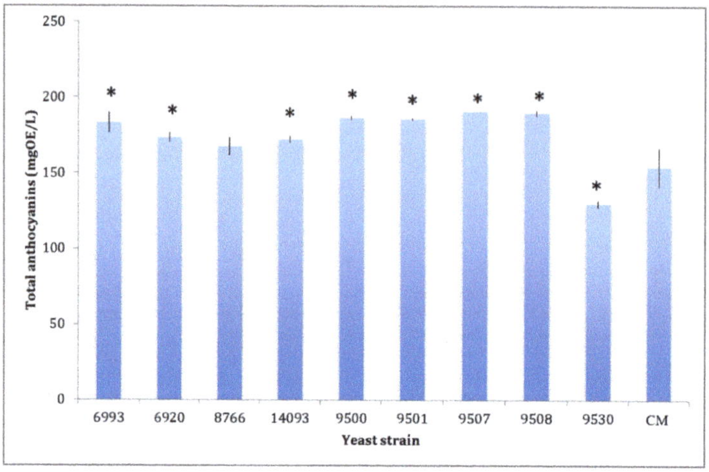

Figure 1. Comparison of total anthocyanin content in wines obtained by the nine-selected yeast and the commercial strain (CM). Data are mean ± S.D. and are representative of three different assays performed. OE = Oenin equivalent. Data were submitted to one-way analysis of variance (ANOVA), Tukey's post hoc method was applied to establish differences between each yeast strain versus control. * indicates statistically significant differences ($p < 0.05$) between each yeast strain versus control.

HPLC data were subjected to Principal Component Analysis (PCA) based on Pearson correlation ($n - 1$) (Figure 2). The two principal components described the 66.33% of the total variance (36.98% and 29.35% for PC1 and PC2) of the phenolic acids data matrix: The analysis permitted the separation of yeast strains on the plane defined by two principal PCs. ITEM 9501, CM, and 9530, located to the left of PC2 result anti-correlated with the identified molecules, showing low concentrations. ITEM 9500 and ITEM 8766 cluster along the positive component of PC2 and show good correlation with trans-resveratrol, oenins, and quercetins. Along the positive component of PC1, ITEM 6993 is positively associated with trans-coutaric, caftaric, gallic acids and catechin. Finally, in the plane bounded by the positive components of the two PCs, we locate the cluster formed by ITEM 9507, 6920, 9508 and 9500 united by values closer to the mean value of all the molecules analyzed, showing profiles that are more balanced.

Based on the processing of the obtained data, strains ITEM6920, ITEM9500, ITEM9507, and ITEM9508 fulfilled the fundamental parameters required, in that: the grafted alcoholic fermentations had a consistent progression and duration (<10 days). FT-IR analysis confirmed the attainment of the expected ethanol content, with sugars being completely consumed (residual < 2 g/L) in the produced wines. Notably, these four demonstrated the absence of undesirable metabolites, particularly showing very low production of acetic acid (volatile acidity < 0.2 g/L). The wines produced showcased a reduction in the concentration of total polyphenols (≤20% TPC reduction) and a reduction in the antioxidant power (≤20% TEAC reduction). As a result, these four strains emerged as promising candidate starters for subsequent pilot and industrial-scale experiments set to be carried out in the winery. In evaluating the fermentation performance of these starter candidate strains, secondary fermentation products were detected and quantified in the nine wines by SPME-GC-MS analysis (Table 5).

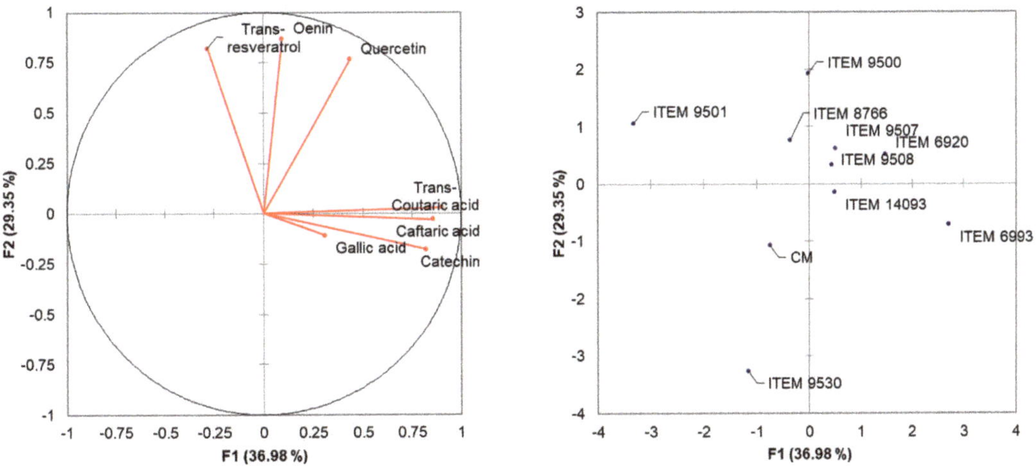

Figure 2. Principal Component Analysis (PCA) performed employing the data obtained by the HPLC analysis of the wines obtained using the nine selected strains.

A total of 23 volatile substances belonging to the classes of alcohols, esters, aldehydes, phenols, and volatile acids were identified. Within the alcohol class, all samples presented higher values, ranging from 26.86 mg/L (ITEM 9530) to 51.57 mg/L (ITEM 9508). Notably, 2+3-methyl-1-butanol and phenylethanol emerged as the most abundant molecules among alcohols. The second class of molecules that quantitatively affect the complete volatile profile of the different samples is that of esters associated with fruity notes. The highest values were detected in ITEM 9507 (7.42 mg/L), ITEM 9500 (7.15 mg/L) ITEM 9501 (6.04 mg/L), ITEM 8766 (5.37 mg/L), and finally ITEM 6920 (5.03 mg/L). The presence of acetic acid was not detected, while an acidic component was identified, in quantities varying from 2.05 mg/L (CM) to 4.64 mg/L (ITEM 9508), due to the presence of butanoic, hexanoic, octanoic, and decanoic which contribute positively with notes of freshness. Furthermore, in order to identify the yeast strains that produce wines with the best volatile profiles, principal component analysis (PCA) was performed on the Pearson correlation matrix, on the concentrations of molecules detected by GC-MS (Figure 3).

The two bi-plots PC1 vs. PC2 showed the projection of the variables considered on the plane defined by the first and second principal components that explain 49.76% of the total variance.

Along the positive component of PC1 (29.15%), the strains ITEM 9500, ITEM 9501, ITEM 9507, and ITEM 9508 cluster in the same group for a positive correlation with n-decanoic acid, isoamyl acetate, monoethyl succinate, ethyl lactate, ethyl hexanoate, 3-hexen-1-ol (E), methyonol. On the other hand, along the positive component of PC2 (20.61), strain ITEM 6920 differs from the others for a greater content of phenylethanol, ethyl butanoate, octanoic acid, 2+3-methyl-1-butanol, ethyl octanoate and 3-hexen-1-ol (Z). All other strains were characterized by low concentrations of all identified volatiles.

Table 5. Volatile Organic Compounds detected in wine samples.

Volatiles (mg/L)	ITEM 6993	±SD	ITEM 6920	±SD	ITEM 8766	±SD	ITEM 14093	±SD	ITEM 9500	±SD	ITEM 9501	±SD	ITEM 9507	±SD	ITEM 9508	±SD	ITEM 9530	±SD	CM	±SD	Statistical Significance
Esters																					
Ethyl butanoate	0.110	0.050	0.715	0.210	0.142	0.060	0.150	0.06	0.760	0.180	0.450	0.170	0.650	0.170	0.611	0.017	0.204	0.042	0.940	0.140	***
Isoamyl acetate	0.722	0.210	1.442	0.620	1.197	0.340	0.170	0.05	1.820	0.640	1.080	0.023	2.050	0.820	0.610	0.032	1.450	0.260	0.514	0.095	*
Ethyl hexanoate	0.149	0.060	0.158	0.050	0.492	0.080	0.127	0.04	0.514	0.120	0.430	0.110	0.610	0.210	0.450	0.120	0.340	0.070	0.110	0.040	**
Hexyl acetate	nd		nd		0.093	0.014	0.076	0.012	0.095	0.014	0.056	0.014	0.066	0.014	0.053	0.011	0.087	0.015	0.250	0.074	***
Ethyl lactate	nd		0.092	0.014	0.235	0.092	0.217	0.051	0.460	0.080	0.560	0.080	0.411	0.130	0.405	0.210	0.326	0.066	0.042	0.011	**
Ethyl octanoate	0.079	0.020	0.570	0.070	0.214	0.014	0.156	0.07	0.440	0.130	0.380	0.110	0.278	0.080	0.316	0.080	0.320	0.080	0.147	0.040	**
Diethyl succinate	0.123	0.060	0.750	0.130	0.341	0.032	0.188	0.012	0.470	0.170	0.270	0.070	0.310	0.070	0.310	0.040	0.167	0.030	0.122	0.040	***
Phenyl acetate	0.353	0.140	0.207	0.080	1.269	0.510	1.160	0.33	1.650	0.910	1.770	0.640	2.110	0.940	0.850	0.170	0.950	0.210	0.281	0.104	*
Mono ethyl succinate	0.654	0.210	1.096	0.340	1.390	0.660	0.860	0.24	0.94	0.18	1.04	0.15	0.930	0.360	0.910	0.310	0.870	0.140	0.142	0.080	ns
Total Esters	**2.19**		**5.03**		**5.37**		**3.10**		**7.15**		**6.04**		**7.42**		**4.52**		**4.71**		**2.55**		
Alcohols																					
2-Methyl-1-propanol	0.20	0.06	0.76	0.12	2.54	0.57	0.95	0.2	1.25	0.04	0.92	0.17	1.62	0.35	0.47	0.06	0.88	0.20	0.55	0.14	***
3-Methyl-1-butanol	27.15	2.35	28.72	5.04	16.12	4.05	18.93	5.1	24.76	5.11	16.90	4.32	21.88	4.15	31.16	4.28	11.87	3.17	14.22	2.07	**
1-Hexanol	1.06	0.04	0.54	0.08	0.82	0.12	0.95	0.18	1.37	0.34	1.78	0.65	1.96	0.62	1.26	0.33	2.87	0.64	1.25	0.04	**
3-Hexen-1-ol (E)	0.17	0.02	0.33	0.03	0.18	0.04	0.26	0.08	0.27	0.07	0.44	0.10	0.22	0.07	0.31	0.01	0.35	0.08	0.28	0.11	ns
3-Hexen-1-ol (Z)	0.17	0.05	0.55	0.08	0.19	0.06	0.25	0.06	0.18	0.06	0.34	0.07	0.21	0.04	0.17	0.05	0.12	0.04	0.27	0.15	**
Methyonol	0.29	0.07	0.36	0.10	0.60	0.17	0.17	0.03	0.55	0.10	0.37	0.08	0.35	0.03	0.22	0.04	0.28	0.06	0.150	0.04	**
Phenylethanol	19.23	3.520	18.607	4.070	11.156	3.070	9.450	1.51	19.400	3.610	8.550	2.170	20.140	5.140	17.980	3.910	10.500	2.180	10.240	1.82	*
Total Alcohols	**48.27**		**49.87**		**31.61**		**30.96**		**47.78**		**29.30**		**46.38**		**51.57**		**26.86**		**26.96**		
Aldehydes																					
Benzaldehyde	0.10	0.0300	0.330	0.0700	0.420	0.1100	0.92	0.24	0.56	0.14	1.10	0.03	1.15	0.07	0.85	0.18	0.870	0.1700	0.25	0.06	***
Volatiles phenol																					
4-Vinylguaiacol	1.480	0.550	nd		4.45	0.95	1.76	0.61	2.55	0.61	2.87	0.51	3.98	0.72	5.38	0.94	4.10	0.92	0.65	0.17	***
Tyrosol	0.995	0.370	nd		5.53	1.04	nd		0.36	0.08	0.77	0.18	0.95	0.18	nd		1.76	0.34	1.040	0.320	***
Total Volatile Phenols	**2.47**				**9.97**		**1.76**		**2.91**		**3.64**		**4.93**		**5.38**		**5.86**		**1.69**		
Volatile acids																					
Butanoic acid	nd		nd		0.26	0.06	0.41	0.07	0.55	0.11	0.63	0.25	0.41	0.07	0.35	0.08	0.36	0.07	nd		***
Hexanoic acid	1.63	0.27	0.83	0.18	1.18	0.04	0.87	0.18	0.77	0.21	0.93	0.18	0.88	0.13	1.34	0.33	0.95	0.16	0.880	0.1700	*
Octanoic acid	1.74	0.14	1.71	0.42	0.63	0.17	0.94	0.24	1.45	0.43	1.76	0.34	1.88	1.04	1.33	0.17	1.36	0.35	0.620	0.2300	ns
n-Decanoic acid	0.43	0.08	0.52	0.17	0.34	0.05	nd		0.56	0.07	0.87	0.21	0.55	0.14	1.62	0.54	0.41	0.11	0.550	0.1600	**
Total Volatile Acids	**3.79**		**3.05**		**2.41**		**2.22**		**3.33**		**4.19**		**3.72**		**4.64**		**3.08**		**2.05**		

nd: not detected; sd: standard deviation; significant differences * $p < 0.05$; ** $p < 0.01$; *** $p < 0.001$.

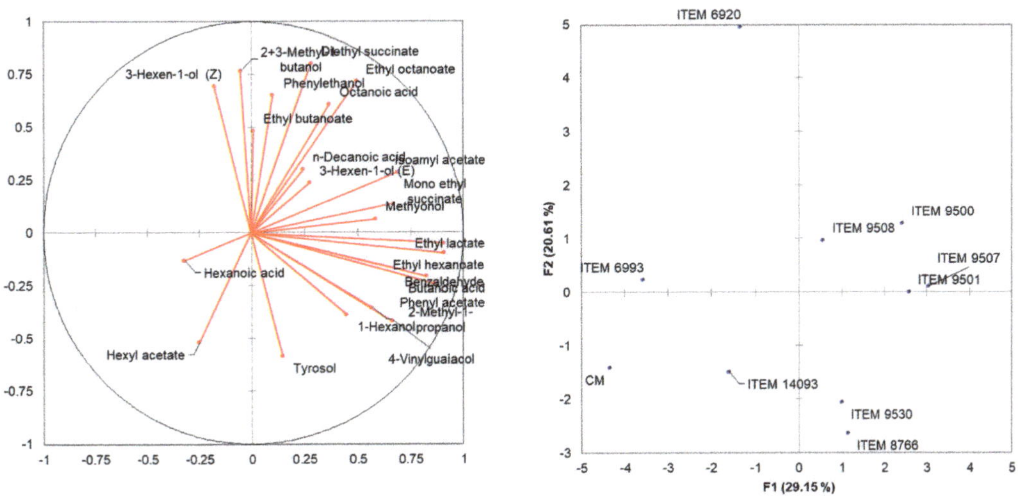

(A) (B)

Figure 3. Two-dimensional Principal Component Analysis (PCA). Scoresplot (**A**) for the nine wines and (**B**) loading plot for Volatile Organic Compounds.

4. Discussion

This investigation marks the first instance of analyzing an autochthonous starter culture collection, with the aim of selecting and characterizing strains exhibiting a low capacity to adsorb polyphenols onto their cell walls. This study establishes selection criteria based on the final concentration of total polyphenols and antioxidant capacity in wine samples obtained at the end of the various experimental phases. Previous studies analyzed polyphenolic variation was by evaluating parameters such as color intensity and shade of the yeast colonies post-fermentation, quantified using empirical criteria [37]. We developed a selection protocol based on the evaluation of the residual polyphenolic component and antioxidant activity in wines fermented with the strains under analysis, employing Folin–Ciocalteu and TEAC assays.

A total of 136 *S. cerevisiae* starter strains were chosen from the CNR ISPA microbial collection, and their capacity to enhance the polyphenolic content in produced wines was evaluated by laboratory-scale fermentation analysis.

The results indicated that all strains decreased the concentration of polyphenolic compounds and, consequently, the antioxidant power. The top-performing strains, demonstrating a lower capacity to adsorb polyphenols, were selected and characterized throughout the various phases of the study. It is known from the literature that yeast's enzymes (pectinase and glycosidases) may affect polyphenolic extraction, modifying the chemical structure of glycosylated phenolic molecules [38]. Moreover, polyphenols interact with yeast by binding to cell wall mannoprotein [9]. The observed decrease in TPC and antioxidant activity is likely related to these interaction mechanisms between polyphenols and yeasts. These interaction mechanisms are related also to the variation of the polyphenolic profile described in Table 4 as hydrolysis process during winemaking alters the rate between free and glycoside forms of phenolic compounds.

Our results confirmed the data obtained by Brandolini et al. [39], who investigated the properties of wines produced by separate inoculation of *S. cerevisiae* strains into the same must. The study highlighted strain-specific abilities to differentially adsorb polyphenols during the vinification process. Similar results were obtained by Kostadinović et al. [40] on Vranec and Merlot wines in Macedonia, emphasizing the strain-specific influence on the concentration of trans-resveratrol and antioxidant activities. The use of different yeast strains also demonstrated varying polyphenolic content in Pinot Noir wines [13]. Indeed, five different yeast starters were tested in several vinifications, in which the *S. cerevisiae* strain RC212 was able to significantly increase the concentrations of total pigments, anthocyanins and tannins. Carrascosa and associates [41] demonstrated that different yeast strains were able to produce Albariño wines characterized by a specific polyphenolic composition. The above results were further confirmed by a recent report [21], in which an unequivocal correlation between the yeast used to promote the fermentation process and the chemical profile of the wine was recognized, thus underlining the strain-specific abilities of yeasts to modify the color and polyphenolic composition of the final product.

Furthermore, a recent study identified yeast starter cultures capable of improving the quality of wine produced from the Italian red cultivar 'Gaglioppo', a cultivar with reduced anthocyanin synthesis [42]. Again, the evidence obtained further highlighted the specific ability of some strains to modify the final amounts of total anthocyanin, total polyphenols and total tannins.

Recently, Grieco et al. [43] highlighted a positive role of indigenous yeast cultures in improving polyphenol content across the industrial production of Negroamaro and Primitivo wines. Statistical analysis showed that the use of indigenous strains increased the concentrations of several classes of polyphenols in the wines produced compared to wines made with a commercial strain.

It can be asserted that the different ability shown by our yeast strains in complexing phenolic compounds on their cell wall is a strain-specific property [44,45]. The evidence that the four yeast starter-culture identified during this study showed the capacity to adsorb the least amount of all types of anthocyanin is consistent with the evidence demonstrated by previous studies in Spain [10], France [45] and Italy [46].

Therefore, it is imperative to consider the adsorption ability of phenolic compounds during selective procedures for yeasts starter culture in wine production [9].

Nevertheless, it is worthy to note the effect of scale comparing the differences between the performance of the selected strains in reducing TPC between the micro-scale and laboratory-scale vinification. This finding is in agreement with our previous investigation [47], where we explored the oenological significance of winemaking scales in evaluating the contribution of new starter cultures to the chemical profile of produced wine.

As regards the volatolomic aspect that influences the sensorial quality, our results confirm that the ITEM 6920-9500-9507-9508 strains contribute to a balanced volatile profile, mainly characterized by secondary fermentation products. In particular, ethyl ester concentrations were influenced by the yeast strain, fermentation temperature, degree of aeration, and sugar content. Both ethyl esters and acetate esters have a key importance in the overall aroma of wine, contributing positive sensory notes like sweet-fruity, grape smell, and sweet balsamic [48,49]. Concerning esters, the mentioned strains showed an increased ester production ofisoamyl acetate, ethyl hexanoate, diethyl succinate, phenyl acetate and mono ethyl succinate. Similarly, among higher alcohols, 2-phenylethanol contributes a floral (pink) aroma [50], but an excess concentration above 300 mg/L would impart a strong and pungent odor and taste [51]. In our study, the use of selected yeast strains allowed us to obtain higher alcohols in concentrations lower than the critical threshold value, with a positive contribution to the sensorial profile, and phenylethanol values higher than the perception threshold (10 mg/L) [52,53] except for ITEM 9501 and 14093. Finally, the contribution of the volatile acid fraction also appears positive. Indeed, fatty acids, produced during fermentation, constitute an important group of aromatic compounds that can provide fruity, cheesy, fatty, and rancid notes. In this case, the quantified fatty acids had levels lower than their perception threshold.

By processing the data obtained at the end of the various experimental phases, we were able to identify four strains (ITEM6920, ITEM9500, ITEM9507, and ITEM9508) capable of producing wine characterized by a higher concentration of total polyphenols, an enhanced antioxidant capacity, and the absence of undesirable metabolites.

5. Conclusions

The results of this study suggest that the exploitation of autochthonous yeast strains can enhance the antioxidant activity and the amount of the phenolic compounds in the produced wine. These autochthonous microbial resources can be denoted as "antioxidant positive strain" [43] since they were able to increase both the health promoting [1] and aromatic properties of wine in synergy with the use of innovative technological processes [54]. Taken together, our finding emphasize the relevance of developing and applying innovative biotechnological approaches to enhance the presence in wine of molecules with potential benefits to human health, thus improving the 'functional parameters' and the overall quality of the final product. Ongoing studies are exploring the industrial application of these four autochthonous strains as starter cultures for the large-scale production of typical red wines.

Supplementary Materials: The following supporting information can be downloaded at: https://www.mdpi.com/article/10.3390/foods13020312/s1, Table S1: List of yeast strains used.

Author Contributions: Conceptualization, F.G. and M.L.; methodology, F.G., C.G. and M.T. (Maria Tufariello); validation C.G. and M.T. (Maria Tufariello); investigation, G.R., M.T. (Maria Tufariello), C.G. and M.T. (Marco Taurino); writing—original draft preparation F.G., M.T. (Marco Taurino), G.R. and M.T. (Maria Tufariello); writing—review and editing, F.G. and M.L. All authors have read and agreed to the published version of the manuscript.

Funding: This research received no external funding.

Institutional Review Board Statement: Not applicable.

Informed Consent Statement: Not applicable.

Data Availability Statement: Data is contained within the article or Supplementary Materials.

Acknowledgments: We thank Leone D'Amico and Annamaria Pascali for the technical assistance.

Conflicts of Interest: The authors declare no conflicts of interest.

References

1. Giovinazzo, G.; Grieco, F. Functional Properties of Grape and Wine Polyphenols. *Plant Foods Hum. Nutr.* **2015**, *70*, 454–462. [CrossRef] [PubMed]
2. Buljeta, I.; Pichler, A.; Šimunović, J.; Kopjar, M. Beneficial Effects of Red Wine Polyphenols on Human Health: Comprehensive Review. *Curr. Issues Mol. Biol.* **2023**, *45*, 52. [CrossRef] [PubMed]
3. Carluccio, M.A.; Siculella, L.; Ancora, M.A.; Massaro, M.; Scoditti, E.; Storelli, C.; Visioli, F.; Distante, A.; De Caterina, R. Olive Oil and Red Wine Antioxidant Polyphenols Inhibit Endothelial Activation. *Arterioscler. Thromb. Vasc. Biol.* **2003**, *23*, 622–629. [CrossRef] [PubMed]
4. Cuevas, A.M.; Guasch, V.; Castillo, O.; Irribarra, V.; Mizon, C.; Martin, A.S.; Strobel, P.; Perez, D.; Germain, A.M.; Leighton, F. A High-Fat Diet Induces and Red Wine Counteracts Endothelial Dysfunction in Human Volunteers. *Lipids* **2000**, *35*, 143–148. [CrossRef]
5. Csiszar, A.; Smith, K.; Labinskyy, N.; Orosz, Z.; Rivera, A.; Ungvari, Z. Resveratrol Attenuates TNF-α-Induced Activation of Coronary Arterial Endothelial Cells: Role of NF-κB Inhibition. *Am. J. Physiol. Heart Circ. Physiol.* **2006**, *291*, H1694–H1699. [CrossRef] [PubMed]
6. Kaminski, B.M.; Steinhilber, D.; Stein, J.M.; Ulrich, S. Phytochemicals Resveratrol and Sulforaphane as Potential Agents for Enhancing the Anti-Tumor Activities of Conventional Cancer Therapies. *Curr. Pharm. Biotechnol.* **2012**, *13*, 137–146. [CrossRef]
7. Calabriso, N.; Scoditti, E.; Massaro, M.; Pellegrino, M.; Storelli, C.; Ingrosso, I.; Giovinazzo, G.; Carluccio, M.A. Multiple Anti-Inflammatory and Anti-Atherosclerotic Properties of Red Wine Polyphenolic Extracts: Differential Role of Hydroxycinnamic Acids, Flavonols and Stilbenes on Endothelial Inflammatory Gene Expression. *Eur. J. Nutr.* **2016**, *55*, 477–489. [CrossRef]
8. Pretorius, I.S. Tailoring Wine Yeast for the New Millennium: Novel Approaches to the Ancient Art of Winemaking. *Yeast* **2000**, *16*, 675–729. [CrossRef]
9. Zhang, P.; Ma, W.; Meng, Y.; Zhang, Y.; Jin, G.; Fang, Z. Wine Phenolic Profile Altered by Yeast: Mechanisms and Influences. *Compr. Rev. Food Sci. Food Saf.* **2021**, *20*, 3579–3619. [CrossRef] [PubMed]
10. Morata, A.; Gómez-Cordovés, M.C.; Suberviola, J.; Bartolomé, B.; Colomo, B.; Suárez, J.A. Adsorption of Anthocyanins by Yeast Cell Walls during the Fermentation of Red Wines. *J. Agric. Food Chem.* **2003**, *51*, 4084–4088. [CrossRef] [PubMed]
11. Villena, M.A.; Iranzo, J.F.Ú.; Pérez, A.I.B. β-Glucosidase Activity in Wine Yeasts: Application in Enology. *Enzym. Microb. Technol.* **2007**, *40*, 420–425. [CrossRef]

12. Manzanares, P.; Rojas, V.; Genovés, S.; Vallés, S. A Preliminary Search for anthocyanin-β-D-glucosidase Activity in non-Saccharomyces Wine Yeasts. *Int. J. Food Sci. Technol.* **2000**, *35*, 95–103. [CrossRef]
13. Carew, A.L.; Smith, P.; Close, D.C.; Curtin, C.; Dambergs, R.G. Yeast Effects on Pinot Noir Wine Phenolics, Color, and Tannin Composition. *J. Agric. Food Chem.* **2013**, *61*, 9892–9898. [CrossRef] [PubMed]
14. Rodriguez-Nogales, J.M.; Fernández-Fernández, E.; Gómez, M.; Vila-Crespo, J. Antioxidant Properties of Sparkling Wines Produced with β-Glucanases and Commercial Yeast Preparations. *J. Food Sci.* **2012**, *77*, C1005–C1010. [CrossRef] [PubMed]
15. Garofalo, C.; Berbegal, C.; Grieco, F.; Tufariello, M.; Spano, G.; Capozzi, V. Selection of Indigenous Yeast Strains for the Production of Sparkling Wines from Native Apulian Grape Varieties. *Int. J. Food Microbiol.* **2018**, *285*, 7–17. [CrossRef] [PubMed]
16. Capozzi, V.; Garofalo, C.; Chiriatti, M.A.; Grieco, F.; Spano, G. Microbial Terroir and Food Innovation: The Case of Yeast Biodiversity in Wine. *Microbiol. Res.* **2015**, *181*, 75–83. [CrossRef] [PubMed]
17. De Benedictis, M.; Bleve, G.; Grieco, F.; Tristezza, M.; Tufariello, M.; Grieco, F. An Optimized Procedure for the Enological Selection of Non-Saccharomyces Starter Cultures. *Antonie van Leeuwenhoek* **2011**, *99*, 189–200. [CrossRef] [PubMed]
18. Tristezza, M.; Vetrano, C.; Bleve, G.; Grieco, F.; Tufariello, M.; Quarta, A.; Mita, G.; Spano, G.; Grieco, F. Autochthonous Fermentation Starters for the Industrial Production of Negroamaro Wines. *J. Ind. Microbiol. Biotechnol.* **2012**, *39*, 81–92. [CrossRef] [PubMed]
19. Tristezza, M.; Vetrano, C.; Bleve, G.; Spano, G.; Capozzi, V.; Logrieco, A.; Mita, G.; Grieco, F. Biodiversity and Safety Aspects of Yeast Strains Characterized from Vineyards and Spontaneous Fermentations in the Apulia Region, Italy. *Food Microbiol.* **2013**, *36*, 335–342. [CrossRef] [PubMed]
20. Tristezza, M.; Fantastico, L.; Vetrano, C.; Bleve, G.; Corallo, D.; Grieco, F.; Mita, G.; Grieco, F. Molecular and Technological Characterization of *Saccharomyces cerevisiae* Strains Isolated from Natural Fermentation of Susumaniello Grape Must in Apulia, Southern Italy. *Int. J. Microbiol.* **2014**, *2014*, 897428. [CrossRef]
21. Tufariello, M.; Chiriatti, M.A.; Grieco, F.; Perrotta, C.; Capone, S.; Rampino, P.; Tristezza, M.; Mita, G. Influence of Autochthonous *Saccharomyces cerevisiae* Strains on Volatile Profile of Negroamaro Wines. *LWT Food Sci. Technol.* **2014**, *58*, 35–48.
22. Garofalo, C.; El Khoury, M.; Lucas, P.; Bely, M.; Russo, P.; Spano, G.; Capozzi, V. Autochthonous Starter Cultures and Indigenous Grape Variety for Regional Wine Production. *J. Appl. Microbiol.* **2015**, *118*, 1395–1408. [CrossRef]
23. Garofalo, C.; Russo, P.; Beneduce, L.; Massa, S.; Spano, G.; Capozzi, V. Non-Saccharomyces Biodiversity in Wine and the 'Microbial Terroir': A Survey on Nero Di Troia Wine from the Apulian Region, Italy. *Ann. Microbiol.* **2016**, *66*, 143–150. [CrossRef]
24. Roscini, L.; Tristezza, M.; Corte, L.; Colabella, C.; Perrotta, C.; Rampino, P.; Robert, V.; Vu, D.; Cardinali, G.; Grieco, F. Early Ongoing Speciation of Ogataea Uvarum Sp. Nov. Within the Grape Ecosystem Revealed by the Internal Variability Among the rDNA Operon Repeats. *Front. Microbiol.* **2018**, *9*, 1687. [CrossRef] [PubMed]
25. Tufariello, M.; Maiorano, G.; Rampino, P.; Spano, G.; Grieco, F.; Perrotta, C.; Capozzi, V.; Grieco, F. Selection of an Autochthonous Yeast Starter Culture for Industrial Production of Primitivo "Gioia Del Colle" PDO/DOC in Apulia (Southern Italy). *LWT* **2019**, *99*, 188–196. [CrossRef]
26. Tristezza, M.; Gerardi, C.; Logrieco, A.; Grieco, F. An Optimized Protocol for the Production of Interdelta Markers in *Saccharomyces cerevisiae* by Using Capillary Electrophoresis. *J. Microbiol. Methods* **2009**, *78*, 286–291. [CrossRef]
27. Grieco, F.; Tristezza, M.; Vetrano, C.; Bleve, G.; Panico, E.; Mita, G.; Logrieco, A. Exploitation of Autochthonous Micro-Organism Potential to Enhance the Quality of Apulian Wines. *Ann. Microbiol.* **2011**, *61*, 67–73. [CrossRef]
28. Magalhães, L.M.; Santos, F.; Segundo, M.A.; Reis, S.; Lima, J.L.F.C. Rapid Microplate High-Throughput Methodology for Assessment of Folin-Ciocalteu Reducing Capacity. *Talanta* **2010**, *83*, 441–447. [CrossRef] [PubMed]
29. Lee, J.; Durst, R.W.; Wrolstad, R.E.; Eisele, T.; Giusti, M.M.; Hach, J.; Hofsommer, H.; Koswig, S.; Krueger, D.A.; Kupina, S.; et al. Determination of Total Monomeric Anthocyanin Pigment Content of Fruit Juices, Beverages, Natural Colorants, and Wines by the pH Differential Method: Collaborative Study. *J. AOAC Int.* **2005**, *88*, 1269–1278. [CrossRef] [PubMed]
30. Scarano, A.; Gerardi, C.; D'Amico, L.; Accogli, R.; Santino, A. Phytochemical Analysis and Antioxidant Properties in Colored Tiggiano Carrots. *Agriculture* **2018**, *8*, 102. [CrossRef]
31. Re, R.; Pellegrini, N.; Proteggente, A.; Pannala, A.; Yang, M.; Rice-Evans, C. Antioxidant Activity Applying an Improved ABTS Radical Cation Decolorization Assay. *Free Radic. Biol. Med.* **1999**, *26*, 1231–1237. [CrossRef]
32. Gerardi, C.; D'amico, L.; Migoni, D.; Santino, A.; Salomone, A.; Carluccio, M.A.; Giovinazzo, G. Strategies for Reuse of Skins Separated from Grape Pomace as Ingredient of Functional Beverages. *Front. Bioeng. Biotechnol.* **2020**, *8*, 645. [CrossRef] [PubMed]
33. Palombi, L.; Tufariello, M.; Durante, M.; Fiore, A.; Baiano, A.; Grieco, F. Assessment of the Impact of Unmalted Cereals, Hops, and Yeast Strains on Volatolomic and Olfactory Profiles of Blanche Craft Beers: A Chemometric Approach. *Food Chem.* **2023**, *416*, 135783. [CrossRef] [PubMed]
34. Tufariello, M.; Pati, S.; D'Amico, L.; Bleve, G.; Losito, I.; Grieco, F. Quantitative Issues Related to the Headspace-SPME-GC/MS Analysis of Volatile Compounds in Wines: The Case of Maresco Sparkling Wine. *LWT* **2019**, *108*, 268–276. [CrossRef]
35. Kang, W.; Xu, Y.; Qin, L.; Wang, Y. Effects of Different β-D-Glycosidases on Bound Aroma Compounds in Muscat Grape Determined by HS-SPME and GC-MS. *J. Inst. Brew.* **2010**, *116*, 70–77. [CrossRef]
36. Perestrelo, R.; Silva, C.; Silva, P.; Câmara, J.S. Unraveling Vitis Vinifera L. Grape Maturity Markers Based on Integration of Terpenic Pattern and Chemometric Methods. *Microchem. J.* **2018**, *142*, 367–376. [CrossRef]
37. Caridi, A. Improved Screening Method for the Selection of Wine Yeasts Based on Their Pigment Adsorption Activity. *Food Technol. Biotechnol.* **2013**, *51*, 137–144.

38. Nogales-Bueno, J.; Baca-Bocanegra, B.; Heredia, F.J.; Hernández-Hierro, J.M. Phenolic Compounds Extraction in Enzymatic Macerations of Grape Skins Identified as Low-Level Extractable Total Anthocyanin Content. *J. Food Sci.* **2020**, *85*, 324–331. [CrossRef] [PubMed]
39. Brandolini, V.; Fiore, C.; Maietti, A.; Tedeschi, P.; Romano, P. Influence of *Saccharomyces cerevisiae* Strains on Wine Total Antioxidant Capacity Evaluated by Photochemiluminescence. *World J. Microbiol. Biotechnol.* **2007**, *23*, 581–586. [CrossRef]
40. Kostadinović, S.; Wilkens, A.; Stefova, M.; Ivanova, V.; Vojnoski, B.; Mirhosseini, H.; Winterhalter, P. Stilbene Levels and Antioxidant Activity of Vranec and Merlot Wines from Macedonia: Effect of Variety and Enological Practices. *Food Chem.* **2012**, *135*, 3003–3009. [CrossRef] [PubMed]
41. Carrascosa, A.V.; Bartolome, B.; Robredo, S.; Leon, A.; Cebollero, E.; Juega, M.; Nunez, Y.P.; Martinez, M.C.; Martinez-Rodriguez, A.J. Influence of Locally-Selected Yeast on the Chemical and Sensorial Properties of Albariño White Wines. *LWT Food Sci. Technol.* **2012**, *46*, 319–325. [CrossRef]
42. Caridi, A.; De Bruno, A.; Salvo, E.; Piscopo, A.; Poiana, M.; Sidari, R. Selected Yeasts to Enhance Phenolic Content and Quality in Red Wine from Low Pigmented Grapes. *Eur. Food Res. Technol.* **2017**, *243*, 367–378. [CrossRef]
43. Grieco, F.; Carluccio, M.A.; Giovinazzo, G. Autochthonous *Saccharomyces cerevisiae* Starter Cultures Enhance Polyphenols Content, Antioxidant Activity, and Anti-Inflammatory Response of Apulian Red Wines. *Foods* **2019**, *8*, 453. [CrossRef] [PubMed]
44. Morata, A.; Loira, I.; Heras, J.M.; Callejo, M.J.; Tesfaye, W.; González, C.; Suárez-Lepe, J.A. Yeast Influence on the Formation of Stable Pigments in Red Winemaking. *Food Chem.* **2016**, *197*, 686–691. [CrossRef] [PubMed]
45. Vernhet, A.; Carrillo, S.; Rattier, A.; Verbaere, A.; Cheynier, V.; Nguela, J.M. Fate of Anthocyanins and Proanthocyanidins during the Alcoholic Fermentation of Thermovinified Red Musts by Different *Saccharomyces cerevisiae* Strains. *J. Agric. Food Chem.* **2020**, *68*, 3615–3625. [CrossRef] [PubMed]
46. Pulvirenti, A.; De Vero, L.; Blaiotta, G.; Sidari, R.; Iosca, G.; Gullo, M.; Caridi, A. Selection of Wine *Saccharomyces cerevisiae* Strains and Their Screening for the Adsorption Activity of Pigments, Phenolics and Ochratoxin A. *Fermentation* **2020**, *6*, 80. [CrossRef]
47. Tufariello, M.; Capozzi, V.; Spano, G.; Cantele, G.; Venerito, P.; Mita, G.; Grieco, F. Effect of Co-Inoculation of *Candida zemplinina*, *Saccharomyces cerevisiae* and *Lactobacillus plantarum* for the Industrial Production of Negroamaro Wine in Apulia (Southern Italy). *Microorganisms* **2020**, *8*, 726. [CrossRef] [PubMed]
48. Rapp, A. Natural Flavours of Wine: Correlation between Instrumental Analysis and Sensory Perception. *Fresenius' J. Anal. Chem.* **1990**, *337*, 777–785. [CrossRef]
49. Swiegers, J.H.; Pretorius, I.S. Yeast Modulation of Wine Flavor. In *Advances in Applied Microbiology*; Academic Press: Cambridge, MA, USA, 2005; Volume 57, pp. 131–175, ISBN 978-0-12-002659-3.
50. Swiegers, J.H.; Bartowski, E.J.; Henschke, P.A.; Pretorius, I.S. Yeast and Bacterial Modulation of Wine Aroma and Flavour. *Aust. J. Grape Wine Res.* **2005**, *11*, 139–173. [CrossRef]
51. Moreira, N.; Pina, C.; Mendes, F.; Couto, J.A.; Hogg, T.; Vasconcelos, I. Volatile Compounds Contribution of *Hanseniaspora guilliermondii* and *Hanseniaspora uvarum* during Red Wine Vinifications. *Food Control* **2011**, *22*, 662–667. [CrossRef]
52. Moreira, N.; Mendes, F.; Guedes de Pinho, P.; Hogg, T.; Vasconcelos, I. Heavy Sulphur Compounds, Higher Alcohols and Esters Production Profile of *Hanseniaspora uvarum* and *Hanseniaspora guilliermondii* Grown as Pure and Mixed Cultures in Grape Must. *Int. J. Food Microbiol.* **2008**, *124*, 231–238. [CrossRef] [PubMed]
53. Medina, K.; Boido, E.; Fariña, L.; Gioia, O.; Gomez, M.E.; Barquet, M.; Gaggero, C.; Dellacassa, E.; Carrau, F. Increased Flavour Diversity of Chardonnay Wines by Spontaneous Fermentation and Co-Fermentation with *Hanseniaspora vineae*. *Food Chem.* **2013**, *141*, 2513–2521. [CrossRef] [PubMed]
54. Madžgalj, V.; Petrović, A.; Čakar, U.; Maraš, V.; Sofrenić, I.; Tešević, V. The influence of different enzymatic preparations and skin contact time on aromatic profile of wines produced from autochthonous grape varieties Krstač and Žižak. *J. Serb. Chem. Soc.* **2023**, *88*, 11–23. [CrossRef]

Disclaimer/Publisher's Note: The statements, opinions and data contained in all publications are solely those of the individual author(s) and contributor(s) and not of MDPI and/or the editor(s). MDPI and/or the editor(s) disclaim responsibility for any injury to people or property resulting from any ideas, methods, instructions or products referred to in the content.

foods

MDPI

Article

Exploring the Phytochemical Composition and the Bioactive Properties of Malbec and Torrontés Wine Pomaces from the Calchaquíes Valleys (Argentina) for Their Sustainable Exploitation

Pablo Ezequiel Tapia [1], Ana Margarida Silva [2], Cristina Delerue-Matos [2], Manuela Moreira [2], Francisca Rodrigues [2], Romina Torres Carro [1], María Daniela Santi [3], María Gabriela Ortega [3], María Amparo Blázquez [4], Mario Eduardo Arena [1,5,*,†] and María Rosa Alberto [1,5,†]

1 Instituto de Biotecnología Farmacéutica y Alimentaria (INBIOFAL) CONICET–UNT, Avenida N Kirchner 1900, San Miguel de Tucumán CP 4000, Argentina; pablotapia2108@gmail.com (P.E.T.); rominatc87@gmail.com (R.T.C.); maria.alberto@fbqf.unt.edu.ar (M.R.A.)

2 REQUIMTE/LAQV, ISEP, Polytechnic of Porto, Rua Dr. António Bernardino de Almeida 431, 4249-015 Porto, Portugal; ana.silva@graq.isep.ipp.pt (A.M.S.); cmm@isep.ipp.pt (C.D.-M.); manuela.moreira@graq.isep.ipp.pt (M.M.); francisca.rodrigues@graq.isep.ipp.pt (F.R.)

3 Farmacognosia, Instituto Multidisciplinario de Biología Vegetal (IMBIV), CONICET and Departamento de Ciencias Farmacéuticas, Facultad de Ciencias Químicas, Universidad Nacional de Córdoba, Ciudad Universitaria, Córdoba X5000HUA, Argentina; mdaniela0505@gmail.com (M.D.S.); maria.gabriela.ortega@unc.edu.ar (M.G.O.)

4 Departament de Farmacologia, Facultat de Farmàcia i Ciències de l'Alimentació, Universitat de València, Avd. Vicent Andrés Es-tellés s/n, 46100 Burjasot, Valencia, Spain; amparo.blazquez@uv.es

5 Facultad de Bioquímica, Química y Farmacia, Universidad Nacional de Tucumán (UNT), Ayacucho 471, San Miguel de Tucumán CP 4000, Argentina

* Correspondence: mario.arena@fbqf.unt.edu.ar

† These authors contributed equally to this work.

Abstract: Hydroalcoholic extracts from Malbec and Torrontés wine pomaces (*Vitis vinifera* L.) originating from the high-altitude vineyards of Argentina's Calchaquí Valleys were characterized. Total phenolics, hydroxycinnamic acids, orthodiphenols, anthocyanins, non-flavonoid phenolics, total flavonoids, flavones/flavonols, flavanones/dihydroflavonols, and tannins were quantified through spectrophotometric methods, with the Malbec extract exhibiting higher concentrations in most of phytochemical groups when compared to Torrontés. HPLC-DAD identified more than 30 phenolic compounds in both extracts. Malbec displayed superior antiradical activity (ABTS cation, nitric oxide, and superoxide anion radicals), reduction power (iron, copper, and phosphomolybdenum), hypochlorite scavenging, and iron chelating ability compared to Torrontés. The cytotoxicity assessments revealed that Torrontés affected the viability of HT29-MTX and Caco-2 colon cancer cells by 70% and 50%, respectively, at the highest tested concentration (1 mg/mL). At the same time, both extracts did not demonstrate acute toxicity in *Artemia salina* or in red blood cell assays at 500 μg/mL. Both extracts inhibited the lipoxygenase enzyme (IC_{50}: 154.7 and 784.7 μg/mL for Malbec and Torrontés), with Malbec also reducing the tyrosinase activity (IC_{50}: 89.9 μg/mL), and neither inhibited the xanthine oxidase. The substantial phenolic content and diverse biological activities in the Calchaquí Valleys' pomaces underline their potentialities to be valorized for pharmaceutical, cosmetic, and food industries.

Keywords: *Vitis vinifera*; antioxidant; cancer cell cytotoxicity; tyrosinase; lipoxygenase; xanthine oxidase

1. Introduction

Agro-industries are currently valued not only due to their productive and economic performance in each country, but also for their relationship and actions with the environment. Although most agro-industrial companies generate wastes whose management or final disposal has been highly questioned in recent decades, they seek to obtain benefits by taking advantage of them, achieving pollution reduction, and generating economic profits derived from the added value that may generate [1–3]. In recent years, the food industry has embraced green technologies, clean labeling, and the incorporation of natural ingredients to benefit consumers and the industry itself. In the wine industry, grapes are one of the most produced crops worldwide, with an estimated production of more than 78 million tons in 2020 [2]. It is estimated that for every 100 L of wine, about 30 kg of pomace is generated, mainly consisting of skins, pulp, and stems [4,5]. Although grape pomace can be used for animal feed and compost, among other uses, only a small amount is reused, and its disposal poses environmental problems [6]. Leveraging wine pomace for sustainable purposes represents an effective strategy to mitigate environmental pollution and serves as a substitute to minimize carbon emissions across a winery's manufacturing process. The biological properties of wine pomace residues are of interest to several industries such as food, cosmetic, and pharmaceutical ones [5].

In previous studies, it was observed that the secondary metabolites present in wine pomace could be used as natural additives due to their antioxidant capacity and ability to improve microbial stability and inhibit the growth of pathogenic microorganisms [5]. Grape pomace extracts offer numerous health benefits due to their antioxidant, anti-inflammatory, anticancer, and hypoglycemic properties [2,5]. These benefits are attributed to the rich contents of nutrients and polyphenols, which underscore their potential as novel pharmacological agents for the treatment and prevention of various diseases [7]. These diseases include cancer, metabolic syndrome, neurological disorders, liver and cardiovascular diseases, and other conditions related to oxidative stress [2,5,7]. Grape pomace is a raw material for the production of dietary supplements (powders, tablets, and capsules), providing an auxiliary source of polyphenols that avoids wine consumption [8]. Among polyphenolic, flavonoids (anthocyanins, flavan-3-ols, and flavonols, among others), tannins and non-flavonoids, such as phenolic acids and stilbenes, are mainly responsible for the well-established antioxidant and anti-inflammatory effects. Due to these properties, phenolic compounds can be used in the cosmetic industry [7,9], inhibiting the enzyme tyrosinase, which is over expressed in skin features such as melasma, freckles, and senile lentigines [10].

According to Wani et al. [7], about 70% of phenolic compounds remain in grape pomace after fermentation-maceration. Therefore, its use and valorization through the extraction of phenolic compounds is an attractive strategy that aims to recover compounds while reducing the environmental impact of their byproducts [11]. However, the biological activity of grape pomace extracts depends on the grape variety, geographical origin, climate, vineyard soil conditions, and the winemaking process [2].

Considering the wide range of possible uses for pomace extracts and the recognition that the polyphenolic composition depends on the plant material's origin and winemaking techniques, it is of special interest to explore the chemical and functional properties of the pomace of each region and cultivar with the aim of finding industrial applications.

In previous studies, white and red wine pomace extracts from the Calchaquí Valleys were found to inhibit the virulence of pathogenic bacteria (*Pseudomonas aeruginosa* and *Staphylococcus aureus*) [12,13]. These extracts demonstrated an inhibition of *P. aeruginosa*'s swarming motility, biofilm production, and metabolic activity in a biofilm environment. The antibiofilm activities showed a positive correlation with the polyphenol content of the extracts [12]. The Calchaquís Valleys' wine pomaces also exhibited efficacy against *Leishmania amazonensis*, the agent responsible for American tegumentary leishmaniasis. The extracts also displayed significant anticholinesterase activity, suggesting potentialities for the palliative treatment of Alzheimer's disease [14].

Given the limited information on the phenolic composition and biological properties of Torrontés and Malbec wine pomaces from this geographical area, the present study aims to investigate the polyphenolic composition, antioxidant capacity, cytotoxicity, and enzyme inhibition potential (tyrosinase, lipoxygenase, and xanthine oxidase) of Torrontés and Malbec pomaces from the Calchaquí Valleys.

2. Materials and Methods

2.1. Chemicals

For HPLC analysis, methanol and formic acid were HPLC graded from Merck (Darmstadt, Germany). The different standards of phenolic compounds, namely the phenolic acids: gallic acid (≥99%), protocatechuic acid (99.63%), neochlorogenic acid (≥98%), caftaric acid (≥97%), chlorogenic acid (>95%), 4-caffeyolquinic acid (≥98%), va-nillic acid (≥97%), caffeic acid (≥98%), syringic acid (≥98%), p-coumaric acid (≥98%), trans-ferulic acid (≥99%), sinapic acid (≥99%), 3,5-di-O-caffeyolquinic acid (≥95%), ellag-ic acid (≥95%), 4,5-di-O-caffeyolquinic acid (≥90%), cinnamic acid (≥99%); flavonoids: (+)-catechin (≥98%), (-)-epicatechin (≥90%), naringin (≥95%), quecetin-3-O-galactoside (≥97%), quercetin-3-O-glucopyranoside (≥99%), rutin hydrate (≥94%), myricetin (≥96%), quercitrin (≥97%), kaempferol-3-O-glucoside (≥95%), kaempferol-3-O-rutinoside (≥98%), isorhamnetin-3-O-glucoside (≥98%), isorhamnetin-3-O-rutinoside (≥99%), naringenin (98%), quercetin (95%), kaempferol (≥98%), apigenin (≥99%), chrysin (≥99%), tiliroside (≥98%); chalcones: phloridzin dehydrate (99%) and phloretin (≥98.5%); and stilbenoids: trans-polydatin (≥98%), trans-epsilon viniferin (≥95%) and resveratrol (≥99%) were purchased from Sigma-Aldrich (Steinhemin, Germany) and Extrasynthese (Genay, Cedex, France). Their stock solutions were prepared in methanol at concentration levels ranging from 1 to 5 g/L and stored at −20 °C.

Dimethylsulfoxide (DMSO), Triton X-100, 2,2′-Azinobis-(3-ethylbenzothiazoline-6-sulfonic acid) cation radical ($ABTS^{\cdot}+$), Trolox (6-hydroxy-2,5,7,8-tetramethylchroman-2-carboxylic acid), Butylated hydroxytoluene (BHT), EDTA, Folin–Ciocalteau reagent, Tyrosinase from mushroom, kojic acid, L-tyrosine, xanthine oxidase from bovine milk, xanthine, nitroblue tetrazolium chloride (NBT), phenazine methosulphate (PMS), β-nicotinamide adenine dinucleotide (NADH), dihydrorhodamine 123 (DHR), and 3-(4,5-dimethylthiazol-2-yl)-2,5-diphenyltetrazolium bromide (MTT) were purchased from Sigma-Aldrich (St. Louis, MO, USA). Other chemicals were obtained from local commercial sources, and were of analytical-grade quality.

Caco-2 cells (clone type C2Bbe1) were acquired from the American Type Culture Collection (ATCC, Manassas, VA, USA), and HT29-MTX was offered from Dr. T. Lesuffleur (INSER-MU178, Villejuif, France). Dulbecco's Modified Eagle Medium (DMEM), Fetal Bovine Serum (FBS), Hank's Balanced Salt Solution (HBSS), non-essential amino acids, penicillin, streptomycin, and trypsin–EDTA were obtained from the Invitrogen Corporation (Life Technologies, S.A., Madrid, Spain).

2.2. Material Collection

Samples of Malbec and Torrontés grape pomaces were gathered from Albarossa Winery, situated in Tafí del Valle, Tucumán, Argentina, during the 2019 and 2020 harvests. Notably, the vineyard cultivation practices at this winery refrain from the use of chemical substances. The grapes exhibited robust health, with no signs of *Botrytis cinerea* detected. The assessment for the presence of *B. cinerea* was conducted visually, and its absence can be attributed to the windy and arid conditions characteristic of the Calchaquí Valleys region.

The Calchaquí Valleys, located in the northwest region of Argentina, have a temperate climate with notable thermal amplitudes and, occasionally, experience prolonged late frosts in spring. Their high-altitude vineyards range from 1700 m to 2400 m. The grapes were grown on soils characterized as sandy loam or sandy with a high proportion of fine sand. The soil profile is deep, with a somewhat rocky subsoil ensuring excellent permeability and the leaching of salts [15].

The regional white wines correspond to the Torrontés varietal (*Vitis vinifera* L.), which has become the emblematic variety of the region. This grape is optimal to produce aromatic wines and adapts very well to the entire area, becoming the most cultivated one in the region [16]. On the other hand, the Malbec wine varietal (Argentina's most exploited purple grape variety) from the Calchaquí Valleys shows different particularities from other country wine regions due to the terroir characteristics described above.

Notably, Torrontés pomace was obtained from a white wine-making process, meaning that grapes were not subjected to ethanolic fermentation. In contrast, Malbec pomace was acquired from a red wine-making process, where grapes are entirely involved in fermentation. The collected specimens were stored at −80 °C before being oven-dried at temperatures below 45 °C for 48 h, ensuring the preservation of phenolic compounds until a constant weight was attained. Moisture was calculated using the weight difference before and after drying. The dried material was crushed with a grinder until reaching a particle size < 250 μm (sieve mesh 60). The extract was prepared from this pomace powder.

2.3. Preparation of Extracts

The active principles were extracted using hydroalcoholic maceration (ethanol:water, 50:50 v/v) in a solid–liquid ratio of 1/4, w/v. Two successive extractions were performed by shaking at room temperature (150 rpm/min) for 3 h. The extracts were vacuum-filtered using Whatman No. 4 filters and evaporated to dryness through vacuum evaporation and subsequent lyophilization. The extracts obtained during the processes were stored in the dark at 4 °C until use. The extraction yields were expressed in mg of soluble principle per gram of dry pomace (mg/g DP) and were calculated as follows: the weight of extract obtained (mg)/the initial weight of the plant matter to be extracted (mg).

2.4. Phytochemical Analysis

2.4.1. Quantification of Different Phenolic Groups

For this assay, stock solutions of 2.5 mg/mL of each extract were used. Total extractable phenols and nonflavonoid compounds were determined colorimetrically using Folin–Ciocalteu's reagent at 765 nm [17]. A standard curve was performed with gallic acid (2–20 μg/mL) as the standard, and the results were expressed in mg of gallic acid equivalents (GAE) per g of dry pomace extract (DPE) (mg GAE/g DPE) and g of dry pomace (mg GAE/g DP) ($R^2 = 0.9965$, $p \leq 0.05$). To eliminate the potential sugar interference, the Torrontés extract was purified using solid-phase extraction prior to Folin–Ciocalteu's analysis. The C18 cartridge (Waters' Sep-Pak Cartridges) was activated with 2 mL of methanol and then 5 mL of deionized water. Then, 9 mL of the extract (2.5 mg/mL) was injected and eluted using 9 mL of acidic deionized water (0.1% formic acid). Finally, 9 mL of acidic methanol (0.1% formic acid) was used to recover the polyphenol compounds [18].

The total flavonoid content was determined using sodium nitrite (5%) and aluminum chloride (10%). The technique is based on the formation of specific colored complexes between the flavonoids and the reagents ($NaNO_2$ and $AlCl_3$), with the color intensity measured spectrophotometrically at 510 nm [19]. For quantification, a quercetin standard curve (4–80 μg/mL) was used and the results were expressed as mg of quercetin equivalents (QE) per g of extract (mg QE/g DPE) ($R^2 = 0.9939$, $p \leq 0.05$)

The flavone and flavonol contents were evaluated spectrophotometrically at 425 nm with aluminum chloride (5%) [17,20]. This technique is based on the formation of a complex between the aluminum ion, Al (III), and the carbonyl and hydroxyl groups of the flavonoid. The standard curve was performed with quercetin (5–40 μg/mL), and the results were expressed in mg of quercetin equivalents (QE) per g of extract (mg QE/g DPE) ($R^2 = 0.9941$, $p \leq 0.05$).

The flavanone and dihydroflavonol contents were measured at 495 nm [17,20]. This technique is based on the reaction of these compounds with 2,4-dinitrophenylhydrazine (DNP) in an acidic medium to form colored phenylhydrazones. The standard curve was performed with naringenin (20–200 μg/mL), and the results were expressed in mg of naringenin equivalents (NE) per g of extract (mg NE/g DPE) ($R^2 = 0.9998$, $p \leq 0.05$).

Orthodiphenols were assessed by UV-Visible spectrophotometry using the sodium molybdate method at a wavelength of 370 nm [21]. A standard curve was performed with caffeic acid (2–20 µg/mL), and the results were expressed as mg of caffeic acid equivalents (CAE) per g of extract (mg CAE/g DPE) (R^2 = 0.9981, $p \leq 0.05$).

Hydroxycinnamic derivates were evaluated at 320 nm using caffeic acid (0.5–5 µg/mL) as the standard, and the results were expressed as mg of caffeic acid equivalents (CAE) per g of extract (mg CAE/g DPE) (R^2 = 0.9972, $p \leq 0.05$) [22].

The assessment of the total anthocyanin content was carried out using the pH differential method, and the results were expressed as mg of cyanidin-3-glucoside equivalents per g of extract (mg C3GLE/g DPE) according to Carullo et al. [23]. Results were determined by means of the following formula:

$$\text{mg C3GLE/L DPE} = \left[\frac{A \times MW \times DF \times 1000}{\varepsilon \times 1}\right] \quad (1)$$

where:

A = (A520 nm − A700 nm)pH = 1 − (A520 nm − A700 nm)pH = 4.5; MW (molecular weights of cyanidin-3-glucoside) = 449.2 g/mol; DF = dilution factor; ε = (molar extinction coefficient) = 26,900 L/mol cm and 1000 = conversion factor from g to mg.

The determination of the tannin content of the extracts was performed as reported previously by Bouabid et al. [24] using the vanillin assay. A standard curve was performed with catechin (5–25 µg/mL), and the results were expressed as mg of catechin equivalents per g of extract (mg CE/g DPE).

2.4.2. Identification of Phenolic Compounds through HPLC-DAD Analysis

The HPLC analyses were carried out in a Shimadzu HPLC system (Shimadzu Corporation, Kyoto, Japan) equipped with a LC-20AD prominence pump, a DGU-20AS prominence degasser, a CTO-10AS VP column oven, a SIL-20A HT prominence autosampler, and an SPD-M20A photodiode array detector.

The phenolic profile of the obtained extracts was analyzed according to the method described by Moreira et al. [25] with slight modifications. Chromatographic analyses were performed using a Shimadzu HPLC system, and polyphenol separation was achieved on a Gemini C18 column (250 × 4.6 mm, 5 µm) from Phenomenex at 25 °C. The solvent system used, pumped at a flow rate of 1 mL/min, was methanol (eluent A) and water (eluent B) with both acidulated with formic acid (0.1%), and the following gradient was employed: 0–5 min: 20–24% A; 5–7 min: 24–25% A; 7–10 min: 25–26% A; 10–11 min: 26–26.5% A; 11–18 min: 26.5% A; 18–25 min: 26.5–30% A; 25–50 min: 30–45% A; 50–60 min: 45–50% A; 60–70 min: 50–55% A; 70–90 min: 55–70% A; 90–100 min: 70–100% A, followed by 100% A for 5 min, back to 20% A in 10 min and 5 min of reconditioning before the next injection. Individual phenolic compounds were identified by comparing the samples' retention time and UV-Vis spectra with those from pure standards. Chromatograms were recorded at 280, 320, and 360 nm depending on the maximum absorption of the phenolic compound identified. Before injection, the dried extract was resuspended in methanol/water (20:80) and filtered through a 0.22 µm PTFE syringe filter. The quantification of phenolic compounds was made based on calibration curves of the pure standards, and results were expressed as mg of compound per 100 g of DPE.

2.5. *Antioxidant Capacity Assays*

At least three independent experiments were performed for each method. For the samples and positive controls, six concentrations were analyzed in duplicate. Prior to the assays, the absorption of the extract was studied at the proper wavelengths. GraphPad Prism 7 software (GraphPad, La Jolla, CA, USA) was used to calculate the results based on the curves of the inhibition percentage versus the antioxidant concentration. For chelating, scavenging, and reducing capacities, the concentration of the extracts necessary to chelate, scavenge, or reduce 50% of the radicals or iron ions (IC_{50}) was determined through linear regression analysis. If the IC_{50} was not reached, the results were expressed as the percentage of inhibition at the highest concentration tested.

2.5.1. Phosphomolybdenum-Reducing Capacity

The phosphomolybdenum method described by Carullo et al. was used [23]. The absorbance of the green-colored complex was measured spectrophotometrically at 695 nm. The standard curve was performed with ascorbic acid (5–50 μg/mL) as the standard, and the results were expressed in μg of ascorbic acid equivalents (AAE) per mg of DPE (μg AAE/mg).

2.5.2. Metal-Chelating Capacity

The metal chelating capacity is based on the formation of colored Fe^{2+} complexes whose concentrations can be determined spectrophotometrically. Molecules present in a sample that can chelate iron will compete with ferrozine, decreasing the reaction's coloration (absorbance) [17]. The absorbance was measured at 562 nm. EDTA (5–20 μg/mL) was the positive control used.

2.5.3. ABTS Cation Radical-Scavenging Capacity

The assay to determine the ability of the extracts to scavenge the 2,2′-Azinobis-(3-ethylbenzothiazoline-6-sulfonic acid) cation radical ($ABTS^{\cdot}+$) was carried out according to a method previously described [17]. The absorbance was measured at 750 nm, and the percent purification was calculated at 6 min. Trolox (2–7 μg/mL) was used as a positive control.

2.5.4. Nitric Oxide-Scavenging Capacity

The method described by Torres-Carro et al. [17] was used to determine the nitric oxide depurating capacity of the extracts. This technique uses a Griess reagent to give a pink-colored azo complex with a maximum absorption at 550 nm. Ascorbic acid (25–200 μg/mL) was used as a positive control.

2.5.5. Iron-Reducing Power

The ability of the residue extracts to reduce potassium ferricyanide (Fe^{3+}) to potassium ferrocyanide (Fe^{2+}), which forms a Prussian blue complex, was detected spectrophotometrically at 700 nm. The absorbance values were used to determine the concentration required to reduce 50% of the Fe^{3+} (RC_{50}). BHT (3–13 μg/mL) was used as a positive control [17].

2.5.6. Copper-Reducing Power

The CUPRAC assay was performed utilizing the copper(II)-neocuproine (Cu(II)-Nc) reagent as the chromogenic oxidant [26]. The standard curve was performed with gallic acid (0.5–5 μg/mL), and the results were expressed in μg of gallic acid equivalents (GAE) per mg of DPE (μg GAE/mg).

2.5.7. Superoxide Anion Radical-Scavenging Assay

The quenching ability of superoxide anion radical ($O_2^{\bullet-}$) was assessed based on the NBT reduction into purple-colored diformazan [27]. $O_2^{\bullet-}$ was produced by the non-enzymatic NADH/PMS/O2 system. The absorbance was measured at 560 nm at 37 °C for 5 min. Results were expressed as the inhibition, in IC_{50}, of the NBT reduction to diformazan. As positive controls, catechin (10–100 μg/mL) and gallic acid (10–100 μg/mL) were employed.

2.5.8. Hypochlorous Acid-Scavenging Assay

The quenching abilities of the samples and the positive controls against hypochlorous acid (HOCl) were determined through a procedure previously described [27]. DHR, used as a fluorescence probe, was oxidized to rhodamine by HOCl. A 1% (*m*/*v*) NaOCl solution was used after adjusting the pH to 6.2. The inhibition of the HOCl-induced oxidation of DHR was determined. Catechin (0.05–0.5 μg/mL) and gallic acid (0.5–2.0 μg/mL) were employed as positive controls.

2.5.9. Saccharomyces Cerevisiae Survival Assay

Yeast cells were exposed to oxidative stress induced by 2 mM of H_2O_2 in the presence and absence of an extract. Two controls were used: yeast exposed to the vehicle of the extract (DMSO) and yeast exposed to extracts without the addition of H_2O_2. Cell viability was analyzed by determining the CFU/mL in a solid medium [28]. The results are expressed as a percentage of survival. One-hundred-percent survival is defined as the CFU/mL observed on the control plate, which contains yeast exposed to an extract vehicle, without extracts or hydrogen peroxide.

2.6. Toxicity Trials

2.6.1. Artemia Salina Test

The acute toxicity levels of Torrontés and Malbec pomace extracts, with concentrations from 250 to 500 µg/mL, were evaluated using the brine shrimp lethality test [17]. The experiments for each concentration were conducted in triplicate. The negative control wells contained DMSO to a final concentration lower than 0.3%, and the positive control potassium dichromate (10–40 µg/mL). Survival percentages were calculated by comparing the number of survivors in the test wells with respect to the negative control.

2.6.2. Hemolysis

The hemolytic impact of the pomace extracts was assessed spectrophotometrically at 550 nm following the method described by Torres-Carro et al. [17]. Extracts ranging from 200 to 1000 µg DPE/mL, or the vehicle (serving as a 0% hemolysis control), were brought into contact with a 10% suspension of human red blood cells (HRBC). A number of 100% hemolysis controls were established by exposing the 10% HRBC suspension to deionized water and to 1% (w/v) Triton X-100.

2.6.3. Cell Viability Assay

The 3-(4,5-dimethylthiazol-2-yl)-2,5-diphenyltetrazolium bromide (MTT) assay was performed to evaluate the effect of the different extract concentrations (0.1–1000 µg DPE/mL) on the intestinal cell lines. Passages 69–71 and 31–32 were used for Caco-2 and HT29-MTX, respectively. Briefly, cells (25×10^3 cells/mL) were incubated for 24 h with fresh medium in the absence or presence of the extracts dissolved in a cell culture medium.

Following the extracts' removal from each well, cells were washed with HBSS. The number of viable cells was determined by adding MTT reagent and incubating for 3 h at 37 °C. DMSO was used to solubilize the crystals. The positive control was DMEM, and the negative control was 1% (w/v) Triton X-100. Cells were grown according to the methodology described by Pinto et al. [27]. The absorbance was read at 590 nm with background subtraction at 630 nm. Results were expressed as percentages of cell viability.

2.7. Enzyme Inhibitions

The enzyme inhibition was calculated as a percent as follows: % inhibition = [(Abs control − Abs sample)/Abs control] × 100, where Abs control is the absorbance of the control solution and Abs sample is the absorbance of the sample solution. When possible, the concentration responsible for inhibiting 50% of enzyme activity (IC_{50}) was established through regression analysis, employing a concentration–inhibition response curve.

2.7.1. Tyrosinase

The tyrosinase inhibitory capacity of Torrontés and Malbec pomace extracts was performed as previously described by Matos et al. [6]. The absorbance was measured at 475 nm after 20 min of incubation. Kojic acid (0.1–5 µg/mL) was used as a positive control agent.

2.7.2. Xanthine Oxidase

The assay was conducted as previously described Quy and Xuan [29]. The absorbance was measured at 290 nm in a spectrophotometer. The reference inhibitor Allopurinol (0.05–5 μg/mL) was used as a positive control.

2.7.3. Lipoxygenase

The assessment of LOX activity followed the methodology outlined by Torres-Carro et al. [17]. Soybean LOX was exposed to different concentrations of the extracts or the vehicle and its substrate, linoleic acid. The inhibitory potential was gauged by computing the percentage of hydroperoxide production inhibition at 234 nm. Quercetin (40–70 μg/mL) and gallic acid (30–80 μg/mL) were used as positive controls.

2.8. Statistical Analysis

All data were expressed as the mean ± standard deviation from at least three independent experiments. The HPLC statistical analysis was conducted using IBM SPSS Statistics 26.0 software (SPSS Inc., Chicago, IL, USA). For the other assays, the statistical analysis was performed using INFOSTAT Analytical Software version 2020e (Universidad Nacional de Córdoba, Córdoba, Argentina). Differences in mean values were evaluated using Student's t-test for independent samples. In all analyses, p-values < 0.05 were considered statistically significant.

3. Results and Discussion

3.1. Chemical Composition

Vitis vinifera byproducts are an important source of phytochemicals with potential health-promoting properties and biotechnological interests. The chemical compositions and health benefits of several grape pomaces have been previously reported [5]. However, little is known about the phytochemical profiles of Malbec and Torrontés wine pomaces from the Calchaquí Valleys.

In the present study, the Malbec and Torrontés pomace samples had moistures between 65 and 70%, and the oven-drying over freeze-drying was selected due to its potential applicability within the wine industry and cost-effectiveness [30]. After drying and milling into flour, the grape pomaces were extracted with the use of non-toxic solvents (ethanol/water) and dried. Green solvents were used to acquire and characterize polyphenolic extracts suitable for human consumption or utilization in medical, cosmetic, or pharmaceutical sectors. Additionally, it is noteworthy that ethanol is a byproduct of the wine industry. The extraction yields of soluble compounds from the Malbec and Torrontés pomaces were 161.6 ± 16.7 and 735.2 ± 51.2 mg of extract/g of flour, respectively (16% and 73%). The extraction yields, and the total phenolic contents align with the results obtained using the same solvent (ethanol 50%) for other red wine pomaces, with values ranging between 5.3% and 16.1% [31,32]. Furthermore, the Torrontés pomace yields were significantly higher than those reported for other white wine pomaces, which ranged between 4.9% and 7.4% [32], having probably a higher content of extractable compounds than polyphenols. It is worth noting that this yields discrepancy may be attributed to the fact that, in the cited study, the authors chose to wash the pomace with water before drying to eliminate residual sugar, a step that was not carried out in the present work.

The analysis of the phenolic metabolites identified in the pomace extracts is detailed in Table 1. A comparison between the two analyzed pomace varieties reveals that the red Malbec variety exhibited a total phenol content 6.5 times higher than that of the Torrontés. Non-flavonoid phenolic compounds accounted for 19.5% of the total polyphenol content in the Malbec pomace extract, while in the Torrontés, these compounds represented 40.2%. The total flavonoid compound content in the Malbec extract surpassed that of the Torrontés 7.5 times. Interestingly, the tannin content was almost similar in both varieties. It should be noted that no anthocyanins were detected in the Torrontés extract (<LOQ).

Table 1. The polyphenol composition of the extracts.

Phytochemical Group	Malbec	Torrontés
	mg/g DPE	mg/g DPE
Total Phenolics (GAE)	156.01 ± 3.49 [a]	19.91 ± 1.21 [b]
Hydroxycinnamic acids (CAE)	11.39 ± 0.32 [a]	0.47 ± 0.01 [b]
Orthodiphenols (CAE)	31.79 ± 0.62 [a]	4.71 ± 0.28 [b]
Anthocyanins (C3GE)	6.30 ± 0.49 [a]	<LOQ
Non-flavonoid phenolics (GAE)	30.49 ± 1.15 [a]	8.01 ± 0.94 [b]
Tannins (CE)	23.20 ± 1.30 [a]	17.37 ± 0.47 [b]
Total flavonoids (QE)	327.25 ± 6.30 [a]	43.67 ± 1.39 [b]
Flavones/Flavonols (QE)	73.56 ± 1.41 [a]	1.17 ± 0.07 [b]
Flavanones/Dihydroflavonols (NE)	63.65 ± 5.56 [a]	10.10 ± 0.85 [b]

DPE: dry pomace extract. DP: dry pomace. GAE: gallic acid equivalents. CAE: caffeic acid equivalents. QE: quercetin equivalents. NE: naringenin equivalents. CE: catechin equivalents. C3GE: cyanidin-3-glucoside equivalents. LOQ: limit of quantification. Results are expressed as means ± standard deviations. Different letters in the same row indicate significant differences between samples ($p < 0.05$) according to Student's t test.

The concentration of total polyphenols in the Malbec dry extract (156.01 mg GAE/g DPE) was very similar to the findings in a Malbec pomace from Mendoza, Argentina (the Cuyo region), which reported 196.2 mg of GAE/g of DPE) [31]. The present findings align with the total polyphenol levels (ranging between 127 and 298 mg/g of DPE) and the total flavonoid content (ranging between 137 and 322 mg/g of DPE) observed in four other red skin pomaces derived from Italian cultivars (Barbera, Grignolino, Pinot Noir, and Nebbiolo). Nevertheless, in the present study, the tannin content was lower [32]. Furthermore, the results achieved exceed those reported for Merlot pomace from Brazil [33] in terms of total polyphenols, total flavonoids, and anthocyanins. It is worth noting that all these studies utilized extraction conditions like the present research. Higher values of total polyphenols (523 mg/g DPE), but lower levels of anthocyanins (1.67 mg/g DPE), were reported for Cabernet Sauvignon grape pomace (Mexico) that was defatted with hexane prior to extraction with ethanol at 60% [30].

Likewise, Torrontés pomace from Galicia (extracted with 65% methanol) exhibited a lower concentration of total phenolic compounds (22 mg GAE/g DPE) as reported by Alvarez-Casas et al. [34], which is consistent with the present findings (19.9 mg GAE/g DPE). The sugars present in phenolic-extracted Torrontés grape marc lead to a reduced polyphenol richness per mass of extract when compared to the red pomace extract. Guaita et al. [32] documented higher values of total phenolics (144–208 mg/g DPE), total flavonoids (108–206 mg/g DPE), and tannins (64–108 mg/g DPE) for three other white-skinned pomaces of Italy (Muscat blanc, Arneis, Cortese) extracted with 50% ethanol. Variations in extraction methods, grape types, and agroclimatic conditions contribute to disparities in the measured total polyphenols, explaining the divergent values reported in the literature. Moreover, the genotype is the main factor that influences the relative concentrations of the different phenolic compounds [35].

Through HPLC-DAD analysis, 35 individual phenolic compounds were identified in the extracts, including 14 phenolic acids, 18 flavonoids, and 3 stilbenoids (Table 2). Most of the compounds identified were found in both pomaces. In the Torrontés extract, the main polyphenols were 4,5-di-O-caffeoylquinic acid > 4-O-caffeyolquinic acid > kaempferol-3-O-glucoside > isorhamnetin-3-O-glucoside > caftaric acid > sinapic acid > gallic acid > ferulic acid > protocatechuic acid > (+)-catechin. Regarding the Malbec extract, the main phenolic compounds were protocatechuic acid > (+)-catechin > gallic acid > 4-O-caffeyolquinic acid > sinapic acid > 4,5-di-O-caffeoylquinic acid.

Table 2. The phenolic compounds quantified in the wine pomace samples through HPLC-DAD.

Phenolic Compound	Retention Time (min)	mg/100 g DPE		mg/100 g DP	
		Malbec	Torrontés	Malbec	Torrontés
Phenolic acids					
Gallic acid	5.618	245.0 ± 12.0 [a]	89.3 ± 4.5 [b]	39.2 ± 1.9 [a]	65.2 ± 3.3 [b]
Protocatechuic acid	9.935	700.0 ± 35.0 [a]	71.2 ± 3.6 [b]	112.0 ± 5.6 [a]	51.9 ± 2.6 [b]
Neochlorogenic acid	10.219	6.1 ± 0.3 [a]	1.9 ± 0.1 [b]	0.9 ± 0.1 [a]	1.4 ± 0.1 [b]
Caftaric acid	15.436	<LOQ	90.6 ± 4.5	<LOQ	66.1 ± 3.3 [b]
Chlorogenic acid	17.869	6.6 ± 0.3 [a]	6.8 ± 0.3 [a]	1.1 ± 0.1 [a]	5.0 ± 0.2 [b]
4-*O*-caffeoylquinic acid	19.897	130.0 ± 6.0 [a]	118.0 ± 6.0 [a]	20.8 ± 1.0 [a]	86.1 ± 4.4 [b]
Vanillic acid	20.748	ND	ND	ND	ND
Caffeic acid	21.224	6.7 ± 0.31 [a]	18.7 ± 0.9 [b]	0.9 ± 0.1 [a]	13.7 ± 0.7 [b]
Syringic acid	22.283	38.6 ± 1.9 [a]	27.6 ± 1.4 [b]	6.2 ± 0.3 [a]	20.2 ± 1.0 [b]
p-Coumaric acid	33.758	28.2 ± 1.4 [a]	36.7 ± 1.8 [b]	4.5 ± 0.2 [a]	26.8 ± 1.3 [b]
trans-Ferulic acid	37.289	45.7 ± 2.3 [a]	74.2 ± 3.7 [b]	7.3 ± 0.4 [a]	54.2 ± 2.7 [b]
Sinapic acid	37.662	108.0 ± 5.0 [a]	89.6 ± 4.5 [b]	17.3 ± 0.8 [a]	65.4 ± 3.3 [b]
3,5-di-*O*-caffeoylquinic acid	50.127	7.2 ± 0.4 [a]	45.0 ± 2.2 [b]	1.2 ± 0.1 [a]	32.9 ± 1.6 [b]
Ellagic acid	55.284	33.4 ± 1.7	ND	5.3 ± 0.3	ND
4,5-di-*O*-caffeoylquinic acid	56.781	50.7 ± 2.5 [a]	258.0 ± 13.0 [b]	8.1 ± 0.4 [a]	188.3 ± 9.5 [b]
Cinnamic acid	58.47	ND	ND	ND	ND
Flavonoids					
(+)-Catechin	14.143	618.0 ± 31.0 [a]	70.1 ± 3.5 [b]	98.9 ± 5.0 [a]	51.2 ± 2.6 [b]
(−)-Epicatechin	23.294	43.2 ± 2.2 [a]	17.2 ± 0.9 [b]	6.9 ± 0.4 [a]	12.6 ± 0.7 [b]
Naringin	49.847	31.9 ± 1.6 [a]	54.6 ± 2.7 [b]	5.1 ± 0.3 [a]	39.9 ± 2.0 [b]
Quercetin-3-*O*-galactoside	52.177	25.2 ± 1.3 [a]	43.8 ± 2.2 [b]	4.0 ± 0.2 [a]	32.0 ± 1.6 [b]
Quercetin-3-*O*-glucopyranoside	52.735	ND	ND	ND	ND
Rutin	53.284	19.1 ± 1.0 [a]	16.4 ± 0.8 [b]	3.1 ± 0.2 [a]	12.0 ± 0.6 [b]
Phloridzin	54.355	32.3 ± 1.6 [a]	ND	5.2 ± 0.3	ND
Myricetin	57.943	17.6 ± 0.9 [a]	18.9 ± 0.9 [a]	2.8 ± 0.1 [a]	13.8 ± 0.7 [b]
Quercitrin	59.07	ND	ND	ND	ND
Kaempferol-3-*O*-glucoside	59.466	ND	116.0 ± 6.0	ND	84.7 ± 4.4
Kaempferol-3-*O*-rutinoside	60.01	<LOD	66.6 ± 3.3	<LOD	48.6 ± 2.4
Isorhamnetin-3-*O*-glucoside	60.277	ND	95.6 ± 4.8	ND	69.8 ± 3.5
Isorhamnetin-3-*O*-rutinoside	61.568	24.2 ± 1.2 [a]	10.2 ± 0.5 [b]	3.9 ± 0.2 [a]	7.4 ± 0.4 [b]
Naringenin	68.149	<LOD	<LOD	<LOD	<LOD
Quercetin	71.031	18.6 ± 0.9 [a]	11.4 ± 0.6 [b]	3.0 ± 0.1 [a]	8.3 ± 0.4 [b]
Phloretin	72.269	<LOQ	<LOD	<LOQ	<LOD
Tiliroside	76.233	22.7 ± 1.1	<LOQ	3.6 ± 0.2	<LOQ
Kaempferol	79.854	3.4 ± 0.2 [a]	4.4 ± 0.2 [b]	0.6 ± 0.1 [a]	3.2 ± 0.2 [b]
Apigenin	81.44	<LOD	<LOD	<LOD	<LOD
Chrysin	90.832	<LOD	<LOD	<LOD	<LOD
Stilbenoids and others					
trans-Polydatin	39.182	18.9 ± 0.9 [a]	3.6 ± 0.2 [b]	3.0 ± 0.1 [a]	2.6 ± 0.1 [b]
Resveratrol	52.507	ND	ND	ND	ND
trans-Epsilon viniferin	69.158	6.5 ± 0.3 [a]	6.2 ± 0.3 [a]	1.0 ± 0.1 [a]	4.5 ± 0.2 [b]

DPE: dry pomace extract. DP: dry pomace. Results are expressed as means ± standard deviations. LOQ: limit of quantification; LOD: limit of detection; ND: not detected. Different letters in the same row indicate significant differences between samples ($p < 0.05$) according to Student's *t* test.

The phenolic patterns among the varieties exhibited notable quantitative differences. For instance, in the Torrontés extract, the content of 4,5-di-O-caffeoylquinic acid is five times higher compared than that of the Malbec extract. Conversely, the Malbec extract displayed significantly elevated levels of (+)-catechin, protocatechuic acid, and gallic acid, which were 8.8, 9.8, and 2.7 times higher, respectively, than those found in the Torrontés extract. Additionally, in the pomace of another red grape variety, Cabernet Sauvignon, quercetin, catechin, epicatechin, and syringic acid emerged as dominant phenolic compounds [36].

To contextualize our findings against other published data for red and white whole pomaces (skin and seeds) extracted with hydroalcoholic solvents, the results were presented in the milligrams of the compound per 100 gram of dry pomace extract (mg/ 100 g DPE) or per 100 gram of dry pomace (mg/100 g DP) (Table 2). The results attested that gallic acid (2.74–73 mg/g DPE and 6.4–14.07 mg/g DP), protocatechuic acid ($\leq$0.29 mg/g DPE and 0.6–2.03 mg/g DP), chlorogenic acid ($\leq$0.06 mg/g DPE and $\leq$0.13–0.23 mg/g DP), and ferulic acid (2.4 mg/g DPE and <0.13 mg/g DP) are present in concentrations much higher than the values reported for red wine pomaces [14,30,31,33,37–39]. Furthermore, the phenolic acids: gallic (3.3–11.12 mg/g DP), protocatechuic (0.39–7 mg/g DP), caffeic (falling within the range of 0.15 mg/g DP), caftaric (1.9–7.9 mg/g DP), and ferulic (0.13–0.22 mg/g DP); and the flavonoids quercetin-3-O-galactoside (approx. 0.96 mg/g DP) and kaempferol-3-O-glucoside (approx. 0.21 mg/g DP) are found in much higher concentrations than those reported in the literature for white wine pomaces [14,34,39]. Although resveratrol has not been identified, the presence of derivatives such as polydatin, also known as piceid (a resveratrol derivative with improved bioavailability), and trans-epsilon viniferin (a resveratrol dimer) has been detected. To the best of our knowledge, this study reports for the first time the presence of neochlorogenic acid, 4-O-caffeoylquinic acid, 3,5-di-O-caffeoylquinic acid, 4,5-di-O-caffeoylquinic acid, sinapic acid, and isorhamnetin-3-O-rutinoside in hydroalcoholic extracts of wine pomace.

Malbec wine pomace from Argentina's Cuyo region exhibited lower levels of catechin (338 mg/100 g DPE) and gallic acid (25 mg/100 g DPE) compared to the present findings for the Malbec pomace from the Calchaquí Valleys. In contrast, Cuyo's Malbec wine pomace displayed higher concentrations of epicatechin (176 mg/100 g DPE) and syringic acid (173 mg/100 g DPE). It is interesting to note that protocatechuic acid, the main phenolis acid observed in this study, was not reported in the Malbec pomace from Cuyo [31]. In another study [14] that assessed the polyphenol content in a concentrated chromatographic fraction obtained from a methanolic extract of red pomace from the Calchaquí Valleys (of an unknown variety), C6-C1 phenolic acids, gallic acid, and syringic acid were identified as the main phenolics. In agreement with the present results, Teixeira et al. [35] reported protocatechuic acid as the most abundant hydroxybenzoic acid in pomace from red varieties.

Regarding white wine pomace from the Calchaquí Valleys (of an unknown variety), Salazar et al. [14] highlighted gallic acid as the predominant phenolic acid. In contrast, the results obtained in this work reveal a distinct composition, with an emphasis on the prevalence of derivatives of caffeoylquinic acid, contrasting with Salazar's observations. In Galician Torrontés pomace, phenolic acids such as gallic, protocatechuic, and caftaric acids, and the flavonoids epicatechin, quercetin, and quercetin derivatives were identified [34]. In the present work, a greater diversity of phenolic compounds was explored, which explains why other authors do not report the caffeoylquinic acid-derived compounds which quantitatively were some of the most important in the present research.

A literature review [35] described notable distinctions in the hydroxycinnamic acid content between red and white grape skins. Specifically, white grape skins exhibited elevated levels of cis-coutaric acid and trans-caftaric acid, whereas red grape skins were predominantly composed of chlorogenic acid (3-O-caffeoylquinic acid). This study observed similar chlorogenic acid contents in both varietals, while caftaric acid was only present in the Torrontés pomace.

3.2. *Biological Activity*

3.2.1. Antioxidant

Reactive species, particularly reactive oxygen species (ROS) and reactive nitrogen species (RNS), play a significant role in various physiological processes, including cell signaling, inflammatory cascades, and homeostasis. Consequently, the evaluation of the scavenging capacity of an extract against ROS and RNS becomes more intriguing due to their pivotal functions in living tissues. The antioxidant and antiradical activities of hydroalcoholic extracts were evaluated in nine different assays (Table 3 and Figure 1). In the ABTS cation radical, nitric oxide, superoxide anion, and hypochlorite assays, the Malbec extract demonstrated the best scavenging efficiency, as indicated by its lower IC_{50} values. In the iron-chelating assay, the Malbec extract could chelate 42% of the metal at 1000 μg/mL (Table 3). Additionally, the Malbec extract exhibited superior Fe^{3+}- and Cu^{2+}-reducing power when compared to the Torrontés extract (Table 3). To assess the bioactivity of wine pomace in safeguarding Saccharomyces cerevisiae cells from induced oxidative damage, cell viability was measured in the presence or absence of extracts such as chemoprotectors. As depicted in Figure 1, the selected concentrations from the samples (12.5–150 μg/mL) exhibited non-cytotoxicity to *S. cerevisiae*. Upon the induction of oxidative stress, yeast cells demonstrated sensitivity to H_2O_2, with only 50% surviving the oxidative insult. Malbec grape pomace extract (12.5 μg/mL) rescues 24% of yeast from oxidative stress induced by H_2O_2, while the Torrontés pomace did not show a protective effect.

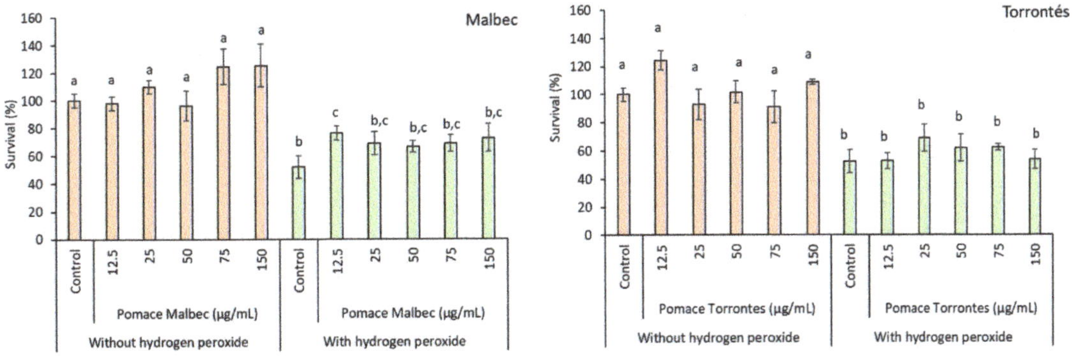

Figure 1. *Saccharomyces cerevisiae* survival in the presence of different pomace extract concentrations (from 12.5 to 150 μg DPE/mL). Different letters indicate significant differences between treatments ($p < 0.05$), according to Tukey's test.

In summary, the Malbec grape pomace demonstrated significant scavenging activity against ROS and RNS, potentially attributed to its higher phenolic content when compared to the Torrontés grape pomace, particularly with concerns to catechin, epicatechin, gallic acid, and protocatechuic acid whose free radical-scavenging potential has been extensively demonstrated [2]. This distinction could account for the heightened antioxidant activity observed in this particular varietal.

The divergence in the antioxidant and antiradical activities between red and white pomace extracts is attributed to their distinct phenolic profiles, significantly influenced by the grape variety and extraction methods [40]. The free radical-scavenging activity and the ferric-reducing power of grape pomace are correlated with specific phenolic acids (gallic and caffeic), flavan-3-ols (catechin, epicatechin, and galloylcatechin), and flavonols (quercetin) [18,38].

The antioxidant activity results obtained in this study surpassed those reported by other authors for hydroalcoholic extracts of red wine pomace (Syrah, Petit Verdot, and Romy) and white wine pomace (Chenin Blanc and Banaty): ABTS scavenging (IC_{50} 56.22 μg/mL for Romy and 78.47 μg/mL for Banaty), iron-reducing ability (RC_{50} 160.97 μg/mL for Romy and 141.5 μg/mL for Banaty), iron-chelating capacity (CC_{50} 262.67 μg/mL for Romy and 248.35 μg/mL for Banaty), superoxide anion scavenging (IC_{50} 190 μg/mL for Petit Verdot, 240 μg/mL for Syrah, and 2160 μg/mL for Chenin Blanc), and hypochlorite scavenging (IC_{50} 17 μg/mL for Petit Verdot, 31 μg/mL for Syrah, and 128 μg/mL for Chenin Blanc) [41,42]. No reports were found regarding nitric oxide-scavenging capacity. Pomace extracts from Syrah, Merlot, and Cabernet Sauvignon boosted the *S. cerevisiae* survival rate by 8% to 16% compared to cells exposed to H_2O_2 [28].

Table 3. The antioxidant capacities of the wine pomace hydroalcoholic extracts.

Sample	Phosphomolybdenum Reducing Capacity (µgAAE/mg DPE)	Cupric Reducing Capacity (µgGAE/mg DPE)	ABTS˙+ Scavenging IC_{50} (µg/mL)	NO Scavenging IC_{50} (µg/mL)	Fe^{3+} Reducing RC_{50} (µg/mL)	Iron Chelating CC_{50} (µg/mL)	$O_2^{\bullet-}$ Scavenging IC_{50} (µg/mL)	HOCl Scavenging IC_{50} (µg/mL)
Malbec	178.57 ± 4.99 [a]	171.18 ± 2.2 [a]	7.79 ± 0.17 [a]	414.19 ± 5.79 [b]	10.22 ± 0.16 [a]	41.82% ± 0.48% *	74.17 ± 4.12 [a,b]	6.71 ± 0.36 [b]
Torrontés	4.74 ± 0.16 [b]	26.12 ± 0.52 [b]	49.5 ± 1.46 [b]	15.34% ± 0.78% **	84.62 ± 0.95 [b]	11.28% ± 1.05% *	874.61 ± 15.71 [c]	27.40 ± 0.19 [c]
Controls								
BHT	-	-	-	-	11.37 ± 0.13 [a]	-	-	-
Trolox	-	-	3.74 ± 0.06 [a]	-	-	-	-	-
Ascorbic acid	-	-	-	36.13 ± 6.01 [a]	-	-	-	-
EDTA	-	-	-	-	-	13.97 ± 0.06	-	-
Catechin	-	-	-	-	-	-	99.21 ± 0.85 [b]	0.095 ± 0.006 [a]
Gallic acid	-	-	-	-	-	-	52.49 ± 1.58 [a]	0.82 ± 0.06 [a]

Different letters in the same column show significant differences among each treated group, according to a Tukey test ($p \leq 0.05$). IC_{50} = the concentration required to demonstrate a decrease of 50% in the reactivity of the reactive species in the tested media (mean ± standard error of the mean). The phosphomolybdenum-reducing capacity is expressed as micrograms of ascorbic acid equivalents per milligram of DPE (dry pomace extract) (µg AAE/mg DPE). The cupric-reducing antioxidant capacity is expressed as micrograms of gallic acid equivalents (µg GAE/mg DPE). The Fe^{3+}-reducing capacity (RC), ABTS radical cation (ABTS˙+)-, nitric oxide (NO)-, superoxide anion radical ($O_2^{\bullet-}$)-, and hypochlorous acid (HOCl)-scavenging capacities (IC). The iron-chelating capacity (CC) determined through linear regression analysis. * An inhibition percentage with the concentration of 1000 µg/mL. ** An inhibition percentage with the concentration of 250 µg/mL.

Protocatechuic acid, catechin, gallic acid, ferulic acid, and chlorogenic acid, present in the Calchaquíes Valleys' wine pomaces, exhibit various pharmacological activities. These include antioxidant, anti-inflammatory, neuroprotective, antibacterial, antiviral, anticancer, antiosteoporotic, analgesic, antiaging, antihypertensive, anti-diabetic, antihyperlipidemic, anticoagulant, antiulcer, cardioprotective, and hepatoprotective properties [43–48]. The antioxidant properties of the pomace extracts observed, particularly those of the Malbec, are partially attributed to the high content of these phenolic compounds.

Additionally, accumulated evidence demonstrated that caffeoylquinic acids present in high concentrations in the Torrontés pomace extract have a wide range of biological activities, such as antiparasitic, antioxidation, antiviral, antibacterial, anti-inflammatory, anticancer, neuroprotective, and anti-diabetic effects [48]. Di-O-caffeoylquinic acids protect bone marrow-derived mesenchymal stem cells from •OH-induced damage and the antioxidant mechanisms include electron transfer, H^+ transfer, and Fe^{2+} chelation [49,50].

Likewise, polydatin has demonstrated countless pharmacological properties, primarily including anticancer, cardioprotective, anti-diabetic, gastroprotective, hepatoprotective, neuroprotective, and antimicrobial effects, along with health-promoting roles in the renal system, the respiratory system, rheumatoid diseases, the skeletal system, and women's health. This resveratrol derivative has higher antioxidant and anti-inflammatory activity than resveratrol [51].

The proven antioxidant capacity of the hydroalcoholic extracts derived from Malbec and Torrontés pomaces, coupled with their demonstrated antivirulence activity against pathogenic bacteria (*P. aeruginosa* and *S. aureus*) as indicated by other authors [12,13], implies a promising prospect for their use as natural preservatives in the food industry. This potential application could play a crucial role in mitigating oxidative processes.

This collective evidence elucidates the biological activity of the Calchaquíes Valleys' wine pomaces, underscoring the substantial health-promoting benefits associated with the presence of these diverse and potent phenolic compounds.

3.2.2. Cytotoxicity

The cytotoxic assay performed on HT29 and Caco-2 colon cancer cells is shown in Table 4. The Torrontés extract significantly reduced cell viability in a dose-dependent manner, achieving 67% and 48% viability for HT29 and Caco-2 cells, respectively, at the highest tested concentration (1 mg/mL). In contrast, the Malbec pomace at the same concentration reduced the viability of Caco-2 cells by 20%. These findings align with previous research indicating that grape pomace can inhibit the proliferation of colon adenocarcinoma cells (Caco-2, HT-29) in a dose-dependent manner, and that white grape pomace is more active than red [52].

Table 4. The effects of the wine pomace samples' exposure on the viability of HT29-MTX and Caco-2 cells at different concentrations as measured via an MTT assay ($n = 3$).

	Cell Viability (%)			
Concentration (µg DPE/mL)	**HT29-MTX Cells**		**Caco-2 Cells**	
	Malbec	**Torrontés**	**Malbec**	**Torrontés**
0.1	132.25 ± 10.18 [a]	109.33 ± 19.90 [a]	120.67 ± 6.92 [a]	70.29 ± 10.03 [b]
1.0	133.03 ± 12.93 [a]	110.34 ± 9.63 [a]	79.08 ± 6.94 [b]	65.14 ± 14.27 [b]
10	131.85 ± 14.80 [a]	103.57 ± 16.69 [a]	82.82 ± 11.80 [b]	66.85 ± 7.36 [b]
100	122.19 ± 16.94 [a]	66.87 ± 13.06 [b]	88.48 ± 12.45 [b]	66.85 ± 10.52 [b]
1000	104.75 ± 23.03 [b]	33.56 ± 4.38 [c]	81.26 ± 12.83 [b]	52.51 ± 1.53 [c]
Medium	100.00 ± 8.86 [a]			
Triton X-100	0.00 ± 0.00			

Different letters mean significant differences between concentrations of the same extract ($p < 0.05$), according to Tukey's test.

The anti-proliferative effects of Zalema (white grape) pomace hydroalcoholic extract (methanol 75%), as well as specific phenolic standards like catechin, epicatechin, quercetin, and gallic acid on Caco-2 cells, have been previously documented [53]. Additionally, the anti-proliferative impact of caffeoylquinic acids on human colon cancer cells has been established [49,54]. The observed effects in this study could be attributed, at least in part, to the high content of these acids in the Torrontés pomace. On the other hand, the synergistic effect of different polyphenolic compounds as chemopreventive agents is well-documented [53].

This study represents the first report on the cytotoxic evaluation of pomaces from the Calchaquí Valleys on human colon carcinoma cells and the first description of the anti-proliferative capacity of Torrontés wine pomace.

3.2.3. Toxicity

Considering that the analyzed extracts are derived from byproducts of the wine industry and may harbor potential health benefits, it becomes crucial to assess their toxicity. This assessment employs two distinct experimental models: a holistic organism model utilizing *Artemia salina* and a eukaryotic cell model involving human red blood cells. Notably, at concentrations up to 1000 μg/mL, none of the extracts demonstrated toxicity towards red blood cells. In the brine shrimp test, the Malbec and Torrontés pomace extracts exhibited no toxicity up to 500 μg/mL (Figure 2). Aligning with these results, polar extracts (both aqueous and alcoholic) from red grape pomaces of Malbec and Syrah varieties in the Cuyo region of Argentina (Mendoza) did not induce toxicity in fish (*Danio rerio*) and crustaceans (*Artemia salina*) up to a concentration of 500 μg/mL. Similarly, no toxicity was observed in a murine macrophage cell line (RAW 264.7) up to a concentration of 1000 μg/mL [55]. In the case of hydroalcoholic extracts derived from white grape pomace (Falanghina) and red grape pomace (Tintilia and Vernaccia Nera di Serrapetrona) from Italy, the toxicity limits surpassed 40 mg/mL. In contrast, for the Sagrantino red variety, the toxic threshold was determined to be greater than 2 mg/mL in the brine shrimp test [56].

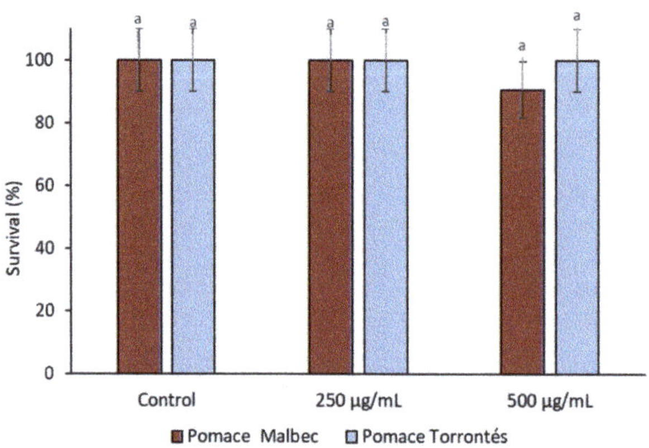

Figure 2. *Artemia salina* survival in the absence and presence of different pomace hydroalcoholic extract concentrations (μg DPE/mL). Results are expressed as means ± standard deviations. Different letters indicate significant differences between samples ($p < 0.05$) according to Student's *t* test.

3.2.4. Enzyme Inhibition

Skin aging often leads to pigmentation disorders, prompting the cosmetic industry to seek anti-hyperpigmentation compounds. These disorders occur due to melanin accumulation, which is influenced by various factors. Therefore, inhibiting melanin production by blocking the enzyme tyrosinase is a key strategy for skin-whitening products, making flavonoids promising compounds for this field [57]. Exploiting the bioactives contained in grape pomaces to obtain high-value cosmetics may support the growth of innovative start-ups and expand the value chain of grapes [9]. In this study, Torrontés demonstrated no impact on this enzyme activity at up to 200 μg/mL concentrations. Conversely, Malbec pomace displayed a dose-dependent inhibitory effect, leading to an approximately 90% decrease in tyrosinase activity at the same concentration (200 μg/mL) (Figure 3). The IC_{50} value for the Malbec extract was 89.9 ± 2.1 μg/mL. For kojic acid, a well-established tyrosinase inhibitor used as a reference, the IC_{50} was 0.37 ± 0.03 μg/mL. In comparison, the IC_{50} reported for Tempranillo (red variety) pomace extract (ethanol 50%) from Valladolid (Spain) was 4000 μg/mL [6]. Additionally, an ethanol extract (0.3 mL aliquots) of a mixture of white grape (Trebbiano and Verdicchio) pomaces demonstrated a 79% reduction in tyrosinase activity [58]. Extracts from the red grape stems of six grape varieties (1 mg/mL) inhibited the tyrosinase enzyme's activity from 41.47% to 53.83%, with the Syrah variety exhibiting the highest activity [59]. Furthermore, it is noteworthy that certain flavonoids, including kaempferol, catechin, myricetin, and quercetin, among others, have been identified as inhibitors of tyrosinase in previous studies [6,58]. The higher content of flavonoid-type compounds in the Malbec extract could elucidate its superior ability to inhibit this enzyme. These results suggest that Malbec pomace from the Calchaquí Valleys is a valuable source of natural ingredients for cosmeceutical applications.

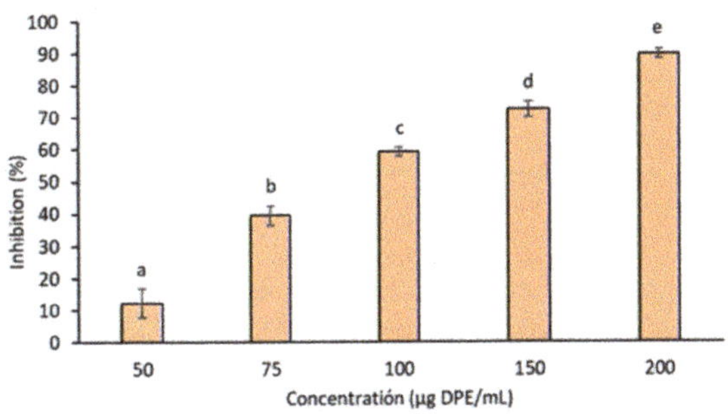

Figure 3. The tyrosinase inhibition present in the Malbec pomace hydroalcoholic extract. Results are expressed as means ± standard deviations. Different letters mean significant differences between concentrations ($p < 0.05$), according to Tukey's test.

5-lipoxygenase (5-LOX) governs a critical biosynthetic pathway in generating eicosanoids. The ultimate product of the 5-LOX pathway, leukotriene B4 (LTB4), is a mediator in several diseases. Inhibitors of 5-LOX exhibit therapeutic potential for various inflammatory conditions, including asthma, allergies, and atherosclerosis [60]. The pivotal role of LOX and its derivatives in tumor initiation and cancer metastasis is well established. Heightened levels of 5-LOX have been discerned in diverse cancer cell lines, encompassing those linked to prostate, lung, colon, breast, and other cancers [61]. Thus, identifying compounds capable of inhibiting the LOX enzyme is a compelling focal point in the quest for bioactive natural products with potential anticancer properties.

Figure 4 illustrates the kinetics of hydroperoxide production by the LOX enzyme at varying concentrations of pomace extracts. Both compounds exhibit a dose-dependent inhibition of LOX activity. Based on the dose–response curves, the extract concentration resulting in a 50% activity reduction (IC_{50}) was estimated to be 154.7 ± 4.1 and 784.7 ± 18.2 μg DPE/mL for Malbec and Torrontés, respectively, indicating that the Malbec extract is a more potent inhibitor of LOX than the Torrontés extract. The IC_{50} values for the positive control employed, namely quercetin and gallic acid, were 48.4 ± 1.9 and 54.1 ± 0.5 μg/mL, respectively. Grape pomace extracts from Montepulciano d'Abruzzo (Italy) red wine displayed dose-dependent LOX inhibitions ranging from 23% to 47% at tested concentrations of 10 and 100 μg/mL, respectively [62]. Previous studies have indicated a positive correlation between LOX inhibition and the concentrations of gallic acid, vanillic acid, p-coumaric acid, catechin, epicatechin, and rutin in pomaces [63]. In the present work, it was observed that the Malbec pomace exhibited higher levels of most of these compounds, potentially accounting for its fivefold greater capacity to inhibit this enzyme.

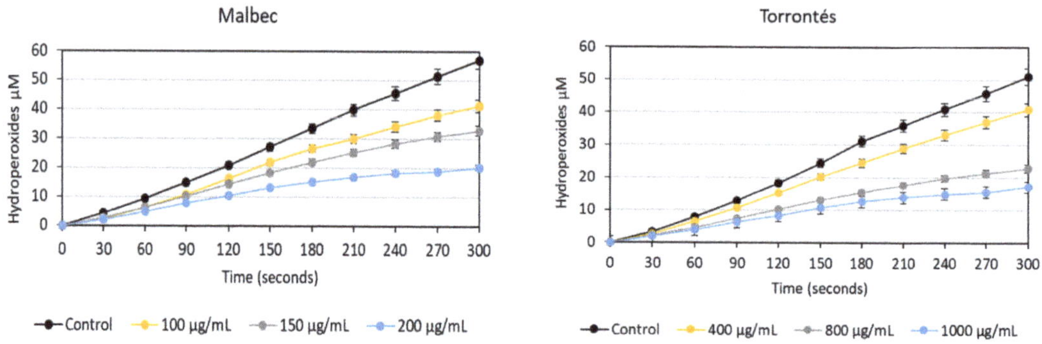

Figure 4. The lipoxygenase inhibition at different concentrations (µg DPE/mL) of hydroalcoholic extracts of Malbec and Torrontés pomaces. Results are expressed as means ± standard deviations.

The xanthine oxidase inhibitors have potential as anti-gout agents. Hyperuricemia, a metabolic anomaly characterized by the excess production of uric acid or its insufficient excretion, often leads to gout, marked by elevated serum urate and the simultaneous accumulation of urate crystals in bones or joints. Xanthine oxidase converts hypoxanthine to xanthine and, subsequently, produces uric acid. Effective strategies for gout prevention and recovery include xanthine oxidase inhibitors. Grape polyphenols, including cinnamic acid derivatives, syringic acid, ellagic acid, caffeic acid, and ferulic acid, are considered to contribute to the inhibition of xanthine oxidase activity, as mentioned in a review of several in vitro studies [64]. Despite the potential of phenolic compounds and flavonoids in mitigating the conditions associated with hyperuricemia, in the trials performed, none of the pomace extracts were able to inhibit the enzyme up to a concentration of 250 µg/mL. This study marks the first report on the evaluation of wine pomaces from the Calchaquí Valleys concerning LOX, tyrosinase, and xanthine oxidase enzymes.

4. Conclusions

In this study, we identified 35 phenolic compounds in the hydroalcoholic extracts derived from the pomace of Malbec and Torrontés wines from the Calchaquí Valleys of Argentina. Both extracts exhibited significant antioxidant activity, with Malbec showing superior antiradical activity, reducing power, hypochlorite scavenging, and iron-chelating capacity. Additionally, Malbec extract demonstrated the ability to inhibit tyrosinase, an enzyme affecting the skin, and the pro-inflammatory enzyme lipoxygenase. However, the Torrontés pomace extract was more effective against colon cancer cells.

These findings support the idea that both extracts could be considered potential functional ingredients with preventive properties against diseases. These extracts have potential applications in pharmacology, cosmetics, and the food industry. Therefore, it is imperative to conduct further comprehensive studies to validate their practical utilization. The exploitation of this byproduct has the potential to impact local economies significantly.

Author Contributions: P.E.T.: Performed the experiments, Analyzed the data, Wrote the paper. A.M.S.: Performed the experiments, Analyzed the data. C.D.-M.: Performed the experiments, Analyzed the data. M.M.: Performed the experiments, Analyzed the data. F.R.: Conceived and designed the experiments, Analyzed the data, Wrote the paper. R.T.C.: Performed the experiments; M.D.S.: Analyzed the data, M.G.O.: Conceived and designed the experiments, Analyzed the data; M.A.B.: Analyzed the data. M.E.A.: Conceived and designed the experiments, Analyzed the data,

Wrote the paper, M.R.A.: Conceived and designed the experiments, Analyzed the data, Wrote the paper, Conceived and initiated the project. All authors have read and agreed to the published version of the manuscript.

Funding: The authors acknowledge the financial support from the SCAIT-UNT (PIUNT D702 y D708), the Agencia Nacional de Promoción Científica y Técnica ANPCyT (PICT 2018-02071 and 02514, PICT 2021-0060, PICT-2021-I-A-00439 I), The Consejo Nacional de Investigaciones Científicas y Técnicas, CONICET (PUE 0021), ANPCyT BID-PICT 1576 and 2417 and SECYT-Universidad Nacional de Córdoba (05/C375). This work also received financial support from project UIDB/50006/2020 and UIDP/50006/2020 through national funds.

Institutional Review Board Statement: Not applicable.

Informed Consent Statement: Not applicable.

Data Availability Statement: The data presented in this study are available on request from the corresponding author. The data are not publicly available due to privacy restrictions.

Acknowledgments: Ana Margarida Silva is thankful for the Ph.D. grant (SFRH/BD/144994/2019) financed by POPH–QREN and subsidized by the European Science Foundation and Ministério da Ciência, Tecnologia e Ensino Superior. Manuela M. Moreira (CEECIND/02702/2017) and Francisca Rodrigues (CEECIND/01886/2020) are thankful for their contracts financed by the FCT/MCTES—CEEC Individual Program Contract. The authors also are grateful to Albarossa cellar for providing the wine pomace.

Conflicts of Interest: The authors declare no conflict of interest.

References

1. Alimenti, C.; Lianza, M.; Antognoni, F.; Giusti, L.; Bistoni, O.; Liotta, L.; Angeloni, C.; Lupidi, G.; Beghelli, D. Characterization and biological activities of in vitro digested olive pomace polyphenols evaluated on ex vivo human immune blood cells. *Molecules* **2023**, *28*, 2122. [CrossRef]
2. Chedea, V.S.; Macovei, Ș.O.; Bocșan, I.C.; Măgureanu, D.C.; Levai, A.M.; Buzoianu, A.D.; Pop, R.M. Grape pomace polyphenols as a source of compounds for management of oxidative stress and inflammation—A possible alternative for non-steroidal anti-inflammatory drugs? *Molecules* **2022**, *27*, 6826. [CrossRef]
3. Gueboudji, Z.; Kadi, K.; Mahmoudi, M.; Hannachi, H.; Nagaz, K.; Addad, D.; Yahya, L.B.; Lachehib, B.; Hessini, K. Maceration and liquid–liquid extractions of phenolic compounds and antioxidants from algerian olive oil mill wastewater. *Environ. Sci. Pollut. Res.* **2023**, *30*, 3432–3439. [CrossRef] [PubMed]
4. Abouelenein, D.; Mustafa, A.M.; Caprioli, G.; Ricciutelli, M.; Sagratini, G.; Vittori, S. Phenolic and nutritional profiles, and antioxidant activity of grape pomaces and seeds from Lacrima Di Morro d'Alba and Verdicchio Varieties. *Food Biosci.* **2023**, *53*, 102808. [CrossRef]
5. Ferrer-Gallego, R.; Silva, P. The wine industry by-products: Applications for food industry and health benefits. *Antioxidants* **2022**, *11*, 2025. [CrossRef]
6. Matos, M.S.; Romero-Díez, R.; Álvarez, A.; Bronze, M.R.; Rodríguez-Rojo, S.; Mato, R.B.; Cocero, M.J.; Matias, A.A. Polyphenol-rich extracts obtained from winemaking waste streams as natural ingredients with cosmeceutical potential. *Antioxidants* **2019**, *8*, 355. [CrossRef]
7. Wani, T.A.; Majid, D.; Dar, B.N.; Makroo, H.A.; Allai, F.M. Utilization of novel techniques in extraction of polyphenols from grape pomace and their therapeutic potential: A review. *J. Food Meas. Charact.* **2023**, *17*, 5412–5425. [CrossRef]
8. Szabó, É.; Marosvölgyi, T.; Szilágyi, G.; Kőrösi, L.; Schmidt, J.; Csepregi, K.; Márk, L.; Bóna, Á. Correlations between total antioxidant capacity, polyphenol and fatty acid content of native grape seed and pomace of four different grape varieties in Hungary. *Antioxidants* **2021**, *10*, 1101. [CrossRef]
9. Hoss, I.; Rajha, H.N.; El Khoury, R.; Youssef, S.; Manca, M.L.; Manconi, M.; Louka, N.; Maroun, R.G. Valorization of wine-making by-products' extracts in cosmetics. *Cosmetics* **2021**, *8*, 109. [CrossRef]
10. Zeng, H.-J.; Li, Q.-Y.; Ma, J.; Yang, R.; Qu, L.-B. A Comparative study on the effects of resveratrol and oxyresveratrol against tyrosinase activity and their inhibitory mechanism. *Spectrochim. Acta Part A Mol. Biomol. Spectrosc.* **2021**, *251*, 119405. [CrossRef] [PubMed]
11. Soceanu, A.; Dobrinas, S.; Sirbu, A.; Manea, N.; Popescu, V. Economic Aspects of Waste Recovery in the Wine Industry. A Multidisciplinary Approach. *Sci. Total Environ.* **2021**, *759*, 143543. [CrossRef] [PubMed]
12. Viola, C.M.; Torres-Carro, R.; Cartagena, E.; Isla, M.I.; Alberto, M.R.; Arena, M.E. Effect of wine wastes extracts on the viability and biofilm formation of *Pseudomonas aeruginosa* and *Staphylococcus aureus* strains. *Evid. Based Complement. Alternat. Med.* **2018**, *2018*, 9526878. [CrossRef] [PubMed]
13. Viola, C.M.; Cartagena, E.; Arena, M.E. Desechos de vinificación como inhibidores de biopelículas de *Staphylococcus aureus*. *Nereis Interdiscip. Ibero-Am. J. Methods Model. Simul.* **2021**, *13*, 135–146. [CrossRef]

14. Salazar, P.B.; Fanzone, M.; Zabala, B.A.; Rodriguez Vaquero, M.J.; Cilli, E.; Barroso, P.A.; Minahk, C.; Acuña, L. A Byproduct from the valles calchaquíes vineyards (Argentina) rich in phenolic compounds: A tool against endemic leishmania dissemination. *Environ. Sci. Pollut. Res.* **2023**, *30*, 97377–97385. [CrossRef] [PubMed]
15. Instituto Nacional de Vitivinicultura. 2018 Regiones Vitivinícolas Argentinas Noroeste. Available online: https://www.argentina.gob.ar/sites/default/files/region_noroeste_18.pdf (accessed on 1 May 2024).
16. Instituto Nacional de Vitivinicultura. Informe Variedad Torrontés. 2021. Available online: https://www.argentina.gob.ar/sites/default/files/2018/10/01-Torront%C3%A9s_2020.pdf (accessed on 1 May 2024).
17. Torres-Carro, R.; Ana, Y.; Rojas-Márquez, J.D.; Stempin, C.C.; Zampini, I.C.; Isla, M.I.; Alberto, M.R. Argentinean puna plants with in vitro antioxidant and anti-inflammatory activities as a potential nutraceutical. *J. Food Sci.* **2019**, *84*, 3352–3363. [CrossRef] [PubMed]
18. Negro, C.; Aprile, A.; Luvisi, A.; De Bellis, L.; Miceli, A. Antioxidant activity and polyphenols characterization of four monovarietal grape pomaces from Salento (Apulia, Italy). *Antioxidants* **2021**, *10*, 1406. [CrossRef] [PubMed]
19. Singh, P.K.; Singh, J.; Medhi, T.; Kumar, A. Phytochemical screening, quantification, FT-IR analysis, and *in silico* characterization of potential bioactive compounds identified in hr-lc/ms analysis of the polyherbal formulation from Northeast India. *ACS Omega* **2022**, *7*, 33067–33078. [CrossRef] [PubMed]
20. Popova, M.; Bankova, V.; Butovska, D.; Petkov, V.; Nikolova-Damyanova, B.; Sabatini, A.G.; Marcazzan, G.L.; Bogdanov, S. Validated methods for the quantification of biologically active constituents of poplar-type propolis. *Phytochem. Anal.* **2004**, *15*, 235–240. [CrossRef]
21. Ben Hmida, R.; Frikha, N.; Bouguerra Neji, S.; Kit, G.; Medina, F.; Bouaziz, M. Synthesis of high added value compounds through catalytic oxidation of 2-phenylethanol: A kinetic study. *Int. J. Chem. Kinet.* **2020**, *52*, 124–133. [CrossRef]
22. Ricco, R.; Agudelo, I.J.; Wagner, M.L. Métodos empleados en el análisis de los polifenoles en un laboratorio de baja complejidad. *Lilloa* **2015**, *52*, 161–174.
23. Carullo, G.; Spizzirri, U.G.; Loizzo, M.R.; Leporini, M.; Sicari, V.; Aiello, F.; Restuccia, D. Valorization of red grape (*Vitis vinifera* cv. sangiovese) pomace as functional food ingredient. *Ital. J. Food Sci.* **2020**, *32*, 367–385. [CrossRef]
24. Bouabid, K.; Lamchouri, F.; Toufik, H.; Faouzi, M.E.A. Phytochemical investigation, in vitro and in vivo antioxidant properties of aqueous and organic extracts of toxic plant: *Atractylis gummifera* L. *J. Ethnopharmacol.* **2020**, *253*, 112640. [CrossRef] [PubMed]
25. Moreira, M.M.; Barroso, M.F.; Boeykens, A.; Withouck, H.; Morais, S.; Delerue-Matos, C. Valorization of apple tree wood residues by polyphenols extraction: Comparison between conventional and microwave-assisted extraction. *Ind. Crops Prod.* **2017**, *104*, 210–220. [CrossRef]
26. Bibi Sadeer, N.; Montesano, D.; Albrizio, S.; Zengin, G.; Mahomoodally, M.F. The versatility of antioxidant assays in food science and safety—Chemistry, applications, strengths, and limitations. *Antioxidants* **2020**, *9*, 709. [CrossRef] [PubMed]
27. Pinto, D.; Vieira, E.F.; Peixoto, A.F.; Freire, C.; Freitas, V.; Costa, P.; Delerue-Matos, C.; Rodrigues, F. Optimizing the extraction of phenolic antioxidants from chestnut shells by subcritical water extraction using response surface methodology. *Food Chem.* **2021**, *334*, 127521. [CrossRef] [PubMed]
28. Lingua, M.S.; Fabani, M.P.; Wunderlin, D.A.; Baroni, M.V. In vivo antioxidant activity of grape, pomace and wine from three red varieties grown in argentina: Its relationship to phenolic profile. *J. Funct. Foods* **2016**, *20*, 332–345. [CrossRef]
29. Quy, T.; Xuan, T. Xanthine Oxidase Inhibitory potential, antioxidant and antibacterial activities of *Cordyceps militaris* (L.) link fruiting Body. *Medicines* **2019**, *6*, 20. [CrossRef] [PubMed]
30. Muñoz-Bernal, Ó.A.; Vazquez-Flores, A.A.; De La Rosa, L.A.; Rodrigo-García, J.; Martínez-Ruiz, N.R.; Alvarez-Parrilla, E. Enriched red wine: Phenolic profile, sensory evaluation and in vitro bioaccessibility of phenolic compounds. *Foods* **2023**, *12*, 1194. [CrossRef] [PubMed]
31. Antoniolli, A.; Fontana, A.R.; Piccoli, P.; Bottini, R. Characterization of polyphenols and evaluation of antioxidant capacity in grape pomace of the Cv. Malbec. *Food Chem.* **2015**, *178*, 172–178. [CrossRef]
32. Guaita, M.; Motta, S.; Messina, S.; Casini, F.; Bosso, A. Polyphenolic profile and antioxidant activity of green extracts from grape pomace skins and seeds of italian cultivars. *Foods* **2023**, *12*, 3880. [CrossRef]
33. Ribeiro, L.F.; Ribani, R.H.; Stafussa, A.P.; Makara, C.N.; Branco, I.G.; Maciel, G.M.; Haminiuk, C.W.I. Exploratory analysis of bioactive compounds and antioxidant potential of grape (*Vitis vinifera*) pomace. *Acta Sci. Technol.* **2022**, *44*, e56934. [CrossRef]
34. Alvarez-Casas, M.; Pájaro, M.; Lores, M.; Garcia-Jares, C. Characterization of grape marcs from native and foreign white varieties grown in northwestern spain by their polyphenolic composition and antioxidant activity. *Eur. Food Res. Technol.* **2016**, *242*, 655–665. [CrossRef]
35. Teixeira, A.; Baenas, N.; Dominguez-Perles, R.; Barros, A.; Rosa, E.; Moreno, D.; Garcia-Viguera, C. Natural bioactive compounds from winery by-products as health promoters: A review. *Int. J. Mol. Sci.* **2014**, *15*, 15638–15678. [CrossRef] [PubMed]
36. Lisov, N.; Čakar, U.; Milenković, D.; Čebela, M.; Vuković, G.; Despotović, S.; Petrović, A. The influence of Cabernet Sauvignon ripeness, healthy state and maceration time on wine and fermented pomace phenolic profile. *Fermentation* **2023**, *9*, 695. [CrossRef]
37. Balea, Ş.S.; Pârvu, A.E.; Pop, N.; Marín, F.Z.; Pârvu, M. Polyphenolic compounds, antioxidant, and cardioprotective effects of pomace extracts from Fetească Neagră cultivar. *Oxid. Med. Cell. Longev.* **2018**, *2018*, 8194721. [CrossRef] [PubMed]
38. Bocsan, I.C.; Măgureanu, D.C.; Pop, R.M.; Levai, A.M.; Macovei, Ș.O.; Pătrașca, I.M.; Chedea, V.S.; Buzoianu, A.D. Antioxidant and anti-inflammatory actions of polyphenols from red and white grape pomace in ischemic heart diseases. *Biomedicines* **2022**, *10*, 2337. [CrossRef] [PubMed]

39. Onache, P.A.; Geana, E.-I.; Ciucure, C.T.; Florea, A.; Sumedrea, D.I.; Ionete, R.E.; Tiţa, O. Bioactive phytochemical composition of grape pomace resulted from different white and red grape Cultivars. *Separations* **2022**, *9*, 395. [CrossRef]
40. Gerardi, C.; Pinto, L.; Baruzzi, F.; Giovinazzo, G. Comparison of antibacterial and antioxidant properties of red (cv. negramaro) and white (cv. fiano) skin pomace extracts. *Molecules* **2021**, *26*, 5918. [CrossRef] [PubMed]
41. Gaafar, A.A.; Asker, M.S. The effectiveness of the functional components of grape (*Vitis vinifera*) pomace as antioxidant, antimicrobial, and antiviral agents. *Jordan J. Biol. Sci.* **2019**, *12*, 625–635.
42. Melo, P.S.; Massarioli, A.P.; Denny, C.; Dos Santos, L.F.; Franchin, M.; Pereira, G.E.; Vieira, T.M.F.D.S.; Rosalen, P.L.; Alencar, S.M.D. Winery By-Products: Extraction optimization, phenolic composition and cytotoxic evaluation to act as a new source of scavenging of reactive oxygen species. *Food Chem.* **2015**, *181*, 160–169. [CrossRef]
43. Bai, J.; Zhang, Y.; Tang, C.; Hou, Y.; Ai, X.; Chen, X.; Zhang, Y.; Wang, X.; Meng, X. Gallic acid: Pharmacological activities and molecular mechanisms involved in inflammation-related diseases. *Biomed. Pharmacother.* **2021**, *133*, 110985. [CrossRef] [PubMed]
44. Ganeshpurkar, A.; Saluja, A. The pharmacological potential of catechin. *Indian J. Biochem. Biophys.* **2020**, *57*, 505–511.
45. Gao, J.; Hu, J.; Hu, D.; Yang, X. A Role of Gallic Acid in Oxidative Damage Diseases: A Comprehensive Review. *Nat. Prod. Commun.* **2019**, *14*, 1934578X1987417. [CrossRef]
46. Li, D.; Rui, Y.; Guo, S.; Luan, F.; Liu, R.; Zeng, N. Ferulic acid: A review of its pharmacology, pharmacokinetics and derivatives. *Life Sci.* **2021**, *284*, 119921. [CrossRef] [PubMed]
47. Singh, S.K.; Thakur, K.; Sharma, V.; Saini, M.; Sharma, D.; Vishwas, S.; Kakoty, V.; Pal, R.S.; Chaitanya, M.V.N.L.; Babu, M.R.; et al. Exploring the multifaceted potential of chlorogenic acid: Journey from nutraceutical to nanomedicine. *S. Afr. J. Bot.* **2023**, *159*, 658–677. [CrossRef]
48. Song, J.; He, Y.; Luo, C.; Feng, B.; Ran, F.; Xu, H.; Ci, Z.; Xu, R.; Han, L.; Zhang, D. New Progress in the Pharmacology of Protocatechuic Acid: A Compound ingested in daily foods and herbs frequently and heavily. *Pharmacol. Res.* **2020**, *161*, 105109. [CrossRef] [PubMed]
49. Liu, W.; Li, J.; Zhang, X.; Zu, Y.; Yang, Y.; Liu, W.; Xu, Z.; Gao, H.; Sun, X.; Jiang, X.; et al. Current advances in naturally occurring caffeoylquinic acids: Structure, bioactivity, and synthesis. *J. Agric. Food Chem.* **2020**, *68*, 10489–10516. [CrossRef] [PubMed]
50. Li, X.; Li, K.; Xie, H.; Xie, Y.; Li, Y.; Zhao, X.; Jiang, X.; Chen, D. Antioxidant and cytoprotective effects of the di-o-caffeoylquinic acid family: The mechanism, structure–activity relationship, and conformational effect. *Molecules* **2018**, *23*, 222. [CrossRef] [PubMed]
51. Karami, A.; Fakhri, S.; Kooshki, L.; Khan, H. Polydatin: Pharmacological mechanisms, therapeutic targets, biological activities, and health benefits. *Molecules* **2022**, *27*, 6474. [CrossRef]
52. Pérez-Ortiz, J.M.; Alguacil, L.F.; Salas, E.; Hermosín-Gutiérrez, I.; Gómez-Alonso, S.; González-Martín, C. Antiproliferative and cytotoxic effects of grape pomace and grape seed extracts on colorectal cancer cell Lines. *Food Sci. Nutr.* **2019**, *7*, 2948–2957. [CrossRef]
53. Jara-Palacios, M.J.; Hernanz, D.; Cifuentes-Gomez, T.; Escudero-Gilete, M.L.; Heredia, F.J.; Spencer, J.P.E. Assessment of white grape pomace from winemaking as source of bioactive compounds, and its antiproliferative activity. *Food Chem.* **2015**, *183*, 78–82. [CrossRef]
54. Cattivelli, A.; Conte, A.; Tagliazucchi, D. Quercetins, chlorogenic acids and their colon metabolites inhibit colon cancer cell proliferation at physiologically relevant concentrations. *Int. J. Mol. Sci.* **2023**, *24*, 12265. [CrossRef] [PubMed]
55. Germanó, M.J.; Muñoz, M.D.; Della-Vedova, M.C.; Feresin, G.E.; Rinaldi-Tosi, M.; Enriz, R.D.; Ramirez, D.C.; Giannini, F.A. Anti-oxidant and anti-inflammatory effect of polar extracts obtained from waste product of wine making. *Nat. Prod. Res.* **2021**, *35*, 4769–4773. [CrossRef] [PubMed]
56. Angelini, P.; Flores, G.A.; Piccirilli, A.; Venanzoni, R.; Acquaviva, A.; Di Simone, S.C.; Libero, M.L.; Tirillini, B.; Zengin, G.; Chiavaroli, A.; et al. Polyphenolic composition and antimicrobial activity of extracts obtained from grape processing by-products: Between green biotechnology and nutraceutical. *Process Biochem.* **2022**, *118*, 84–91. [CrossRef]
57. El-Nashar, H.A.S.; El-Din, M.I.G.; Hritcu, L.; Eldahshan, O.A. Insights on the inhibitory power of flavonoids on tyrosinase activity: A survey from 2016 to 2021. *Molecules* **2021**, *26*, 7546. [CrossRef] [PubMed]
58. Ferri, M.; Rondini, G.; Calabretta, M.M.; Michelini, E.; Vallini, V.; Fava, F.; Roda, A.; Minnucci, G.; Tassoni, A. White grape pomace extracts, obtained by a sequential enzymatic plus ethanol-based extraction, exert antioxidant, anti-tyrosinase and anti-inflammatory activities. *New Biotechnol.* **2017**, *39*, 51–58. [CrossRef]
59. Leal, C.; Gouvinhas, I.; Santos, R.A.; Rosa, E.; Silva, A.M.; Saavedra, M.J.; Barros, A.I.R.N.A. Potential application of grape (*Vitis vinifera* L.) stem extracts in the cosmetic and pharmaceutical industries: Valorization of a by-product. *Ind. Crops Prod.* **2020**, *154*, 112675. [CrossRef]
60. Sinha, S.; Doble, M.; Manju, S.L. 5-Lipoxygenase as a drug target: A review on trends in inhibitors structural design, sar and mechanism based approach. *Bioorg. Med. Chem.* **2019**, *27*, 3745–3759. [CrossRef] [PubMed]
61. Vishnupriya, P.; Aparna, A.; Viswanadha, V.P. Lipoxygenase (lox) pathway: A promising target to combat cancer. *Curr. Pharm. Des.* **2021**, *27*, 3349–3369. [CrossRef]
62. Mollica, A.; Scioli, G.; Della Valle, A.; Cichelli, A.; Novellino, E.; Bauer, M.; Kamysz, W.; Llorent-Martínez, E.J.; Fernández-de Córdova, M.L.; Castillo-López, R.; et al. Phenolic analysis and in vitro biological activity of red wine, pomace and grape seeds oil derived from *Vitis vinifera* L. Cv. Montepulciano d'Abruzzo. *Antioxidants* **2021**, *10*, 1704. [CrossRef]

63. Bucić-Kojić, A.; Fernandes, F.; Silva, T.; Planinić, M.; Tišma, M.; Šelo, G.; Šibalić, D.; Pereira, D.M.; Andrade, P.B. Enhancement of the anti-inflammatory properties of grape pomace treated by *Trametes versicolor*. *Food Funct.* **2020**, *11*, 680–688. [CrossRef]
64. Dwibedi, V.; Jain, S.; Singhal, D.; Mittal, A.; Rath, S.K.; Saxena, S. Inhibitory activities of grape bioactive compounds against enzymes linked with human diseases. *Appl. Microbiol. Biotechnol.* **2022**, *106*, 1399–1417. [CrossRef]

Disclaimer/Publisher's Note: The statements, opinions and data contained in all publications are solely those of the individual author(s) and contributor(s) and not of MDPI and/or the editor(s). MDPI and/or the editor(s) disclaim responsibility for any injury to people or property resulting from any ideas, methods, instructions or products referred to in the content.

foods

MDPI

Article

Sequential Fermentation in Red Wine *cv.* Babić Production: The Influence of *Torulaspora delbrueckii* and *Lachancea thermotolerans* Yeasts on the Aromatic and Sensory Profile

Stipe Ivić [1], Ana Jeromel [2], Bernard Kozina [2], Tihomir Prusina [3], Irena Budić-Leto [1], Ana Boban [1], Višnja Vasilj [3] and Ana-Marija Jagatić Korenika [2,*]

1 Institute for Adriatic Cultures and Karst Reclamation, Put Duilova 11, 21000 Split, Croatia; stipe.ivic@krs.hr (S.I.); irena.budic-leto@krs.hr (I.B.-L.); aboban@krs.hr (A.B.)

2 Faculty of Agriculture, University of Zagreb, Svetošimunska cesta 25, 10000 Zagreb, Croatia; amajdak@agr.hr (A.J.); bkozina@agr.hr (B.K.)

3 Faculty of Agriculture and Food Technology, University of Mostar, Biskupa Čule 10, 88000 Mostar, Bosnia and Herzegovina; tiho@vinarija-citluk.ba (T.P.); visnja.vasilj@aptf.sum.ba (V.V.)

* Correspondence: amjagatic@agr.hr; Tel.: +385-1-239-3834

Abstract: This research aimed to analyze the impact of two different non-*Saccharomyces* yeast species on the aromatic profile of red wines made from the *cv.* Babić (*Vitis vinifera* L.) red grape variety. The grapes were obtained from two positions in the Middle and South of Dalmatia. This study compared a control treatment with the *Saccharomyces cerevisiae* (*Sc*) strain as a type of sequential inoculation treatment with *Lachancea thermotolerans* (*Lt x Sc*) and *Torulaspora delbrueckii* (*Td x Sc*). The focus was on the basic wine parameters and volatile aromatic compound concentrations determined using the SPME-Arrow-GC/MS method. The results revealed significant differences in *cis*-linalool oxide, geraniol, neric acid, and nerol, which contribute to the sensory profile with floral and rose-like aromas; some ethyl esters, such as ethyl furoate, ethyl hexanoate, ethyl lactate, ethyl 2-hydroxy-3-methylbutanoate, ethyl 3-hydroxy butanoate, diethyl glutarate, and diethyl succinate, contribute to the aromatic profile with fruity, buttery, overripe, or aging aromas. A sensory evaluation of wines confirmed that *Td x Sc* treatments exhibited particularly positive aromatic properties together with a more intense fullness, harmony, aftertaste, and overall impression.

Keywords: aroma profile; gas chromatography–mass spectrometry; non-*Saccharomyces* yeast; sensory properties; volatile compounds

1. Introduction

The most recent trend in enology involves the utilization of selected non-*Saccharomyces* (non-*Sc*) yeasts. Non-*Sc* yeasts encompass any yeast species found in wine other than *Sc* that have a beneficial impact on the winemaking process. More attention was paid to them in winemaking due to their influence on wine aroma and polyphenolic composition [1]. In the past, yeast selection for winemaking was limited to the *Saccharomyces cerevisiae* species (*Sc*). However, now there are numerous commercial yeast strains derived through specific genetic methods and selection [2]. These strains can produce varying concentrations of secondary compounds, which contribute to the unique characteristics of wine. While *Sc* remains the dominant species for alcoholic fermentation, non-*Saccharomyces* yeasts have gained interest in recent years [3,4]. Traditionally, non-*Sc* yeasts have been considered as contaminants in winemaking, and measures such as must pasteurization, sulfite addition, and equipment and processing area disinfection have been routinely employed to eliminate them from the fermentation process. Due to their limitations, such as their sensitivity to ethanol and SO_2, non-*Sc* yeasts can only be involved in sequential fermentation with *Sc* [5]. This method closely resembles spontaneous fermentation. The incorporation of non-*Sc* yeasts in the winemaking process yields several advantageous effects on wine quality, including moderate ethanol levels, an increased glycerol content, a higher acidity, and a more intricate aromatic profile. For this study, two commercially available non-*Saccharomyces* yeast strains were used—*Torulaspora delbrueckii* (*Td*), the oldest, and *Lachancea thermotolerans* (*Lt*), the most recently selected [6]. *Lt* is renowned for acidifying musts with low total acidity and high pH values [7]. Nowadays, it is widely employed in sequential inoculation for red wine vinification through the production of significant concentrations of lactic acid. This inherent feature renders it a valuable resource for blending and/or re-equilibrating red wines from warm climates [8] like *cv.* Babić, an autochthonous variety of red grapes, cultivated in the warm, coastal region of Dalmatia, in Croatia. In addition to enhancing freshness and acidity, *Lt* contributes to the aromatic complexity right from the beginning of alcoholic fermentation. *Torulaspora delbrueckii* is also recommended for the modification of the aroma profile of wine. Its metabolic activity facilitates the release of terpene aromas, including α-terpineol and linalool [9]. The utilization of this strain can intensify the fruity characteristics of wines. Furthermore, it can enhance red wine color, reduce ethanol levels, decrease fatty acid concentrations, and increase mannoprotein and glycerol through sequential fermentation with *Sc* [10]. So, sequential fermentation with different yeast strains enhances wine's flavor profile by increasing the presence of diverse volatile compounds such as alcohols, esters, phenols, terpenes, and C13-norisoprenoids [11].

The fermentation process of must or pomace plays a pivotal role in extracting aromatic compounds from grapes. This extraction process alters the aromatic compounds and generates secondary metabolites produced by yeast. Moreover, the presence of other compounds in wine, such as ethanol, phenols, and acids, also impacts the composition of volatile components and thus influences the aroma perceived in a glass of wine [12]. The perception of volatile compounds in wine aroma is closely linked to the orthonasal and retronasal human senses. The complexity of wine aroma arises from the interaction between volatile compounds and other components such as water, ethanol, phenolic compounds, and polysaccharides. Organic acids have various important functions in the context of wine production. They contribute significantly to the wine's overall stability, both from a microbiological and physicochemical standpoint [13]. Furthermore, these acids have a notable impact on the wine's visual perspective such as color intensity, as well as aging potential and flavor balance [14]. Additionally, organic acids influence the oxidation process of compounds found in both the must and the final product, as well as microbial metabolism, protein and polysaccharide solubility, potassium bitartrate solubility, and the efficacy of sulfur dioxide, fining agents, and pectolytic enzymes [14].

This study aimed to examine the impact of two different non-*Sc* yeast species on the aromatic and sensory characteristics of red wines produced from the Babić grape variety planted in two vine-growing positions.

2. Materials and Methods

2.1. Sample Preparation

The *cv.* Babić grapes were cultivated in the warm vine-growing hills of Šibenik (locality Jadrtovac) and Primošten (locality Široke) in the sub-regions (Protected Designation of Origin) of Middle and South Dalmatia. In the Jadrtovac locality, 50 ha of vineyards, at 70 to 100 m above sea level, mainly under the *cv.* Babić, were planted in the period from 2007 to 2012. The direction of the rows is north–south. The trellis system was a modified cordon with two cuttings with two to three buds. The soil is brown and skeletal. The position is extremely windy, and the entire production is organic with a maximum of two treatments per season with sulfur and copper preparations. The Široke locality is located on the southern slopes, between 230 and 250 m above sea level, characterized by natural rock material ingrown into the soil, which must be extracted manually to acquire a small quantity of soil in the lot (cassette) for cultivation. The direction of the rows is west–east. The total area of the vineyard is about 0.23 ha with about 1250 vines, and is more than 50 years old. The cultivation form was 'en goblet'. This form of cultivation produces 4–7 bunches per vine, but it is important to note that the wind is one of the most important characteristics of this terroir and that part of the crop is regularly lost due to the strong winds.

The research specifically concentrated on the vintage 2020. The grape harvest was performed manually during the early morning hours, with each site yielding quantities of 200 kg, grapes were transported in plastic crates with a capacity of 20 kg, and primary processing was immediately performed, involving destemming and crushing. An electric crusher-destemmer was used for the primary processing of the grapes. The experiment was designed with three treatments for each of the two wine-growing positions with each treatment comprising three replications (Table 1).

Table 1. Inoculums for each treatment in the present study.

K	*Lt x Sc*	*Td x Sc*
Control treatment, alcoholic fermentation with the yeast strain *Saccharomyces cerevisiae* (Uvaferm BDX®, Lallemand Montreal, QC, Canada)	Sequential alcoholic fermentation with the yeast strains *Lachancea thermotolerans* (Laktia®, Lallemand, Montreal, QC, Canada) + *Saccharomyces cerevisiae* (Uvaferm BDX®, Lallemand, Montreal, QC, Canada)	Sequential alcoholic fermentation with the yeast strains *Torulaspora delbrueckii* (Biodiva®, Lallemand, Montreal, QC, Canada) + *Saccharomyces cerevisiae* (Lalvin ICV D254®, Lallemand, Montreal, QC, Canada)

2.2. Alcoholic Fermentation Trials

Detailed protocols for alcoholic fermentations are presented in Table 2. The control treatment (K) pomace was treated with 10 g/hL of $K_2S_2O_5$, and 2 g/hL of pectolytic enzyme. Micronutrient-rich inactive yeast was added to the pomace, prepared in rehydration water with selected *Sc* yeast. Next, 72 h into fermentation, a complex yeast nutrient was added.

Table 2. Protocols for the treatments.

K	*Lt x Sc*	*Td x Sc*
- Potassium metabisulfite ($K_2S_2O_5$)—10 g/hL - pectolytic enzyme Lallzyme® OE (2 g/hL, Lallemand, Montreal, QC, Canada) - inactive yeast Go-Ferm Protect® (20 g/hL, Lallemand, Montreal, QC, Canada) - *Saccharomyces cerevisiae* Uvaferm BDX® (25 g/hL, Lallemand, Montreal, QC, Canada) - yeast nutrient Fermaid E® (20 g/hL, Lallemand, Montreal, QC, Canada)	- Potassium metabisulfite ($K_2S_2O_5$)—3 g/hL - pectolytic enzyme Lallzyme® OE (2 g/hL, Lallemand, Montreal, QC, Canada) - *Lachancea thermotolerans* Laktia® (25 g/hL, Lallemand, Montreal, QC, Canada) + after 72 h, *Saccharomyces cerevisiae* Uvaferm BDX® (25 g/hL, Lallemand, Montreal, QC, Canada) - yeast nutrient Fermaid E® (20 g/hL, Lallemand, Montreal, QC, Canada)	- Potassium metabisulfite ($K_2S_2O_5$)—3 g/hL - pectolytic enzyme Lallzyme® OE (2 g/hL, Lallemand, Montreal, QC, Canada) - *Torulaspora delbrueckii* Biodiva® (25 g/hL, Lallemand, Montreal, QC, Canada) + after 72 h, *Saccharomyces cerevisiae* Lalvin ICV D254® (25 g/hL, Lallemand, Montreal, QC, Canada) - yeast nutrient Fermaid E® (20 g/hL, Lallemand, Montreal, QC, Canada)

The *Lt x Sc* and *Td x Sc* musts were treated with 3 g/hL of $K_2S_2O_5$ and 2 g/hL of pectolytic enzyme. Treatment *Lt x Sc* was inoculated with rehydrated *Lt* strains, and yeast nutrients were added. The pomace was sequentially inoculated 72 h into fermentation with the rehydrated *Sc* strain. Treatment *Td x Sc* was inoculated with rehydrated *Td* strains, yeast nutrients were added, and the pomace was inoculated with rehydrated *Sc* yeast strains 72 h into alcoholic fermentation.

All treatments were punched down daily, every eight hours, with the pomace temperature ranging from 22 to 25 °C. After eight days of maceration, the pomace was pressed and the partly fermented must was transferred to 10 L glass carboys (three replications per treatment) at 22 °C. During the alcoholic fermentation process, the degradation of sugar was monitored daily using a refractometer and specific gravity, and the must temperature was measured. One month after the first racking, wines from all replications were sampled for their chemical composition analysis.

2.3. Identification and Quantification of Volatile Compounds

The analysis of volatile compounds in the wine samples was conducted using the SPME-Arrow-GC/MS (gas chromatography–mass spectrometry) [15]. The SPME-Arrow extraction was performed using the RSH Triplus autosampler (Thermo Fisher Scientific Inc., Brookfield, WI, USA). A total of 5 mL of sample and 2.00 g of NaCl were put in 20 mL headspace screw-top vials sealed with PTFE/silicone septum-containing caps. The sorption conditions were as follows: the sample was incubated at 60 °C for 20 min and then SPME-Arrow fiber DVB/CWR/PDMS (120 μm × 20 mm; Thermo Fisher Scientific Inc., Brookfield, WI, USA) was exposed for 49 min. Then, the fiber was inserted into the GC injector port operating in splitless mode and was desorbed at 250 °C for 10 min.

Sample analysis was conducted on a TRACE 1300 Gas Chromatographer coupled to an ISQ 7000 TriPlus quadrupole mass spectrometer (Thermo Fisher Scientific Inc., USA) equipped with a TG-WAXMS A capillary column (60 m × 0.25 mm × 0.25 μm film thickness; Thermo Fisher Scientific, USA). The volatile compounds injected into the inlet were delivered to the column at a splitless mode and helium was used as a carrier gas at a constant flow rate of 1 mL/min. The oven temperature program was as follows: an initial temperature of 40 °C was maintained for 5 min, followed by an increase of 2 °C/min to 210 °C and being held for 10 min. MS spectra were recorded in the electron impact ionization mode (EI) at an ionization energy of 70 eV. Mass spectrometry was performed in full scan mode in the range of 30–300 m/z. The data obtained were processed using the Chromeleon Data System (Thermo Fisher Scientific Inc., USA). Identification was carried out by comparing retention times, retention index, and mass spectra with those of standards and with the data available in the Wiley Registry 12th Edition/NIST Spectral Library. Quantification was carried out using calibration curves. The curves (based on quantification ions) were constructed with Chromeleon 7 Chromatography Data System (CDS) software (version 7.2.10). As an internal standard, 3-methyl-3-pentanol was used at a final concentration of 1 mg/L. For all available standards, six different concentrations were prepared, while for the other compounds, semi-quantitative analysis was performed. Their concentrations were expressed in equivalents of similar compounds, with the assumption that a response factor was equal to one. The parameters of the identification and the calibration of wine volatiles are presented in Supplemental Table S1.

2.4. Determination of Organic Acids

Organic acids (tartaric, malic, lactic, citric, and succinic) were analyzed using High-performance Liquid Chromatography, Agilent 1050 (Palo Alto, CA, USA). The sample was previously filtered using PTFE membrane filters (0.45 μm). Identification and quantification were conducted at a wavelength of λ = 210 nm on Aminex HPX-87H (BioRad, Hercules, CA, USA).

2.5. Physicochemical Analysis

The basic physicochemical parameters were analyzed in must (reducing sugars) and wines (alcohol, total dry extract, total acidity, volatile acidity, pH, and ash) according to methods set by the International Organization of Vine and Wine (OIV, 2021) [16].

2.6. Sensory Analysis

For the 2020 vintage wines assessment, a panel of seven expert tasters participated (four females and three males), who were members of the Committee for Organoleptic Evaluation of Wine and Fruit Wines appointed by the Ministry of Agriculture. The panelists were specialists in the field and were well-experienced based on the evaluations in the

Croatian Agency for Agriculture and Food, accredited according to the HRN EN ISO/IEC 17065 standard [17] for the implementation of the procedure for placing wines with PDO, i.e., certification of wines with a label of origin, on the market. Evaluation was performed in the Laboratory for Sensory Analysis of Agricultural and Food Products, University of Zagreb Faculty of Agriculture, in individual booths under standardized conditions.

All wines included in this study underwent sensory analysis using Quantitative Descriptive Analysis (QDA) six months after the conclusion of alcoholic fermentation. The Babić red wines (20 mL) were served at 15 °C in standard wine tasting glasses (ISO 3591:1977) [18] covered with a watch-glass to reduce the volatility of wine aromas. Blind tasting of coded samples was performed using three random replicates in the experiment. A total of 14 wine attributes for taste and odor (Figures 1 and 2) were selected by the research group and were further developed and evaluated by panelists. The panel evaluated five referent monovarietal *cv.* Babić 2020 wines to achieve a consensus about the attributes describing wine's sensory profiles. The additional training of the panel before the formal evaluation included assessing wine aroma using the aqueous solutions of different selected compounds and an "Aromaster" kit (Vinofil Co., Ltd., Hong Kong) that includes 88 typical wine aromas in vials (Supplemental Table S2). Quantification was performed using a six-point scale, on a paper sheet, as follows: 0–1—weak, 2–3—medium, and 4–5—strongly intensive attribute. Sample differences were graphically presented using radar graphs. Subsequently, the samples were ranked based on the overall quality, with the highest-ranked wine deemed the best, and the lowest-ranked wine identified as the worst.

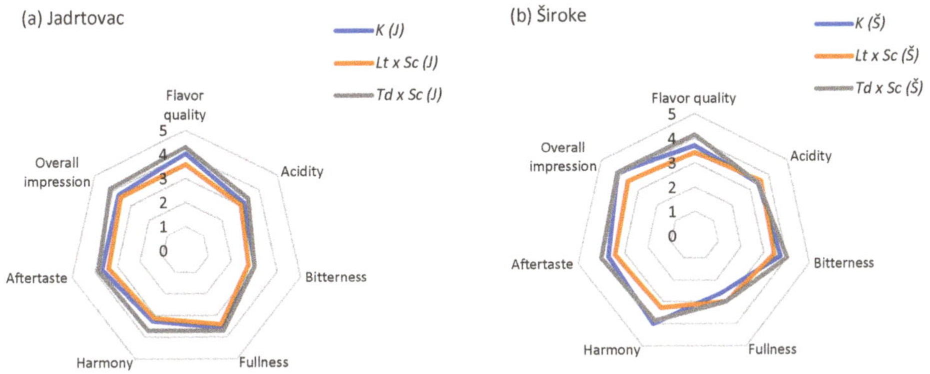

Figure 1. Quantitative descriptive sensory analysis of the taste parameters, cv. Babić red wine 2020, (**a**) Jadrtovac, (**b**) Široke.

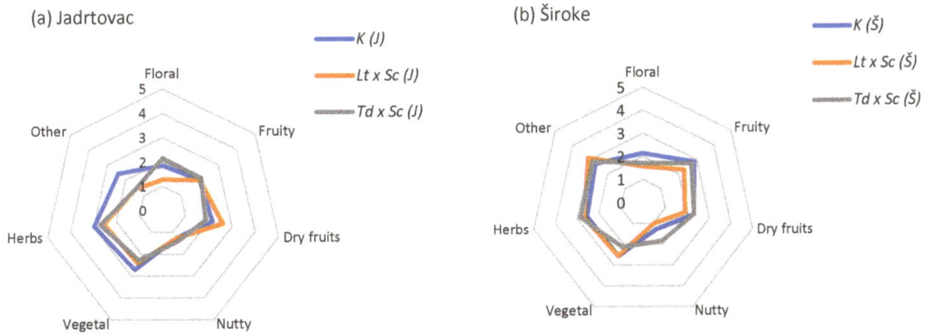

Figure 2. Quantitative descriptive sensory analysis of the odor parameters, cv. Babić red wine 2020, (**a**) Jadrtovac, (**b**) Široke.

2.7. Statistical Analysis

In the study of *cv.* Babić red wines, preconditions for applying ANOVA were examined using the Kolmogorov–Smirnov test for normality and the Bartlett test for variance homogeneity. All dependent variables met the conditions of normality and variance homogeneity, so a one-way analysis of variance (ANOVA) was applied. Differences in the chemical composition of wine (aromatic and phenolic compounds) between treatments and vineyard locations were tested using a two-way analysis of variance (ANOVA) with the SAS 9.4 statistical program (Cary, NC, USA).

3. Results and Discussion

3.1. Physicochemical Parameters

The study of the impact of non-*Saccharomyces* yeasts on the physicochemical parameters of *cv.* Babić wines pointed out significant differences among the treatments (Table 3). Based on the presented results, it can be seen that the prevalence of non-*Sc* yeasts at the start of fermentation impacted the overall composition of wine, which is consistent with prior published research [19]. Compared to the control, a lower alcohol content was determined in wines produced using *Lt x Sc* from the Jadrtovac location. The lower alcohol content can be attributed to the synthesis of lactic acid from sugars within the *Lt* metabolism [20], as confirmed by the results of organic acid analysis (Table 4). Depending on the chosen yeast strain and the conditions of alcoholic fermentation, the alcohol content of wine can be reduced by 1–2% (*v*/*v*) [21,22].

Table 3. Physicochemical parameters of Babić red wines, vintage 2020.

Compound	Location	Treatment		
		K	*Lt x Sc*	*Td x Sc*
Alcohol (vol%)	J	13.60 ± 0.01 c	13.30 ± 0.06 b	13.40 ± 0.02 b
	Š	13.00 ± 0.02 a	13.01 ± 0.03 a	12.93 ± 0.05 a
Total dry extract (g/L)	J	35.40 ± 0.20 c	37.50 ± 0.40 d	35.20 ± 0.20 bc
	Š	34.60 ± 0.20 ab	36.20 ± 0.20 d	34.40 ± 0.20 a
Reducing sugars (g/L)	J	3.80 ± 0.20 abc	4.00 ± 0.10 bc	3.40 ± 0.20 a
	Š	4.20 ± 0.10 c	3.70 ± 0.10 ab	3.40 ± 0.20 a
Total acidity * (g/L)	J	6.30 ± 0.10 a	7.50 ± 0.10 b	6.80 ± 0.20 ab
	Š	7.50 ± 0.10 b	8.20 ± 0.10 c	8.00 ± 0.20 c
Volatile acidity ** (g/L)	J	0.30 ± 0.02 ab	0.40 ± 0.01 ab	0.50 ± 0.01 a
	Š	0.42 ± 0.03 c	0.44 ± 0.04 c	0.40 ± 0.01 c
pH	J	3.62 ± 0.02 b	3.51 ± 0.01 a	3.53 ± 0.02 a
	Š	3.36 ± 0.01 a	3.32 ± 0.02 a	3.32 ± 0.02 a
Ash (g/L)	J	3.35 ± 0.01 a	3.48 ± 0.01 b	3.40 ± 0.10 ab
	Š	3.29 ± 0.01 a	3.39 ± 0.01 ab	3.35 ± 0.02 a
Total phenols (mg/L)	J	1775.00 ± 300.52 a	1632.50 ± 38.89 a	1610.00 ± 162.63 a
	Š	1567.50 ± 24.74 b	1590.00 ± 134.35 b	1520.00 ± 14.14 b

* tartaric acid and ** acetic acid equivalents. Concentrations are expressed as mean ± standard deviation (n = 3). Different letters in the rows represent statistically significant differences between treatments at the significance level of $p < 0.05$, separately for two localities (two-way ANOVA and LSD test). Different letters in the columns represent statistically significant differences between localities of the same treatment at the significance level of $p < 0.05$. J—Jadrtovac, Š—Široke, K—control treatment (*S. cerevisiae*), *Lt x Sc*—*L. thermotolerans x S. cerevisiae*, *Td x Sc*—*T. delbrueckii x S. cerevisiae*.

The total dry extract concentrations in wines were significantly the highest in the *Lt x Sc* treatments. The content of extract, resulting from the action of non-*Saccharomyces* yeasts, has a positive influence on taste properties [23].

Reducing sugar concentrations ranging from 3.40 g/L to 4.00 g/L, obtained through sequential fermentation, indicated the production of dry wines in both non-*Sc* treatments. Significantly, the lowest reducing sugar concentration was in the *Td x Sc* treatments. A higher reducing sugar concentration is attributed to a greater consumption of nutrients by

Lt yeasts [24]. The addition of *Lt* yeasts to the must several days before the addition of *Sc* yeasts resulted in a depletion of nutrients for further activity.

Table 4. Concentration of organic acids (g/L).

Organic Acid (g/L)	Location	Treatment		
		K	*Lt x Sc*	*Td x Sc*
Tartaric	J	2.76 ± 0.01 a	2.83 ± 0.02 b	2.96 ± 0.02 c
	Š	4.12 ± 0.04 d	4.18 ± 0.02 d	4.38 ± 0.02 e
Malic	J	0.29 ± 0.00 d	0.26 ± 0.00 c	0.27 ± 0.00 c
	Š	0.18 ± 0.01 b	0.17 ± 0.01 b	0.15 ± 0.01 a
Lactic	J	0.01 ± 0.01 a	1.58 ± 0.01 d	0.06 ± 0.01 b
	Š	0.01 ± 0.01 a	0.62 ± 0.01 c	0.06 ± 0.01 b
Citric	J	0.49 ± 0.01 c	0.35 ± 0.01 b	0.26 ± 0.00 a
	Š	0.33 ± 0.01 b	0.34 ± 0.02 b	0.27 ± 0.01 a
Succinic	J	0.43 ± 0.00 a	0.71 ± 0.02 d	0.58 ± 0.01 c
	Š	0.54 ± 0.01 b	0.42 ± 0.02 a	0.42 ± 0.01 a

Concentrations are expressed as mean ± standard deviation (n = 3). Different letters in the rows represent statistically significant differences between treatments at the significance level of $p < 0.05$, separately for two localities (two-way ANOVA and LSD test). Different letters in the columns represent statistically significant differences between localities of the same treatment at the significance level of $p < 0.05$. J—Jadrtovac, Š—Široke, K—control treatment (*S. cerevisiae*), *Lt x Sc*—*L. thermotolerans x S. cerevisiae*, *Td x Sc*—*T. delbrueckii x S. cerevisiae*.

This research revealed certain differences in the total acidity. Control treatments had lower concentrations of total acidity compared to the sequential fermentations, where it was higher for 1.20 in the *Lt x Sc* treatment and 0.50 g/L in the *Td x Sc* treatment. Earlier conducted studies [25–27] also pointed out a significant increase in total acidity in the sequential inoculation using *Sc x Lt* and *Sc x Td*. In the warmer climate of southeastern Europe, sequential inoculation can increase total acidity by up to 3.00 g/L [28]. Berbegal et al. stated that the maturity of grapes in warmer regions affects the concentration of total acidity, with tartaric acid being more stable at higher temperatures [29]. The use of *Lt* leads to the synthesis of lactic acid, which is an alternative to traditional malolactic fermentation in the production of red wine [30].

Recent research comparing several strains of *Lt* observed a significant degree of variability in volatile acidity, approximately 50% [31]. In this study, the range was 0.30–0.50 g/L with the highest concentrations in the *Td x Sc* (J) treatment. Comparing different strains of non-*Saccharomyces* yeasts [32] and three *Saccharomyces* yeasts, the reported concentrations were between 0.32 and 0.58 g/L for *Lt*, and 0.37 and 0.63 g/L for *Td* [29].

The pH values in the study ranged from 3.36 to 3.62, with significantly lower concentrations in sequential fermentations in the Jadrtovac locality, which was in line with the highest total acidity concentrations. Lower pH values were observed in all non-*Sc* treatments. It was noted by Porter et al. (2019) [33] that the main reason for the reduction in pH values with the sequential fermentation process using *Lt* yeasts, by as much as 0.5 units at the beginning of fermentation, is related to lactic acid synthesis. According to Morata et al. (2018) [7], lower pH values at lower concentrations of total SO_2 result in increased levels of molecular SO_2, protecting against the effects of yeasts and bacteria such as *Brettanomyces* during aging. The same author noted positive effects in warmer climates affected by climate change, where the pH of the wine naturally decreases during fermentation without acid correction.

3.2. Organic Acids in Wines

Table 4 presents the results of the organic acids analysis. Significant differences were observed in the organic acids between treatments. It is well known that Babić wines have higher concentrations of tartaric acid in comparison to the other red Dalmatian wines. Sequential fermentations resulted in the same or significantly higher tartaric acid concentrations in the *Td x Sc* treatment. The increase in tartaric acid in sequential fermentation is yeast-dependent [34]. Given that tartaric acid is stable and only slightly variable, it is essential to maintain its stability, as noted before [35]. Sequential fermentation resulted in lower concentrations of malic acid at both locations. This is consistent with previous data [21] presenting a decrease in wines produced using the *Td* strain in comparison to the *Sc*.

There were significant differences in the concentrations of lactic acid, with the highest concentrations in the *Lt x Sc* treatments. *Lt* is known for the natural acidification of wine through the synthesis of lactic acid, but its action is effective only with co-inoculation with *Sc* [36,37]. One of the main criteria for selecting *Lt* is its ability to produce lactic acid, with a maximum concentration of 9.60 g/L [38], which exceeds the values obtained in this study. The synthesis of lactic acid influenced the total acidity of the wine, resulting in a significant increase in the *Lt x Sc* (J) treatment, which is in accordance with previously published data [7,31].

Wines also showed significant variation in the concentration of citric acid, and sequential fermentations influenced the reduction. The highest concentration was observed in the K(J), while the lowest was in the *Td x Sc* (J) treatment. Significant differences were also noted in the concentration of succinic acid, with higher concentrations in sequential fermentation treatments (J). Succinic acid is formed only during alcoholic fermentation influenced by various factors such as yeast strain, temperature, nitrogen content, and vitamins. In this case, the fermentation temperatures were consistent across all fermentations, while the yeasts varied, and there may have been differences in the amino acid profile and nitrogen content in the musts, as evidenced by significant variations in succinic acid concentrations at both locations. *Lt x Sc* (J) exhibited the highest succinic acid concentration, and a positive correlation between succinic acid and total acidity was observed. The role of succinic acid in elevating the total acidity of wine is also highlighted [39], which aligns with the findings of this study.

3.3. Volatile Aromatic Compounds in Wines

The results of the analysis of 101 volatile aroma compounds in Babić wines are shown in Table 5. There were no significant differences between all treatments regarding the total fatty acids, which ranged from 1044.00 µg/L to 1773.00 µg/L. The highest concentration was in the *Td x Sc* (J) treatment. Identical reductions during sequential fermentation with *Td* yeast are reported [40], while Belda et al. (2017) reported the unchanged concentration of fatty acids in wine produced by sequential fermentation [41]. The obtained results are consistent with those previously reported [42], and the reduction could result from the formation of smaller amounts of acetate and ethyl esters. A significant difference was observed only for hexanoic acid. Lower concentrations produced by sequential inoculation with Td yeast have been reported earlier [43]. A reduction in the concentrations of medium-chain fatty acids can be considered positive because of their contribution to negative aromas resembling fat, cheese, and even rancidity if present in higher concentrations [44]. Applied treatments did not significantly affect the concentration of total terpenes, which ranged from 122.49 µg/L K(J) to 159.85 µg/L (*Td x Sc* (Š)). The obtained results are in contradiction with a paper reporting the increase in total terpenes in other red grape varieties produced using *Lt x Sc* sequential inoculation [45]. Higher concentrations of linalool in wines produced by sequential fermentation with *Lt* and *Td* [45] were also not confirmed in this study. Significantly higher concentrations of geraniol were observed in treatments *Lt x Sc* and *Td x Sc* (J), which agrees with the influence of sequential inoculation with *Td* [45]. Nerol significantly differed in wines *Lt x Sc* (J) and *Td x Sc* (Š), similar to a previous study [46]. An increase in nerol and hotrienol was detected in sequential fermentation under the influence of *Lt* [29]. Certain strains of Td release conjugated terpenes that characterize specific wine varieties [45]. There is a scientific consensus regarding the positive influence of Td on the aromatic profile of the wine, which is also associated with the release of mannoproteins and the emphasizing of varietal characteristics [40]. A possible reason for this increase is closely related to the activity of glucosidase enzymes [30].

Table 5. Concentrations of volatile aroma compounds (μg/L) in *cv.* Babić red wines.

Compounds (μg/L)	Locality	Treatments		
		K	*Lt x Sc*	*Td x Sc*
Fatty acids				
Propanoic acid	J	3.68 ± 0.28 a	2.86 ± 0.26 a	2.60 ± 0.00 a
	Š	2.70 ± 0.12 a	2.70 ± 0.12 a	3.20 ± 0.93 a
2-Methylpropanoic acid	J	521.71 ± 17.41 a	933.11 ± 558.58 a	1052.60 ± 50.55 a
	Š	535.41 ± 0.79 a	661.98 ± 19.50 a	762.14 ± 86.67 a
Butanoic acid	J	159.07 ± 2.39 a	191.02 ± 26.57 a	264.78 ± 13.21 a
	Š	217.19 ± 4.58 a	209.52 ± 14.63 a	153.61 ± 154.60 a
Isovaleric acid	J	2.63 ± 0.02 a	2.84 ± 0.13 a	2.69 ± 0.09 a
	Š	2.74 ± 0.01 a	2.79 ± 0.16 a	3.29 ± 0.43 a
Hexanoic acid	J	338.12 ± 20.36 a	309.98 ± 43.79 a	431.63 ± 6.05 ab
	Š	599.29 ± 71.17 b	467.86 ± 62.50 ab	543.92 ± 83.35 ab
Heptanoic acid	J	5.35 ± 2.43 a	8.03 ± 0.77 a	5.65 ± 2.92 a
	Š	9.03 ± 0.36 a	8.92 ± 0.68 a	6.02 ± 3.46 a
Nonanoic acid	J	7.95 ± 0.09 a	8.34 ± 0.21 a	8.20 ± 0.15 a
	Š	8.26 ± 0.09 a	8.11 ± 0.38 a	8.43 ± 0.21 a
Decanoic acid	J	5.46 ± 0.47 a	5.01 ± 1.56 a	5.17 ± 0.01 a
	Š	5.04 ± 0.22 a	4.76 ± 0.41 a	5.00 ± 0.19 a
Σ Fatty acids	J	1044.00 ± 37.10 a	1461.00 ± 489.00 a	1773.00 ± 60.70 a
	Š	1380.00 ± 76.70 a	1367.00 ± 97.90 a	1486.00 ± 12.50 a
Terpenes				
Farnesol	J	10.38 ± 4.85 a	8.90 ± 0.37 a	10.57 ± 0.88 a
	Š	6.27 ± 3.66 a	3.01 ± 0.22 a	7.12 ± 5.42 a
Tetrahydrolinalool	J	6.43 ± 0.33 a	11.07 ± 0.02 a	30.10 ± 0.08 b
	Š	12.42 ± 0.67 a	5.76 ± 8.15 a	12.32 ± 1.66 a
Linalyl format	J	0.36 ± 0.14 a	0.42 ± 0.00 a	0.32 ± 0.12
	Š	1.51 ± 0.61 a	1.29 ± 0.24 a	1.27 ± 0.17 a
cis-Linalool oxide, fur.	J	1.60 ± 0.19 a	2.19 ± 0.03 ab	1.59 ± 0.03 a
	Š	3.23 ± 0.24 bc	3.59 ± 0.14 c	3.98 ± 0.55 c
Linalool	J	4.48 ± 0.00 a	4.38 ± 0.74 a	4.43 ± 0.16 a
	Š	4.82 ± 0.85 a	4.39 ± 0.09 a	4.34 ± 0.62 a
Terpinen-4-ol	J	4.14 ± 1.30 a	7.81 ± 0.92 b	5.89 ± 0.22 ab
	Š	4.57 ± 0.09 a	5.58 ± 0.22 ab	4.44 ± 0.17 a
Hotrienol	J	0.59 ± 0.24 a	1.56 ± 1.01 ab	1.55 ± 1.32 ab
	Š	4.54 ± 2.44 a	3.34 ± 0.67 ab	3.12 ± 0.53 a
β-Ionone-5,6-epoxide	J	0.13 ± 0.02 a	0.10 ± 0.05 b	0.16 ± 0.04 ab
	Š	0.14 ± 0.04 a	0.09 ± 0.10 ab	0.15 ± 0.08 a

Table 5. *Cont.*

Compounds (μg/L)	Locality	Treatments		
		K	*Lt x Sc*	*Td x Sc*
cis-β-Farnesene	J	0.56 ± 0.02 a	0.59 ± 0.00 a	0.56 ± 0.01 a
	Š	0.83 ± 0.22 a	0.71 ± 0.24 a	0.55 ± 0.02 a
trans-β-Farnesene	J	0.87 ± 0.04 a	1.05 ± 0.14 a	1.24 ± 0.02 a
	Š	0.90 ± 0.17 a	1.06 ± 0.07 a	0.90 ± 0.49 a
Menthol	J	0.29 ± 0.16 a	0.19 ± 0.24 a	0.12 ± 0.04 a
	Š	0.25 ± 0.17 a	0.25 ± 0.21 a	0.33 ± 0.04 a
Ocimenol	J	0.18 ± 0.20 a	0.20 ± 0.20 a	0.05 ± 0.01 a
	Š	0.40 ± 0.25 a	0.18 ± 0.06 a	0.36 ± 0.18 a
Nerolic acid	J	11.68 ± 1.50 b	2.48 ± 0.14 a	2.39 ± 0.38 a
	Š	2.53 ± 0.02 a	12.59 ± 0.90 b	12.95 ± 0.45 b
2,6-Dimethyl-3,7-octadien-2,6-diol	J	0.19 ± 0.02 a	0.33 ± 0.07 a	0.37 ± 0.00 a
	Š	0.47 ± 0.51 a	0.16 ± 0.02 a	0.32 ± 0.06 a
α-Terpineol	J	1.73 ± 0.51 a	2.23 ± 0.65 a	1.98 ± 0.19 a
	Š	1.75 ± 0.14 a	1.67 ± 0.00 a	2.14 ± 0.48 a
Terpendiol I	J	4.2 ± 0.64 b	4.15 ± 0.72 b	3.32 ± 0.07 ab
	Š	2.48 ± 0.04 ab	1.60 ± 0.12 a	1.86 ± 0.24 a
Citronellol	J	31.77 ± 3.91 a	28.02 ± 4.12 a	31.98 ± 0.60 a
	Š	27.83 ± 0.10 a	21.01 ± 0.51 a	32.52 ± 3.51 a
Nerol	J	1.02 ± 0.05 a	2.55 ± 0.14 b	3.26 ± 0.00 c
	Š	2.52 ± 0.09 b	2.66 ± 0.14 bc	2.35 ± 0.25 b
Geraniol	J	6.85 ± 0.74 a	10.58 ± 1.41 b	10.33 ± 0.09 b
	Š	5.57 ± 0.05 a	5.08 ± 0.09 a	4.50 ± 0.31 a
Terpendiol II	J	0.63 ± 0.77 a	0.16 ± 0.09 a	0.35 ± 0.31 a
	Š	1.14 ± 0.24 a	0.49 ± 0.23 a	0.93 ± 0.84 a
6,7-Dihydro-7-hydroxylinalool	J	10.11 ± 0.53 a	14.01± 0.50 a	13.71 ± 2.72 a
	Š	23.01 ± 0.50 a	20.56 ± 1.84 a	23.18 ± 6.07 a
2,6-Dimethyl-7-octen-2,6-diol	J	9.33 ± 1.65 a	12.59 ± 1.37 ab	12.11 ± 1.38 ab
	Š	21.38 ± 1.35 bc	19.29 ± 1.83 abc	22.86 ± 3.93 c
Nerolidol	J	0.59 ± 0.09 a	0.63 ± 0.03 a	0.89 ± 0.35 a
	Š	0.76 ± 0.02 a	0.76 ± 0.26 a	0.82 ± 0.07 a
1,8-Terpin	J	1.10 ± 0.96 a	1.66 ± 0.41 a	1.17 ± 1.15 a
	Š	1.45 ± 1.68 a	1.10 ± 1.15 a	1.32 ± 1.56 a
Geranyl acetate	J	9.84 ± 0.33 a	11.20 ± 0.06 a	11.79 ± 1.90 a
	Š	14.39 ± 1.34 a	11.69 ± 0.48 a	12.07 ± 1.65 a
8-Hydroksylinalool	J	1.86 ± 1.83 a	1.28 ± 1.18 a	2.15 ± 2.02 a
	Š	8.86 ± 11.59 a	0.89 ± 0.28 a	1.89 ± 1.40 a
Ethyl linalyl acetate	J	1.91 ± 1.21 a	0.33 ± 0.24 a	0.95 ± 0.45 a
	Š	0.87 ± 0.12 a	0.72 ± 0.04 a	1.06 ± 0.16 a

Table 5. *Cont.*

Compounds (μg/L)	Locality	Treatments		
		K	*Lt x Sc*	*Td x Sc*
Σ Terpenes	J	122.49 ± 14.10 a	130.98 ± 9.56 a	153.68 ± 3.55 a
	Š	154.55 ± 0.84 a	129.82 ± 11.70 a	159.85 ± 27.60 a
C_{13}-norisoprenoids				
β-Damascenone	J	2.23 ± 0.19 b	3.75 ± 0.28 c	3.25 ± 0.05 c
	Š	1.18 ± 0.12 a	1.65 ± 0.26 ab	1.53 ± 0.18 ab
TDN	J	n.d.	n.d.	n.d.
	Š	n.d.	n.d.	n.d.
β-Ionone	J	0.12 ± 0.04 a	0.13 ± 0.02 ab	0.12 ± 0.04 a
	Š	0.30 ± 0.02 b	0.20 ± 0.00 ab	0.27 ± 0.04 ab
α-Ionone	J	0.10 ± 0.14 a	0.08 ± 0.12 a	0.22 ± 0.00 a
	Š	0.18 ± 0.01 a	0.20 ± 0.04 a	0.17 ± 0.00 a
Σ C_{13}-norisoprenoids	J	2.45 ± 0.38 a	3.97 ± 0.18 b	3.60 ± 0.00 b
	Š	1.66 ± 0.16 a	2.06 ± 0.21 a	1.98 ± 0.14 a
Higher alcohols				
Isobutanol	J	5130.80 ± 278.80 bc	5415.53 ± 71.36 ab	3941.29 ± 237.07 a
	Š	429.55 ± 467.41 abc	4206.45 ± 69.14 c	4322.40 ± 137.48 abc
1-Butanol	J	139.92 ± 16.40 a	167.33 ± 5.61 a	136.29 ± 2.85 a
	Š	140.27 ± 31.10 a	196.66 ± 4.29 a	170.33 ± 8.69 a
2-Methyl-1-butanol	J	11,114.39 ± 1069.30 a	21,785.33 ± 250.97 a	20,076.06 ± 473.04 a
	Š	20,415.29 ± 628.22 a	20,973.72 ± 51.15 a	21,340.38 ± 131.95 a
Isoamyl alcohol	J	12,793.67 ± 8.23 a	8048.26 ± 9.82 a	8419.33 ± 8.10 a
	Š	10,245.44 ± 3.73 a	8827.16 ± 1.13 a	8914.66 ± 5.97 a
4-Methyl-1-pentanol	J	37.455 ± 1.47 a	32.15 ± 6.20 a	26.20 ± 5.18 a
	Š	27.72 ± 0.75 a	26.13 ± 1.20 a	30.34 ± 5.21 a
1-Octanol	J	0.75 ± 0.94 a	1.00 ± 0.52 a	0.03 ± 0.00 a
	Š	0.13 ± 0.03 a	0.28 ± 0.31 a	0.33 ± 0.33 a
1-Nonanol	J	7.23 ± 1.12 ab	5.53 ± 1.29 ab	4.85 ± 0.07 a
	Š	7.49 ± 0.04 ab	6.06 ± 0.57 ab	8.70 ± 0.53 b
2-Penten-1-ol	J	8.66 ± 0.39 a	8.61 ± 1.34 a	6.46 ± 1.4 a
	Š	5.91 ± 0.48 a	6.01 ± 0.09 a	6.80 ± 1.35 a
1-Hexanol	J	1179.26 ± 47.80 a	1239.97 ± 203.88 a	1412.61 ± 27.33 a
	Š	1649.18 ± 57.89 a	1409.09 ± 30.79 a	1510.73 ± 195.77 a
trans-3-Hexen-1-ol	J	22.65 ± 2.11 a	24.76 ± 4.24 a	22.96 ± 0.19 a
	Š	42.88 ± 0.10 b	41.63 ± 0.94 b	43.26 ± 6.54 b
3-Etoxy-1-propanol	J	7.08 ± 1.35 a	121.48 ± 27.28 b	48.35 ± 2.80 a
	Š	15.13 ± 0.30 a	42.08 ± 2.51 a	7.67 ± 0.72 a
cis-3-Hexen-1-ol	J	12.49 ± 0.16 ab	15.15 ± 2.48 ab	10.46 ± 0.26 a
	Š	18.96 ± 0.16 ab	20.61 ± 0.28 b	20.50 ± 3.76 b

Table 5. *Cont.*

Compounds (μg/L)	Locality	Treatments		
		K	*Lt x Sc*	*Td x Sc*
trans-3-Hexen-1-ol	J	4.83 ± 0.26 a	5.90 ± 0.14 a	3.77 ± 0.16 a
	Š	12.01 ± 0.26 b	11.39 ± 0.41 b	12.40 ± 1.38 b
2-Ethyl-1-hexanol	J	0.11 ± 0.08 a	0.14 ± 0.14 a	0.04 ± 0.00 a
	Š	0.41 ± 0.48 a	0.65 ± 0.07 a	0.25 ± 0.19 a
1-Decanol	J	3.83 ± 0.40 bc	2.10 ± 0.14 a	2.48 ± 0.06 ab
	Š	4.22 ± 0.23 c	2.31 ± 0.33 a	3.87 ± 0.48 bc
Phenylethyl alcohol	J	5847.45 ± 171.74 a	5207.13 ± 28.62 a	5302.40 ± 29.76 a
	Š	4634.85 ± 72.85 a	2768.87 ± 2452.29 a	2721.91 ± 1921.65 a
Σ Higher alcohols	J	36,311.00 ± 236.00 a	42,080.00 ± 592.00 a	39,414.00 ± 565.00 a
	Š	41,516.00 ± 623.00 a	38,539.00 ± 3504.00 a	39,163.00 ± 189.00 a
Esters				
Isobutyl acetate	J	80.59 ± 30.38 a	107.63 ± 0.47 a	64.89 ± 0.49 a
	Š	81.74 ± 1.02 a	81.30 ± 0.17 a	98.57 ± 3.09 a
Ethyl butanoate	J	113.9 ± 6.06 a	113.25 ± 4.53 a	159.915 ± 2.58 ab
	Š	172.53 ± 0.26 ab	176.33 ± 7.00 ab	202.57 ± 34.18 b
Isoamyl acetate	J	333.97 ± 38.13 a	514.63 ± 89.22 a	576.81 ± 38.76 a
	Š	607.65 ± 9.61 a	658.69 ± 40.51 a	688.53 ± 196.17 a
Ethyl hexanoate	J	68.59 ± 15.28 ab	59.06 ± 9.04 a	93.98 ± 4.96 ab
	Š	163.6 ± 0.85 c	128.17 ± 1.45 bc	173.26 ± 25.58 c
Ethyl lactate	J	185.44 ± 1.35 a	1826.19 ± 347.42 b	385.48 ± 27.91 a
	Š	282.66 ± 15.52 a	608.89 ± 8.36 b	402.41 ± 46.77 a
Ethyl 2-hydroxy-3-methylbutanoate	J	1.71 ± 0.04 a	3.45 ± 0.07 b	10.14 ± 0.03 c
	Š	3.88 ± 0.23 b	3.46 ± 0.14 b	3.92 ± 0.70 b
Ethyl octanoate	J	31.08 ± 1.67 ab	19.89 ± 1.35 a	27.32 ± 3.30 ab
	Š	72.50 ± 3.13 ab	47.96 ± 3.36 ab	80.78 ± 29.86 b
Ethyl 3-hydroxybutanoate	J	11.70 ± 0.63 a	11.57 ± 2.31 a	14.03 ± 0.82 a
	Š	18.42 ± 1.01 a	16.81 ± 0.14 a	31.02 ± 3.73 b
Ethyl furoate	J	2.24 ± 0.17 b	2.62 ± 0.31 bc	3.51 ± 0.04 c
	Š	0.28 ± 0.04 a	3.06 ± 0.02 bc	3.82 ± 0.54 c
Diethyl succinate	J	155.10 ± 7.17 b	73.13 ± 7.87 a	184.26 ± 3.63 b
	Š	167.58 ± 1.04 b	149.10 ± 2.34 b	191.31 ± 27.18 b
2-Phenylethyl acetate	J	1.10 ± 0.20 ab	1.86 ± 0.29 b	1.66 ± 0.19 ab
	Š	1.05 ± 0.02 ab	1.04 ± 0.01 a	0.94 ± 0.11 a
Diethyl malate	J	5.62 ± 0.05 a	5.62 ± 0.31 a	10.21 ± 0.82 ab
	Š	8.25 ± 0.50 ab	8.69 ± 0.02 ab	12.90 ± 3.22 b
Ethyl hydrogen succinate	J	0.22 ± 0.19 a	0.50 ± 0.07 a	0.83 ± 0.15 a
	Š	0.77 ± 0.20 a	0.71 ± 0.20 a	0.37 ± 0.07 a
Ethyl linoleate	J	0.16 ± 0.02 a	0.37 ± 0.05 a	0.21 ± 0.01 a
	Š	0.27 ± 0.00 a	0.00 ± 0.25 a	0.32 ± 0.20 a

Table 5. *Cont.*

Compounds (μg/L)	Locality	Treatments		
		K	*Lt x Sc*	*Td x Sc*
Ethyl vanillate	J	5.30 ± 0.73 b	6.09 ± 0.26 bc	6.72 ± 0.24 bc
	Š	7.62 ± 0.08 c	7.16 ± 0.08 c	0.00 ± 0.00 a
Σ Esters	J	995.83 ± 25.90 a	2745.56 ± 273.00 c	1540.36 ± 20.60 ab
	Š	1588.58 ± 25.20 ab	1891.05 ± 31.40 bc	1890.31 ± 366.00 bc
Aldehydes				
2,4-Hexadienal	J	1.22 ± 0.00 a	1.29 ± 0.02 ab	1.27 ± 0.14 ab
	Š	1.56 ± 0.05 b	1.50 ± 0.02 ab	1.52 ± 0.06 ab
Benzaldehyde	J	12.62 ± 1.13 ab	10.18 ± 0.89 a	10.27 ± 0.09 a
	Š	16.15 ± 0.81 b	15.34 ± 1.17 b	12.11 ± 1.30 ab
2,4-Heptadienal (E)	J	8.14 ± 0.65 a	10.26 ± 1.18 a	9.62 ± 0.29 a
	Š	10.27 ± 0.39 a	9.10 ± 0.38 a	11.21 ± 1.61 a
Decanal	J	1.59 ± 0.18 a	2.53 ± 0.44 a	1.81 ± 0.12 a
	Š	3.22 ± 0.79 a	2.38 ± 0.17 a	2.71 ± 0.16 a
Acetylfuran	J	1.13 ± 0.09 a	1.25 ± 0.27 a	1.10 ± 0.07 a
	Š	1.18 ± 0.19 a	1.02 ± 0.12 a	1.43 ± 0.28 a
2,4-Nonadienal	J	1.33 ± 0.19 b	1.32 ± 0.20 b	1.04 ± 0.03 ab
	Š	0.79 ± 0.05 ab	0.52 ± 0.06 a	0.66 ± 0.06 a
2,4-Decadienal	J	0.17 ± 0.22 a	0.07 ± 0.09 a	0.21 ± 0.25 a
	Š	0.12 ± 0.03 a	9.02 ± 12.68 a	2.74 ± 1.24 a
2,4-Heptadienal (Z)	J	0.40 ± 0.00 a	0.11 ± 0.15 a	n.d.
	Š	0.28 ± 0.19 a	0.25 ± 0.00 a	0.35 ± 0.07 a
Σ Aldehydes	J	26.59 ± 0.45 a	27.06 ± 3.59 a	25.33 ± 1.30 a
	Š	33.60 ± 0.60 a	39.15 ± 14.6 a	32.72 ± 2.88 a
Lactones				
γ-Decalactone	J	1.27 ± 0.12 a	3.29 ± 0.03 bc	1.65 ± 0.24 ab
	Š	1.44 ± 0.33 a	3.96 ± 0.76 c	1.72 ± 0.28 ab
γ-Nonalactone	J	33.74 ± 2.25 a	31.44 ± 0.45 a	39.84 ± 5.61 a
	Š	35.64 ± 0.96 a	34.71 ± 0.89 a	49.18 ± 12.77 a
γ-Hexalactone	J	6.30 ± 0.28 a	6.82 ± 0.79 a	6.70 ± 0.37 a
	Š	8.38 ± 0.45 a	9.34 ± 0.04 ab	12.31 ± 1.62 b
γ-Octalactone	J	0.84 ± 0.07 b	1.93 ± 0.02 c	0.53 ± 0.03 ab
	Š	0.53 ± 0.07 ab	0.45 ± 0.01 ab	0.31 ± 0.20 a
δ-Decalactone	J	2.79 ± 0.15 a	2.71 ± 0.07 a	3.71 ± 0.16 b
	Š	2.71 ± 0.14 a	2.50 ± 0.11 a	2.89 ± 0.13 a
γ-Undecalactone	J	0.46 ± 0.04 ab	0.45 ± 0.07 ab	0.43 ± 0.02 a
	Š	0.41 ± 0.06 a	0.43 ± 0.01 a	0.60 ± 0.02 b
γ-Butyrolactone	J	325.83 ± 11.60 ab	543.00 ± 90.58 b	299.10 ± 35.63 a
	Š	201.61 ± 9.75 a	243.26 ± 5.72 a	327.91 ± 77.36 ab

Table 5. *Cont.*

Compounds (μg/L)	Locality	Treatments		
		K	*Lt x Sc*	*Td x Sc*
Σ Lactones	J	371.00 ± 14.40 ab	590.00 ± 91.80 b	352.00 ± 30.00 ab
	Š	251.00 ± 9.45 a	295.00 ± 7.42 a	395.00 ± 91.80 ab
Volatile phenols				
Guaiacol	J	3.51 ± 0.19 b	3.10 ± 0.38 a	3.18 ± 0.00 b
	Š	1.32 ± 0.02 a	1.01 ± 0.01 a	1.48 ± 0.33 a
Homovanillyl alcohol	J	70.75 ± 1.93 a	75.03 ± 2.63 a	79.80 ± 0.96 a
	Š	115.85 ± 1.43 b	112.26 ± 2.77 b	132.01 ± 14.79 b
Eugenol	J	0.60 ± 0.12 ab	0.55 ± 0.04 ab	0.81 ± 0.015 b
	Š	0.30 ± 0.01 a	0.27 ± 0.03 a	0.36 ± 0.11 a
4-Ethylphenol	J	3.25 ± 0.04 b	1.90 ± 0.00 a	3.08 ± 0.14 b
	Š	2.53 ± 0.33 ab	2.52 ± 0.11 ab	3.40 ± 0.36 b
4-Vinylphenol	J	12.09 ± 1.32 a	9.76 ± 0.05 a	9.30 ± 0.73 a
	Š	7.30 ± 0.98 a	9.46 ± 1.35 a	7.60 ± 1.68 a
Vanillin	J	9.29 ± 2.74 a	12.13 ± 1.56 a	15.22 ± 1.48 a
	Š	15.28 ± 0.73 a	15.34 ± 0.48 a	17.58 ± 2.67 a
Σ Volatile phenols	J	99.50 ± 0.77 a	102.00 ± 1.47 ab	111.00 ± 1.70 b
	Š	143.00 ± 0.00 c	141.00 ± 2.06 c	162.00 ± 20.00 d
Other compounds				
2-Pentylfuran	J	245.51 ± 9.17 a	255.59 ± 4.92 a	264.88 ± 14.15 a
	Š	233.42 ± 2.08 a	248.72 ± 23.94 a	246.61 ± 23.49 a
Acetoin	J	12.84 ± 1.32 a	14.23 ± 6.73 a	8.69 ± 3.98 a
	Š	12.96 ± 0.79 a	15.82 ± 6.83 a	18.61 ± 4.90 a
6-Methyl-5-hepten-2-one	J	34.90 ±1.22 a	362.02 ± 67.38 b	73.19 ± 5.53 a
	Š	53.61 ± 4.17 a	117.57 ± 1.58 a	75.73 ± 8.88 a
Furfuryl alcohol	J	0.26 ± 0.24 a	0.47 ± 0.08 a	0.45 ± 0.25 a
	Š	3.78 ± 0.07 bc	3.02 ± 0.16 b	4.95 ± 0.53 c
4-Ethyl-cyclohexanol	J	8.75 ± 0.77 a	11.28 ± 1.40 a	10.51 ± 0.35 a
	Š	11.28 ± 0.46 a	9.89 ± 0.45 a	12.40 ± 1.90 a
Furfural	J	1.69 ± 0.37 a	2.08 ± 0.08 a	1.87 ± 0.67 a
	Š	1.96 ± 0.01 a	1.66 ± 0.31 a	1.33 ± 0.55 a
Benzyl alcohol	J	9.65 ± 0.38 a	8.36 ± 0.26 a	9.84 ± 0.74 a
	Š	7.82 ± 0.03 a	7.03 ± 0.00 a	8.79 ± 1.18 a
Σ Other compounds	J	314.00 ± 4.61 a	653.69 ±72.80 b	369.41 ± 8.25 a
	Š	327.57 ± 7.15 a	404.11 ± 19.00 a	368.83 ±38.70 a

Concentrations are expressed as mean ± standard deviation ($n = 3$). Different letters in the same row represent statistically significant differences between treatments at the significance level of $p < 0.05$, separately for each location (two-way ANOVA and LSD test). Different letters in the column represent statistically significant differences between locations of the same treatment at the significance level of $p < 0.05$. J—Jadrtovac, Š—Široke, *K*—control treatment (*S. cerevisiae*), *Lt x Sc*—*L. thermotolerans x S. cerevisiae*, *Td x Sc*—*T. delbrueckii x S. cerevisiae*, n.d.—not detected.

The concentration of total C_{13}-norisoprenoids ranged from 1.66 μg/L to 3.97 μg/L. A significant difference was determined in the *Lt x Sc* and *Td x Sc* (J) treatments. *β*-damascenone and *β*-ionone are carriers of floral and fruity aromas [32]. β-damascenone significantly differed in the *Lt x Sc* and *Td x Sc* (J) treatments. As reported before [47], a significant increase in *β*-damascenone is noticed when Lt and Td yeast are used compared to *Sc* yeast. According to Korenika et al. (2021) [46], *Lt* did not affect the total concentration of C_{13}-norisoprenoids regardless of the variety. In wines from both locations, TDN was not detected.

Total higher alcohols concentrations ranged from 36,311.00 (K(J)) to 42,080.00 μg/L (*Lt x Sc* (J)), with no significant differences between the treatments. Studies report a decrease in total higher alcohols produced during sequential inoculation with *Lt* yeast [25,28,48]. The *Lt x Sc* (Š) treatment had lower concentrations of total higher alcohols, although without statistical significance. This decrease could be due to strain variability within the *Lt* species and oxygen availability [49,50]. Wines produced using conventional *Sc* yeast have the highest concentrations of isoamyl alcohol compared to the *Lt* yeast [51]. All wines in this study have exceeded the isoamyl alcohol detection threshold of 300.00 mg/L [32,52]. Isoamyl alcohol has a strong sensory effect on wine [53], and sequential inoculation with *Lt* reduces the concentration of isoamyl alcohol in Sangiovese wines compared to *Sc* [25,48]. A significant difference was observed for 1-decanol at both localities.

Esters are volatile compounds produced by yeasts during alcoholic fermentation and contribute to the fruity aroma of wine. The intensity of fruitiness is mainly related to higher concentrations of esters and lower concentrations of alcohols and fatty acids [54]. Total esters in this study have shown significant differences at both localities. The *Lt x Sc* (J) treatment showed the highest concentration. Non-*Sc* yeasts are known to increase ester concentrations, but some studies report a decrease in certain ethyl esters compared to *Sc* [55]. Commercial *Sc* strains are known to produce esters such as isoamyl acetate, hexyl acetate, ethyl hexanoate, and ethyl octanoate, which affect the aromatic profile of wine [32]. Sequential inoculation with non-*Sc* yeasts is one way to increase acetate ester concentration [56]. There was a significant change in 2-phenylethyl acetate concentration in the *Lt x Sc* (J) treatment. A significant increase in ethyl lactate was observed in wines produced using *Lt* yeast and sequential fermentation with *Lt* yeast [47,53]. This corresponds to the highest concentration of ethyl lactate found in the *Lt x Sc* (J) treatment.

Aldehydes are volatile aromatic compounds that can be produced by non-*Sc* yeasts during alcoholic fermentation [4]. Depending on their thickness, certain aldehydes, esters, and terpenes can be adsorbed onto yeast cell walls, leading to a decrease in their concentration [57]. Total aldehydes ranged from 25.33 μg/L (*Td x Sc* (J)) to 39.15 μg/L (*Lt x Sc* (Š)) without a significant difference between treatments. A significant increase in 2-octenal in Babić and Trnjak wines produced throughout sequential inoculation with *Lt* yeasts contradicts the results of this study [46].

Lactones were found in overripe Syrah grapes, and their presence was confirmed in the Riesling variety, contributing to varietal aroma [58,59]. Most individual lactones have a positive effect on the wine aroma [60]. An increase in their concentration is attributed to the dominance of *Td* over *Sc*, resulting in a reduction in the amount of common ethyl esters [61]. This reduction in esters simultaneously leads to an increase in lactone concentration, achieving a better sensory effect on white wines [35]. The *Lt x Sc* (J) treatment resulted in the highest concentration of total lactones (590 μg/L), while the lowest concentration of 251 μg/L was determined in (K (J)). The most represented lactones in wine are butyrolactone and γ-butyrolactone [62]. This study found a significant increase in γ-butyrolactone in *Lt x Sc* (J), which is consistent with previous reports [46,51]. *Td x Sc* (J) wine had significantly higher concentrations of γ-nonalactone and δ-decalactone, which is in accordance with work by Azzolini et al. (2012) [63]. γ-decalactone dominated in wines from *Lt x Sc* treatments on both positions, similar to the study of white wines from warm regions [64].

Volatile phenols are classified as aromatic compounds, and the most significant representatives are vinyl and ethyl derivatives. Especially noteworthy is 4-ethylphenol, responsible for unpleasant odors such as 'horse sweat or 'barnyard'. Concentrations above the sensitivity threshold (0.23 mg/L) have a negative effect on the wine aroma [31]. *Sc* yeasts produce very low concentrations of volatile phenols during alcoholic fermentation due to their low level of hydroxycinnamic acid decarboxylation [65]. This was confirmed by the results for volatile phenol analysis in K (J) (99.50 µg/L) in this study. A possible reason for the lower concentration of volatile phenols is the earlier inoculation of non-*Saccharomyces* yeasts, which blocks the action of decarboxylase enzymes [31]. A significant difference between the control treatment with *Sc* and wines from sequential inoculation treatments with *Lt* and *Td* was observed [66]. Treatments *Td x Sc* resulted in the highest concentrations of total volatile phenols, which is in accordance with the study on Trnjak red wine [46].

During this research, a significant increase in the other compounds was observed in wines from sequential inoculations with *Lt* and *Td*. The only compound that significantly differed was 6-methyl-5-hepten-2-one in the *Lt x Sc* (J) treatment, which is not in accordance with a previous study [51].

3.4. Sensory Analysis

The results of a quantitative descriptive sensory analysis of the aroma and flavor properties of Babić wines from the Jadrtovac and Široke localities are shown in Figures 1 and 2. Sequential fermentation with different yeast strains contributes to the enhanced wine flavor. This was achieved through the formation of various volatile compounds, including fatty acids, alcohols, esters, phenols, terpenes, and C_{13}-norisoprenoids. It can be seen that *Td x Sc* wines were the best ranked in terms of flavor quality, fullness or body, aftertaste, and overall impression for both localities (Figure 1). *Lt x Sc* treatments had the lowest intensity of all taste parameters for both localities, except for the acidity in (Š). The obtained results were in accordance to the ranking method results, where the best-rated Babić red wine was *Td x Sc* (Š), and the lowest ranked wines on both positions were those from *Lt x Sc* treatments.

More pronounced differences in sensory properties between the treatments and positions were presented in odor parameters evaluation (Figure 2). Different treatments resulted in different aroma descriptions regarding the position. The best-ranked wine according to the overall impression, in this study—*Td x Sc* (Š) —showed a stronger intensity of fruity, dry fruits, nutty, and herbal aromas, and the lowest vegetal odors intensity (Figure 2b). More intense fruity aromas in these wines were associated with higher concentrations of ethyl hexanoate, which gives aromas of green apple, orange juice, and grapefruit, detected in the *Td x Sc* (Š) treatment. Additionally, the highest concentrations of ethyl 2-hydroxy-3-methylbutanoate contributed to the fruitiness. *Td x Sc* (Š) wine also had a significantly higher concentration of ethyl butanoate, which emits aromas resembling pineapple and apple. Moreover, the higher concentrations of ethyl hexanoate likely influenced the fruity odor, as the concentrations exceeded the sensory threshold of 0.014 mg/L [60]. *Td x Sc* (Š) treatment demonstrated a significant effect on the terpene composition, particularly on ethyl 3-hydroxybutanoate, which emits fruity and grape scents. The intensity of the rose-like floral scent was similar in the *Lt x Sc* (J) and *Td x Sc* (J) treatments. *Td x Sc* (J) wine exhibited the highest concentration of nerol, which resembles a rose and thyme odor. This study revealed a significant increase in β-damascenone in *Lt x Sc* (J) and *Td x Sc* (J), which emits fruity scents, particularly plum and honey [65].

The *Lt x Sc* (J) treatment exhibited the highest concentrations of ethyl lactate. Sequential fermentation with *Lt* strains significantly reduced the concentration of diethyl succinate, which emits scents reminiscent of ripe and overripe fruit. Wines from the *Lt x Sc* (J) treatment showed a significant decrease in β-ionone-5,6-epoxide. This treatment also displayed a significant decrease in nerol concentration, which resembles lemon-like aromas. *Lt x Sc* (J) treatment showed a significant increase in 3-ethoxy-1-propanol, which resembles aromas of blackcurrant and green pepper [67]. The concentration in this treatment was 15 folds higher than in (K), which could have a positive influence. Among other odor descriptors in *Lt x Sc* (Š) wines, the presence of cheese and buttery notes were detected probably due to higher concentrations of some fatty acids.

4. Conclusions

Based on the presented research, it can be concluded that sequential fermentation using non-*Saccharomyces* yeast species, specifically *Torulaspora delbrueckii x Saccharomyces cerevisiae* (*Td x Sc*) and *Lachancea thermotolerans x Saccharomyces cerevisiae* (*Lt x Sc*), had a significant impact on the chemical composition and sensory properties of cv. Babić red wines from two different vine-growing positions. The sequential fermentation partially resulted in a significant reduction in alcoholic strength and an increase in total acidity. Furthermore, a significant increase in tartaric, citric, and succinic acids was reported. In the *Lt x Sc* treatment, lactic acid synthesis occurred, as expected. *Td x Sc* (J) and *Lt x Sc* (Š) led to an increase in total esters and C_{13}-norisoprenoids concentrations, as well as a significant decrease in total aldehydes, fatty acids, higher alcohols, and volatile phenols.

The *Td x Sc* treatments exhibited particularly positive aromatic properties, together with more intense fullness, harmony, aftertaste, and overall impression, while a more acidic taste was pronounced in the *Lt x Sc* treatment. However, the sensory properties of the *Lt x Sc* Babić red wines were negatively impacted, resulting in the lowest-rated wines regardless of positions. Further research is required to confirm the effects of non-*Sc* yeasts on the chemical composition and sensory properties of warm-climate red wines from other grape varieties and on large-scale production.

Supplementary Materials: The following supporting information can be downloaded at: https://www.mdpi.com/article/10.3390/foods13132000/s1, Supplemental Table S1. Parameters of the identification and calibration of wine volatile compounds. Supplemental Table S2. Sensory attributes and material used for the sensory panel training.

Author Contributions: Conceptualization, S.I. and A.-M.J.K.; methodology, A.J.; software, A.B.; validation, A.J., I.B.-L. and V.V.; formal analysis, A.J. investigation, B.K.; resources, T.P.; writing—original draft preparation, S.I.; writing—review and editing, A.-M.J.K.; visualization, A.-M.J.K.; supervision, A.J.; funding acquisition, A.-M.J.K. All authors have read and agreed to the published version of the manuscript.

Funding: This research received no external funding.

Institutional Review Board Statement: The panelists were specialists in the field and well-experienced based on the evaluations in the Croatian Agency for Agriculture and Food, accredited according to the HRN EN ISO/IEC 17065 standard for the implementation of the procedure for placing wines with PDO, i.e., certification of wines with a label of origin, on the market. Panelists did not receive compensation for the evaluation approved and performed in a Laboratory for Sensory Analysis of Agricultural and Food Products, University of Zagreb Faculty of Agriculture, under standardized conditions. This study does not involve human ethical issues and ethical approval was not required for this type of study.

Informed Consent Statement: Informed consent was obtained from all subjects involved in the study.

Data Availability Statement: The original contributions presented in the study are included in the article/Supplementary Material, further inquiries can be directed to the corresponding author.

Conflicts of Interest: The authors declare no conflicts of interest.

References

1. Maicas, S. The Role of Yeasts in Fermentation Processes. *Microorganisms* **2020**, *28*, 1142. [CrossRef] [PubMed]
2. Suárez-Lepe, J.A.; Morata, A. New trends in yeast selection for winemaking. *Trends Food Sci. Technol.* **2012**, *23*, 39–50. [CrossRef]
3. Ciani, M.; Comitini, F. Non-*Saccharomyces* wine yeasts have a promising role in biotechnological approaches to winemaking. *Ann. Microbiol.* **2011**, *61*, 25–32. [CrossRef]
4. Padilla, B.; Gil, J.V.; Manzanares, P. Past and Future of Non-*Saccharomyces* Yeasts: From Spoilage Microorganisms to Biotechnological Tools for Improving Wine Aroma Complexity. *Front. Microbiol.* **2016**, *7*, 411. [CrossRef] [PubMed]

5. Torres-Díaz, L.L.; Murillo-Peña, R.; Iribarren, M.; Sáenz de Urturi, I.; Marín-San Román, S.; González-Lázaro, M.; Pérez-Álvarez, E.P.; Garde-Cerdán, T. Exploring *Metschnikowia pulcherrima* as a Co-Fermenter with *Saccharomyces cerevisiae*: Influence on Wine Aroma during Fermentation and Ageing. *Beverages* **2024**, *10*, 26. [CrossRef]
6. Albertin, W.; Loira, I.; Tesfaye, W.; Bañuelos, M.A.; González, C.; Suárez Lepe, J.A. *New Insights into Wine Yeast Diversities, Yeasts in the Production of Wine*, 1st ed.; Springer: New York, NY, USA, 2019; pp. 117–163.
7. Morata, A.; Loira, I.; Tesfaye, W.; Bañuelos, M.A.; González, C.; Suárez Lepe, J.A. *Lachancea thermotolerans* Applications in Wine Technology. *Fermentation* **2018**, *4*, 53. [CrossRef]
8. Tzamourani, A.; Paramithiotis, S.; Favier, M.; Coulon, J.; Moine, V.; Paraskevopoulos, I.; Dimopoulou, M. New Insights into the Production of Assyrtiko Wines from the Volcanic Terroir of Santorini Island Using *Lachancea thermotolerans*. *Microorganisms* **2024**, *12*, 786. [CrossRef] [PubMed]
9. Tufariello, M.; Fragasso, M.; Pico, J.; Panighel, A.; Castellarin, S.D.; Flamini, R.; Grieco, F. Influence of Non-*Saccharomyces* on Wine Chemistry: A Focus on Aroma-Related Compounds. *Molecules* **2021**, *26*, 644. [CrossRef] [PubMed]
10. Benito, Á.; Calderón, F.; Benito, S. The Influence of Non-*Saccharomyces* Species on Wine Fermentation Quality Parameters. *Fermentation* **2019**, *5*, 54. [CrossRef]
11. Romano, P.; Braschi, G.; Siesto, G.; Patrignani, F.; Lanciotti, R. Role of Yeasts on the Sensory Component of Wines. *Foods* **2022**, *11*, 1921. [CrossRef]
12. Perez-Jiménez, M.; Esteban-Fernández, A.; Muñoz-González, C.; Pozo-Bayón, M.A. Interactions among Odorants, Phenolic Compounds, and Oral Components and Their Effects on Wine Aroma Volatility. *Molecules* **2020**, *25*, 1701. [CrossRef] [PubMed]
13. Jackson, R. *Wine Science—Principles and Applications*, 4th ed.; Academic Press: London, UK, 2014; pp. 357–359.
14. Boulton, R.B.; Singleton, V.L.; Singleton, V.L.; Bisson, L.F. *Principles and Practices of Winemaking*, 1st ed.; Springer: New York, NY, USA, 1998; pp. 521–538.
15. Jagatić Korenika, A.-M.; Preiner, D.; Tomaz, I.; Skendrović Babojelić, M.; Jeromel, A. Aroma Profile of Monovarietal Pét-Nat Ciders: The Role of Croatian Traditional Apple Varieties. *Horticulturae* **2022**, *8*, 689. [CrossRef]
16. OIV. *Compendium of International Methods of Wine and Must Analysis*; International Organization of Vine and Wine: Paris, France, 2021; Volume 1, ISBN 978-2-85038-033-4.
17. *HRN EN ISO/IEC 17065:2013*; Hrvatski Normativni Document. Croatian Accreditation Agency (HAA): Zagreb, Croatia, 2013.
18. *ISO 3591:1977*; Sensory Analysis–Apparatus–Wine-Tasting Glass. International Organization for Standardization (ISO): Geneva, Switzerland, 1977.
19. Romano, P.; Suzzi, G.; Domizio, P.; Fatichenti, F. Secondary products formation as a tool for discriminating non-*Saccharomyces* wine strains. Strain diversity in non-*Saccharomyces* wine yeasts. *Antonie Van Leeuwenhoek* **1997**, *71*, 239–242. [CrossRef] [PubMed]
20. Morata, A.; Escott, C.; Loira, I.; Del Fresno, J.M.; González, C.; Suárez-Lepe, J.A. Influence of *Saccharomyces* and non-*Saccharomyces* Yeasts in the Formation of Pyranoanthocyanins and Polymeric Pigments during Red Wine Making. *Molecules* **2019**, *24*, 4490. [CrossRef]
21. Contreras, A.; Hidalgo, C.; Henschke, P.A.; Chambers, P.J.; Curtin, C.; Varela, C. Evaluation of non-*Saccharomyces* yeasts for the reduction of alcohol content in wine. *Appl. Environ. Microbiol.* **2014**, *80*, 1670–1678. [CrossRef]
22. Röcker, J.; Schmitt, M.; Pasch, L.; Ebert, K.; Grossmann, M. The use of glucose oxidase and catalase for the enzymatic reduction of the potential ethanol content in wine. *Food Chem.* **2016**, *210*, 660–670. [CrossRef]
23. Dutraive, O.; Benito, S.; Fritsch, S.; Beisert, B.; Patz, C.D.; Rauhut, D. Effect of Sequential Inoculation with Non-*Saccharomyces* and *Saccharomyces* Yeasts on Riesling Wine Chemical Composition. *Fermentation* **2019**, *5*, 79. [CrossRef]
24. Escribano-Viana, R.; Portu, J.; Garijo, P.; López, R.; Santamaría, P.; López-Alfaro, I.; Gutiérrez, A.R.; González-Arenzana, L. Effect of the Sequential Inoculation of Non-*Saccharomyces/Saccharomyces* on the Anthocyans and Stilbenes Composition of Tempranillo Wines. *Front. Microbiol.* **2019**, *10*, 773. [CrossRef]
25. Gobbi, M.; Comitini, F.; Domizio, P.; Romani, C.; Lencioni, L.; Mannazzu, I.; Ciani, M. *Lachancea thermotolerans* and *Saccharomyces cerevisiae* in simultaneous and sequential co-fermentation: A strategy to enhance acidity and improve the overall quality of wine. *Food Microbiol.* **2013**, *33*, 271–281. [CrossRef]
26. Kapsopoulou, K.; Mourtzini, A.; Anthoulas, M.; Nerantzis, E. Biological acidification during grape must fermentation usingmixed cultures of *Kluyveromyces thermotolerans* and *Saccharomyces cerevisiae*. *World J. Microbiol. Biotechnol.* **2007**, *23*, 735–739. [CrossRef]
27. García, M.; Esteve-Zarzoso, B.; Crespo, J.; Cabellos, J.M.; Arroyo, T. Influence of Native *Saccharomyces cerevisiae* Strains from D.O."Vinos de Madrid" in the Volatile Profile of White Wines. *Fermentation* **2019**, *5*, 94. [CrossRef]
28. Benito, Á.; Calderón, F.; Benito, S. Combined Use of *S. pombe* and *L. thermotolerans* in Winemaking. Beneficial Effects Determined Through the Study of Wines' Analytical Characteristics. *Molecules* **2016**, *21*, 1744. [CrossRef] [PubMed]
29. Comitini, F.; Gobbi, M.; Domizio, P.; Romani, C.; Lencioni, L.; Mannazzu, I.; Ciani, M. Selected non-*Saccharomyces* wine yeasts in controlled multistarter fermentations with *Saccharomyces cerevisiae*. *Food Microbiol.* **2011**, *28*, 873–882. [CrossRef]
30. Benito, Á.; Calderón, F.; Palomero, F.; Benito, S. Combine Use of Selected *Schizosaccharomyces pombe and Lachancea thermotolerans* Yeast Strains as an Alternative to the Traditional Malolactic Fermentation in Red Wine Production. *Molecules* **2015**, *20*, 9510–9523. [CrossRef] [PubMed]
31. Chatonnet, P.; Dubourdieu, D.; Boidron, J.N.; Lavigne, V. Synthesis of volatile phenols by *Saccharomyces cerevisiae* in wines. *J. Sci. Food Agric.* **1993**, *62*, 191–202. [CrossRef]

32. Tomasino, E.; Bolman, S. The Potential Effect of β-Ionone and β-Damascenone on Sensory Perception of Pinot Noir Wine Aroma. *Molecules* **2021**, *26*, 1288. [CrossRef] [PubMed]
33. Porter, T.J.; Divol, B.; Setati, M.E. *Lachancea* yeast species: Origin, biochemical characteristics and oenological significance. *Food Res. Int.* **2019**, *119*, 378–389. [CrossRef] [PubMed]
34. Chidi, B.S.; Bauer, F.F.; Rossouw, D. Organic Acid Metabolism and the Impact of Fermentation Practices on Wine Acidity: A Review. *S. Afr. J. Enol. Vitic.* **2018**, *39*, 1–15. [CrossRef]
35. Guise, R.; Filipe-Ribeiro, L.; Nascimento, D.; Bessa, O.; Nunes, F.M.; Cosme, F. Comparison between different types of carboxylmethylcellulose and other oenological additives used for white wine tartaric stabilization. *Food Chem.* **2014**, *156*, 250–257. [CrossRef]
36. Caridi, A.; Cufari, A.; Lovino, R.; Palumbo, R.; Tedesco, I. Influence of yeast on polyphenol composition of wine. *Food Technol. Biotechnol.* **2004**, *42*, 37–40.
37. Palomo, E.S.; González-Viñas, M.A.; Díaz-Maroto, M.C.; Soriano-Pérez, A.; Pérez-Coello, M.S. Aroma potential of Albillo wines and effect of skin-contact treatment. *Food Chem.* **2007**, *103*, 631–640. [CrossRef]
38. Kapsopoulo, K.; Kapaklis, A.; Spyropoulos, H. Growth and Fermentation Characteristics of a Strain of the Wine Yeast *Kluyveromyces thermotolerans* Isolated in Greece. *World J. Microbiol. Biotechnol.* **2005**, *21*, 1599–1602. [CrossRef]
39. Masson, P.; Bertoni, G. Ground cover trials with soil-improving clover. *Obstbau-Weinbau* **1996**, *33*, 7–8.
40. Jolly, N.P.; Varela, C.; Pretorius, I.S. Not your ordinary yeast: Non-*Saccharomyces* yeasts in wine production uncovered. *FEMS Yeast Res.* **2014**, *14*, 215–237. [CrossRef]
41. Belda, I.; Ruiz, J.; Beisert, B.; Navascués, E.; Marquina, D.; Calderón, F.; Rauhut, D.; Benito, S.; Santos, A. Influence of *Torulaspora delbrueckii* in varietal thiol (3-SH and 4-MSP) release in wine sequential fermentations. *Int. J. Food Microbiol.* **2017**, *18*, 183–191. [CrossRef]
42. Hernández-Orte, P.; Cersosimo, M.; Loscos, N.; Cacho, J.; Garcia-Moruno, E.; Ferreira, V. The development of varietal aroma from non-floral grapes by yeasts of different genera. *Food Chem.* **2008**, *107*, 1064–1077. [CrossRef]
43. González-Royo, E.; Pascual, O.; Kontoudakis, N.; Esteruelas, M.; Esteve-Zarzoso, B.; Mas, A.; Canalis, J.M.; Zamora, F. Oenological consequences of sequential inoculation with non-*Saccharomyces yeasts* (*Torulaspora delbrueckii* or *Metschnikowia pulcherrima*) and *Saccharomyces cerevisiae* in base wine for sparkling wine production. *Eur. Food Res. Technol.* **2015**, *240*, 999–1012. [CrossRef]
44. Oliveira, J.M.; Oliveira, P.; Baumes, R.L.; Maia, O. Changes in aromatic characteristics of Loureiro and Alvarinho wines during maturation. *J. Food Compos. Anal.* **2008**, *21*, 695–707. [CrossRef]
45. Korenika, A.-M.J.; Tomaz, I.; Preiner, D.; Lavrić, M.; Šimić, B.; Jeromel, A. Influence of *L. thermotolerans* and *S. cerevisiae* Commercial Yeast Sequential Inoculation on Aroma Composition of Red Wines (*Cv* Trnjak, Babic, Blatina and Frankovka). *Fermentation* **2021**, *7*, 4. [CrossRef]
46. Whitener, M.E.B.; Stanstrup, J.; Carlin, S.; Divol, B.; Du Toit, M.; Vrhovsek, U. Effect of non-*Saccharomyces* yeasts on the volatile chemical profile of Shiraz wine. *Aust. J. Grape Wine Res.* **2017**, *23*, 179–192. [CrossRef]
47. Delač Salopek, D.; Horvat, I.; Hranilović, A.; Plavša, T.; Radeka, S.; Pasković, I.; Lukić, I. Diversity of Volatile Aroma Compound Composition Produced by Non-*Saccharomyces* Yeasts in the Early Phase of Grape Must Fermentation. *Foods* **2022**, *11*, 3088. [CrossRef]
48. Balikci, E.K.; Tanguler, H.; Jolly, N.P.; Erten, H. Influence of *Lachancea thermotolerans* on *cv.* Emir wine fermentation. *Yeast* **2016**, *33*, 313–321. [CrossRef]
49. Escribano, R.; González-Arenzana, L.; Portu, J.; Garijo, P.; López-Alfaro, I.; López, R.; Santamaría, P.; Gutiérrez, A.R. Wine aromatic compound production and fermentative behaviour within different non-*Saccharomyces* species and clones. *J. Appl. Microbiol.* **2018**, *124*, 1521–1531. [CrossRef]
50. Shekhawat, K.; Porter, T.J.; Bauer, F.F.; Setati, M.E. Employing oxygen pulses to modulate *Lachancea thermotolerans–Saccharomyces cerevisiae* Chardonnay fermentations. *Ann. Microbiol.* **2018**, *68*, 93–102. [CrossRef]
51. Mucalo, A.; Budić-Leto, I.; Zdunić, G. Effect of Sequential Fermentation with *Lachancea thermotolerans/S. cerevisiae* on Aromatic and Flavonoid Profiles of Plavac Mali Wine. *Foods* **2023**, *12*, 1912. [CrossRef]
52. Guth, H. Quantitation and Sensory Studies of Character Impact Odorants of Different White Wine Varieties. *J. Agric. Food Chem.* **1997**, *45*, 3027–3032. [CrossRef]
53. Cameleyre, M.; Lytra, G.; Tempere, S.; Barbe, J.C. Olfactory Impact of Higher Alcohols on Red Wine Fruity Ester Aroma Expression in Model Solution. *J. Agric. Food Chem.* **2015**, *63*, 9777–9788. [CrossRef]
54. Medina, K.; Boido, E.; Fariña, L.; Gioia, O.; Gomez, M.E.; Barquet, M.; Gaggero, C.; Dellacassa, E.; Carrau, F. Increased flavour diversity of Chardonnay wines by spontaneous fermentation and co-fermentation with *Hanseniaspora vineae*. *Food Chem.* **2013**, *141*, 2513–2521. [CrossRef]
55. Lilly, M.; Bauer, F.F.; Lambrechts, M.G.; Swiegers, J.H.; Cozzolino, D.; Pretorius, I.S. The effect of increased yeast alcohol acetyltransferase and esterase activity on the flavour profiles of wine and distillates. *Yeast* **2006**, *23*, 641–659. [CrossRef]
56. Rojas, V.; Gil, J.V.; Piñaga, F.; Manzanares, P. Acetate ester formation in wine by mixed cultures in laboratory fermentations. *Int. J. Food Microbiol.* **2003**, *86*, 181–188. [CrossRef]
57. Gallardo-Chacón, J.J.; Vichi, S.; López-Tamames, E.; Buxaderas, S. Changes in the sorption of diverse volatiles by *Saccharomyces cerevisiae* lees during sparkling wine aging. *J. Agric. Food Chem.* **2010**, *58*, 12426–12430. [CrossRef] [PubMed]

58. Ribéreau-Gayon, P.; Glories, Y.; Maujean, A.; Dubourdieu, D. *Handbook of Enology*, 2nd ed.; Wiley: Chichester, UK, 2006; pp. 141–203.
59. Jackson, R.S. *Wine Science*, 2nd ed.; Academic Press: Amsterdam, The Netherland, 2000; pp. 232–280.
60. Ferreira, V.; Lopez, R.; Cacho, J.F. Quantitative determination of the odorants of young red wines from different grape varieties. *J. Sci. Food Agric.* **2000**, *80*, 1659–1667. [CrossRef]
61. Velázquez, R.; Zamora, E.; Álvarez, M.L.; Hernández, L.M.; Ramírez, M. Effects of new *Torulaspora delbrueckii* killer yeasts on the must fermentation kinetics and aroma compounds of white table wine. *Front. Microbiol.* **2015**, *6*, 1222. [CrossRef]
62. Genovese, A.; Caporaso, N.; Moio, L. Influence of Yeast Strain on Odor-Active Compounds in Fiano Wine. *Appl. Sci.* **2021**, *11*, 7767. [CrossRef]
63. Azzolini, M.; Fedrizzi, B.; Tosi, E.; Finato, F.; Vagnoli, P.; Scrinzi, C. Effects of *Torulaspora delbrueckii* and *Saccharomyce cerevisiae* mixed cultures on fermentation and aroma of Amarone wine. *Eur. Food. Res. Technol.* **2012**, *2*, 303–313. [CrossRef]
64. Vaquero, C.; Izquierdo-Cañas, P.M.; Mena-Morales, A.; Marchante-Cuevas, L.; Heras, J.M.; Morata, A. Use *of Lachancea thermotolerans* for Biological vs. Chemical Acidification at Pilot-Scale in White Wines from Warm Areas. *Fermentation* **2021**, *7*, 193. [CrossRef]
65. Bakker, J.; Clarke, R.J. *Wine: Flavour Chemistry*, 1st ed.; John Wiley and Sons: Chichester, UK, 2004; pp. 189–229.
66. Morata, A.; Loira, I.; González, C.; Escott, C. Non-*Saccharomyces* as Biotools to Control the Production of Off-Flavors in Wines. *Molecules* **2021**, *26*, 4571. [CrossRef]
67. Amores-Arrocha, A.; Roldán, A.; Jiménez-Cantizano, A.; Caro, I.; Palacios, V. Evaluation of the use of multiflora bee pollen on the volatile compounds and sensorial profile of Palomino fino and Riesling white young wines. *Food Res. Int.* **2018**, *105*, 197–209. [CrossRef]

Disclaimer/Publisher's Note: The statements, opinions and data contained in all publications are solely those of the individual author(s) and contributor(s) and not of MDPI and/or the editor(s). MDPI and/or the editor(s) disclaim responsibility for any injury to people or property resulting from any ideas, methods, instructions or products referred to in the content.

foods

MDPI

Article

Ethanol Reduction in Montepulciano Wine: *Starmerella bombicola* Sequential Fermentation at Pilot Scale Under Aeration Conditions

Laura Canonico, Silvia Gattucci, Laura Moretti, Alice Agarbati, Francesca Comitini and Maurizio Ciani *

Department of Life and Environmental Sciences, Polytechnic University of Marche, Via Brecce Bianche, 60131 Ancona, Italy; l.canonico@univpm.it (L.C.); s.gattucci@staff.univpm.it (S.G.); laura.moretti@pm.univpm.it (L.M.); a.agarbati@univpm.it (A.A.); f.comitini@univpm.it (F.C.)
* Correspondence: m.ciani@univpm.it

Abstract: One of the most relevant challenges in winemaking is the increase in the alcohol content of wine, mainly due to climate change. The use of selected non-*Saccharomyces* yeasts in sequential fermentation with *Saccharomyces cerevisiae* is one of the effective strategies for dealing with this issue, even if it has been poorly confirmed at the winery level. This work evaluated the use of *Starmerella bombicola* and commercial *S. cerevisiae* strains in sequential fermentation at pilot scale in winery conditions to reduce the ethanol content and obtain a wine with enhanced aroma complexity. The results showed that the sequential *S. bombicola*/*S. cerevisiae* fermentation in aeration conditions (20 mL/L/min for the first three days) resulted in a reduction in ethanol of 0.80% (v/v) compared to pure *S. cerevisiae* fermentation. The aeration conditions of sequential fermentation did not affect the fermentation performance of yeasts. The winery conditions determined, in the sequential fermentation modalities, an enhancement of wild yeasts' presence. At the same time, the inoculation of *S. bombicola* determined an enhancement of glycerol and lactic acid, which positively influences the structure and body of the wine as well as specific aromatic notes. In winery conditions, better control of fermentation is needed to achieve potential ethanol reduction and favorable by-product formation using *S. bombicola*.

Keywords: yeast interactions; alcohol content; red winemaking; oxygen; *Starmerella bombicola*

1. Introduction

Climate change has influenced the winemaking industry from viticultural practices to fermentation processes. Indeed, grapes have been characterized by an increase in sugar content that has led to an increase in wine's ethanol content. In the last two decades, there has been a generalized increase of 2% (v/v) of ethanol in wines over the world [1,2]. The high ethanol content in wine can impact wine quality by increasing the perception of heat, body, viscosity, and, to a lesser extent, sweetness and acidity, and it is also related to the health aspect [1,3–7]. In this context, the wine sector has been trying to develop processes that produce wines with lower alcohol content without compromising their quality [8,9]. Most recently, the EU introduced the categories of "dealcoholized wine", including wines with "actual alcoholic strength no more than 0.5% v/v ethanol", and "partially dealcoholized wine", where "actual alcoholic strength above 0.5% v/v ethanol is below the minimum actual alcoholic strength of the wine category" [6]. Several approaches could be used to reduce the ethanol content in wine, such as vineyard management and winemaking practices including microbiological, physical, and biochemical approaches [10–14]. Among the biological approaches, in addition to the use of genetically modified *Saccharomyces cerevisiae* strains, the use of non-*Saccharomyces* strains has been proposed [15–17]. Several non-*Saccharomyces* yeast species used in combination with *S. cerevisiae* under different fermentation conditions have formed wines with reduced ethanol concentrations [15,18–23].

Several non-*Saccharomyces* yeast species are characterized by respiro-fermentative regulatory mechanisms (Crabtree negative effect) that allow the respiration in high sugar content substrates to be carried out differently from *S. cerevisiae*, a feature that could be used to reduce ethanol content in wine [19,23–26]. In some non-*Saccharomyces* yeast species/strains, the diversion of alcoholic fermentation with an abundant formation of secondary compounds may partly explain the low ethanol yield [21,27]. Both these behaviors may contribute to achieving ethanol reduction [28]. Several non-*Saccharomyces* yeast strains were investigated under aerated or other fermentation conditions with the following ethanol reduction levels: *Metschnikowia pulcherrima* (0.8–1.6%); *Starmerella bombicola* (0.6–1.6%); *Torulaspora delbrueckii* (0.3–1.5%); *Zygosaccharomyces bailii* (1–2%); *Starmerella bacillaris* (0.3–0.8%); *Schizosaccharomyces pombe* (0.4–0.65%); *Schizosaccharomyces japonicus* (1.7–2.4%); *Lachancea thermotolerans* (0.4–1.2%); *Hanseniaspora opuntiae* (0.6–1.3%); and *Hanseniaspora osmophila* (0.8–1.3%), which can decrease ethanol yields through respiration [29–33]. Among them, *S. bombicola* (formerly *Candida stellata*) showed promising applications in winemaking for its by-products, particularly glycerol [30,34]. Following the results of previous works [29,30], here, *S. bombicola*/*S. cerevisiae* sequential fermentation was evaluated under winery conditions at the pilot scale under aeration (20 mL/L/min for the first three days) for ethanol reduction and the overall improvement of the wine profile. The influence of cellar conditions and the red winemaking process on the use of *S. bombicola* for ethanol reduction was also evaluated.

2. Materials and Methods

2.1. Yeast Strains

The non-*Saccharomyces* yeast strain used in this study was *S. bombicola* DiSVA66 (DBVPG # 3827 Industrial Yeast Collection of the University of Perugia). This strain was previously selected and used in laboratory-scale sequential fermentation for ethanol reduction in different fermentation conditions [15,30].

S. cerevisiae commercial strain Lalvin EC1118 (Lallemand Inc., Toulouse, France) was used as a control strain and in sequential fermentation with *S. bombicola*. All the strains were maintained at −80 °C for long-term storage in cryovials supplemented with 40% (*w*/*v*) glycerol as the cryoprotective agent. Subsequently, the strains were cultured on Yeast Peptone Dextrose (YPD) agar medium at 25 °C for 48–72 h and stored at 4 °C.

2.2. Pilot Scale Fermentation

Montepulciano grape juice, a red grape variety provided by the winery Terre Cortesi Moncaro s.r.l., Montecarotto (AN), Italy, was used at pilot scale fermentation. Montepulciano grape juice showed the following characteristics: pH 3.52, sugar content 255.71 g/L, malic acid 0.87 g/L, lactic acid 0.36 g/L, total acidity 4.02 g/L. Modified YPD (yeast extract 0.5%, peptone 0.1%, glucose 2%) was used to obtain biomass for fermentation trials. *S. bombicola* was incubated at 25 °C for 72 h under shaking conditions (150 rpm). Biomass was harvested by centrifugation. Cell concentration was determined through the Thoma-Zeiss Counting Chamber. The fermentation trials were carried out in duplicate in 150 L steel tanks containing 100 L. A uniform batch of Montepulciano grapes (500 Kg) was destemmed and pressed, and grape must with skins was distributed into four steel tanks. Sequential trials were inoculated with *S. bombicola* DiSVA66 at a concentration of 5×10^6 cells/smL at 25 °C. A supplement of aeration was maintained using 20 mL/L/min of airflow during the initial 72 h. After this, no aeration was applied, and *S. cerevisiae* was inoculated (1×10^6 cells/mL). Pure fermentation trials of *S. cerevisiae* (inoculum 1×10^6 cells/mL) were used as a control. The skins were kept in both fermentation trials until the 0 °Babo degree (about 20 g/L of reduced sugars), and the fermentation processes were conducted without skin. The temperature was maintained at 25 ± 1 °C.

2.3. Biomass and Sugar Kinetics

Biomass evolution was evaluated by viable cell count (CFU/mL) on lysine agar selective medium and WL nutrient agar (Oxoid, Hampshire, UK). Wild non-*Saccharomyces* yeasts (WNSs) were easily distinguished by *S. bombicola* through macro- and microscopic characterization of the colony on WL nutrient agar. To confirm the belonging of the isolates to the species *S. bombicola*, some strains underwent DNA extraction and were then identified through ITS1- 5.8S rRNA-ITS2 region analyses using the primer pairs ITS1 (5′-TCCGTAGGT GAACCTCGCG-3′)-ITS4 (5′-TCCTCCGCTTTATTG ATATGC-3′). PCR products were separated by horizontal electrophoresis (Bio-Rad, Hercules, CA, USA) in a 1.5% (*w*/*v*) agarose gel using 0.5×TBE buffer and used for identification by sequencing [35]. The genomic sequences obtained were compared with those already present in the data library using the BLAST program and the GenBank database of the ITS. The fermentations were carried out in duplicate. The glucose and fructose (K-FRUGL) concentrations were determined using specific enzyme kits (Megazyme International, Wicklow, Ireland).

2.4. Analytical Determinations

Total acidity [36], organic acids [37], volatile acidity [38], pH [39], ethanol content [40], sugar content [41], and free and total SO_2 [42] were evaluated according to the use of the standard methods of OIV. Acetaldehyde, ethyl acetate, and higher alcohols were analyzed using a gas chromatograph system (GC-2014; Shimadzu, Kyoto, Japan) using direct injection [15]. Samples were injected into a 30 m × 0.32 mm column with a 0.25 µm film thickness (Zebron ZB-WAXPlus; Phenomenex, Torrance, CA, USA) using 1-pentanol (162 mg/L) as an internal standard. Helium served as carrier gas. A Shimadzu gas chromatograph (Japan) equipped with a flame ionization detector was used. The oven temperature ranged from 40 °C to 200 °C. The oven temperature ranged from 40 °C for 5 min, then 5 °C/min until 200 °C for 10 min, while the injector and detector temperatures were maintained at 220 °C. The main volatile compounds were analyzed using the solid-phase microextraction (HS-SPME) method. Five mL of each sample was placed in a vial containing 1 g NaCl closed with a septum-type cap. HS-SPME was carried out under magnetic stirring for 10 min at 25 °C. After this period, an amount of 3-octanol as the internal standard (1.6 mg/L) was added, and the solution was heated to 40 °C and extracted with a Divinylbenzene/Carboxen/Polydimethylsiloxane (DVB/CAR/PDMS) fiber (Sigma-Aldrich, St. Louis, MI, USA) for 30 min by insertion into the vial headspace. The compounds were desorbed by inserting the fiber into a Shimadzu gas chromatograph GC injector for 5 min. The following glass capillary column was used: 0.25 µm Supelcowax 10 (length, 60 m; internal diameter, 0.32 mm). The fiber was inserted in split–splitless mode. The compounds were identified and quantified by comparisons with calibration curves for each compound.

2.5. Sensorial Analysis

At the end of the fermentation, the wines were decanted and after three months, transferred into filled 750 mL bottles, closed with the crown cap, and maintained at 4 °C until sensory analysis. After this period of refinement, they were subjected to sensory evaluation. A group of 10 testers, 8 males and 2 females aged 25–45 years (80% expert and 20% non-expert), used a score scale of 1 to 10, where 10 was the score that quantitatively represented the best judgment (maximum satisfaction), and 1 was the score to be attributed in case of poor satisfaction. The expert testers were composed of oenologists, sommeliers, and wine producers. The order of presentation was randomized among judges. A list of descriptors related to both the olfactory aromatic notes (ripe fruit, tropical fruit, citrus, honey, spicy, aromatic herbs, herbal, and floral) and the taste features (acidity, bitter, softness, structure, balance, tannicity, and intensity). Their data were combined, and the means were subjected to statistical analysis. The data processed in this way were used to provide information on both the contributions of each descriptor to the overall organoleptic quality of the wines and the significant differences between the wines about each descriptor.

2.6. Statistical Analysis

Analysis of variance (ANOVA) was applied to the experimental data for the main enological characteristics and volatile compounds of the wines. The data were analyzed using STATISTICA 7—version 7, the statistical software. Duncan tests were used to detect significant differences, where significance was associated with p-values < 0.05. The data from the sensory analysis were also subjected to Fisher ANOVA to determine the significant differences ($p < 0.05$).

3. Results

3.1. Fermentation Kinetics and Biomass Evolution

The kinetics of sugar consumption, reported in Figure 1, showed a similar trend among the two fermentation trials. It should be highlighted that the *S. bombicola/S. cerevisiae* sequential fermentations in aeration conditions exhibited higher fermentation kinetics up to the tenth day of fermentation (especially on the third day of fermentation) and then overlapped to *S. cerevisiae* pure culture.

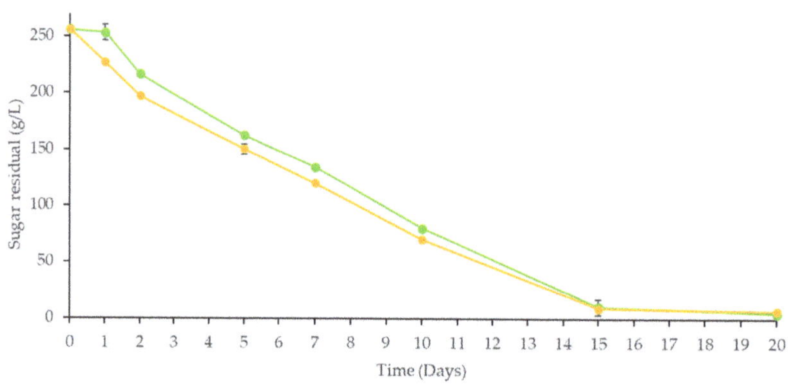

Figure 1. Kinetics of sugar consumption of pure and sequential fermentation are carried out at an industrial level. *S. bombicola/S. cerevisiae* (green line), *S. cerevisiae* pure culture (yellow line).

The biomass evolution of fermentations is reported in Figure 2. *S. cerevisiae* pure culture (Figure 2a) reached the maximum biomass concentration (over 10^7 CFU/mL) on the third day of fermentation and then remained constant until the end of the fermentation process. The same trend was exhibited by wild non-*Saccharomyces* (mainly represented by apiculate yeasts and *Starmerella baciillaris* and occasionally by *Metschnikowia* spp. and *Pichia* spp.) present in the initial grape juice until the fifteenth day of fermentation and then disappeared. *S. bombicola* in sequential fermentation (Figure 2b) achieved the highest biomass concentration on the third day of fermentation and remained constant until the end of fermentation. A similar trend was exhibited by wild non-*Saccharomyces* yeasts, showing, however, a slightly higher concentration. Regarding *S. cerevisiae* inoculated on the third day of fermentation, it showed a limited evolution until the end of fermentation. The high presence of wild non-*Saccharomyces* yeasts may be due to the practice of the red winemaking process that requires the presence of grape skins during fermentation.

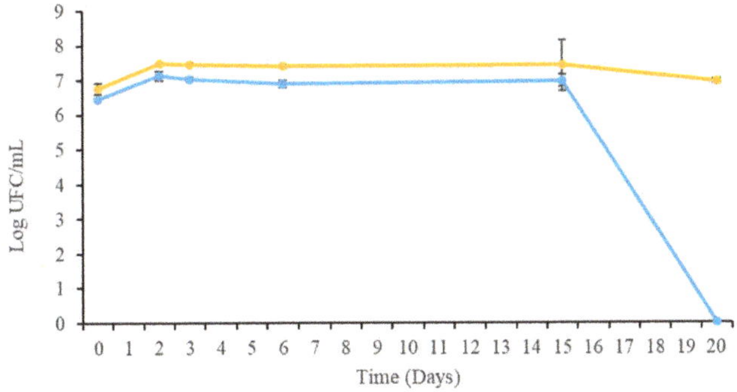

(**a**)

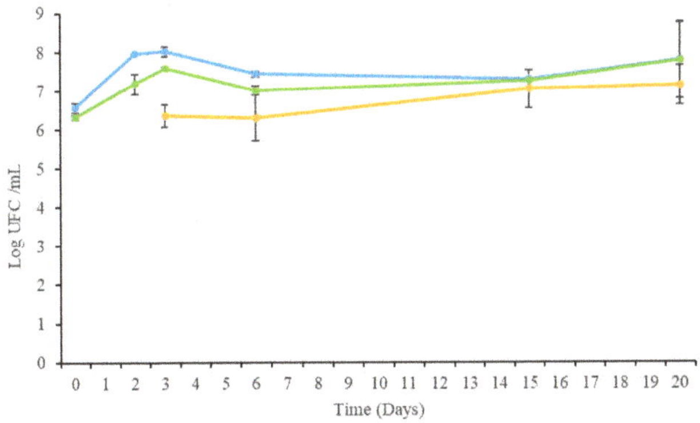

(**b**)

Figure 2. Growth kinetics of *S. cerevisiae* pure culture in Montepulciano grape juice (**a**) and *S. bombicola/S. cerevisiae* sequential fermentation (**b**). *S. cerevisiae* (yellow line), wild yeasts (blue line), *S. bombicola* (green line). Wild yeasts were apiculate yeasts (*Hanseniaspora* spp.), *Starmerella baciillaris*, and occasionally (0th and 2nd day) *Metschnikowia* spp. and *Pichia* spp.

3.2. Main Fermentation Parameters

Table 1 reports the data of the main analytical characters on the third day of fermentation. The only significant difference was in the ethanol content: sequential fermentation *S. bombicola/S. cerevisiae* showed lower ethanol (1.44% *v/v*) compared to *S. cerevisiae* pure culture (3.24% *v/v*).

Table 1. Chemical characterization of wine on the third day of fermentation. Data are means ± standard deviations. Values displaying different superscript letters (^{a,b}) within each column are significantly different according to Duncan's tests ($p < 0.05$).

Sample	Ethanol (% v/v)	Ethanol Yield (wt/vol%)	Total Acidity (as Tartaric Acid g/L)	pH	Volatile Acidity (as Acetic Acid g/L)	Sugar Content (g/L)
S.cerevisiae pure culture	3.24 ± 0.03^{a}	0.54 ± 0.04^{a}	5.92 ± 0.06^{a}	3.50 ± 0.02^{a}	0.21 ± 0.01^{a}	196.01 ± 1.96^{a}
S. bombicola/S. cerevisiae	1.77 ± 0.14^{b}	0.45 ± 0.10^{b}	5.69 ± 0.17^{a}	3.50 ± 0.01^{a}	0.25 ± 0.05^{a}	216.40 ± 7.21^{a}

The low value of the ethanol yield in the first three days of fermentation highlighted the reduced ethanol produced by sequential fermentation trials due to the inoculum of *S. bombicola* and wild yeasts' presence as well as air supplementation.

The data of the main analytical characters of the final wines are reported in Table 2. *S. bombicola* sequential fermentation with *S. cerevisiae* led to wine with a significant reduction in ethanol (0.8% v/v) and total SO_2 and an increase in glycerol compared with *S. cerevisiae* pure culture. No significant differences were shown for the other parameters analyzed, even if an increase in volatile acidity was detected without determining relevant influence on its perception. This behavior may be due to the presence and development of wild yeast during the fermentation process in sequential aeration conditions.

Table 2. Chemical characterization of resulting wine. Data are means ± standard deviations. Values displaying different superscript letters (^{a,b}) within each column are significantly different according to Duncan's tests ($p < 0.05$).

Sample	Ethanol (% v/v)	Ethanol Yield (wt/vol%)	Total Acidity (as Tartaric Acid g/L)	pH	Volatile Acidity (as Acetic Acid g/L)	Free SO_2	Total SO_2	Sugar Content (g/L)	Net Extract (g/L)	Glycerol (g/L)
S. cerevisiae pure culture	15.18 ± 0.02^{b}	0.59 ± 0.02^{b}	6.89 ± 0.03^{a}	3.52 ± 0.01^{a}	0.33 ± 0.01^{a}	12.50 ± 0.71^{a}	41.50 ± 2.12^{a}	2.4 ± 0.1^{a}	27.25 ± 0.01^{a}	8.98 ± 0.85^{b}
S. bombicola/ S. cerevisiae	14.39 ± 0.25^{a}	0.56 ± 0.01^{a}	5.94 ± 0.97^{a}	3.68 ± 0.10^{a}	0.60 ± 0.16^{a}	10.00 ± 1.41^{a}	28.50 ± 0.71^{b}	2.9 ± 0.2^{a}	30.89 ± 0.33^{a}	13.60 ± 0.36^{a}

3.3. The Main Volatile Compounds

The data of the main volatile compounds of *S. cerevisiae* pure cultures and sequential fermentation are reported in Table 3. Regarding ester compounds, *S. bombicola* sequential fermentation led to wine with a significant increase in ethyl acetate and phenyl ethyl acetate, which are responsible for fruity, floral aromas, and the wine was sweeter in comparison to that of the *S. cerevisiae* pure culture. Regarding the other esters, no significant differences were shown among the compounds analyzed. Moreover, the sequential fermentation significantly increased the linalool and isobutanol concentrations compared to the *S. cerevisiae* pure culture. On the other hand, *S. cerevisiae* starter strains exhibited significant increases in nerol and amyl alcohol. No significant differences were shown for the other volatile compounds tested.

Table 3. The main volatile compounds (mg/L) of *S. cerevisiae* pure culture and *S. bombicola* sequential fermentations. OAV = odor activity value. Data are means ± SD from two independent experiments and two repetitions of analysis. Data with different superscript letters ([a,b]) within each column are different according to Duncan's tests (0.05%).

ESTERS	*S. cerevisiae* Pure Culture	OAV	*S. bombicola/S. cerevisiae*	OAV
Ethyl butyrate	1.28 ± 0.72 [a]	3.2	3.71 ± 0.94 [a]	9.28
Ethyl acetate	23.70 ± 4.46 [b]	1.98	95.21 ± 3.07 [a]	7.93
Phenyl ethyl acetate	0.07 ± 0.01 [b]	0.96	0.09 ± 0.00 [a]	1.23
Ethyl octanoate	0.003 ± 0.001 [a]	0.01	0.005 ± 0.001 [a]	0.01
Isoamyl acetate	0.56 ± 0.06 [a]	3.5	0.66 ± 0.05 [a]	4.13
Hexyl acetate	0.002 ± 0.000 [a]	0.00	0.004 ± 0.00 [a]	0.01
Diethyl succinate	0.02 ± 0.01 [a]	0.00	0.07 ± 0.02 [a]	0.00
CARBONYL COMPOUNDS				
Acetaldehyde	1.85 ± 0.67 [a]	3.7	20.59 ± 10.68 [a]	41.18
MONOTERPENS				
Linalool	0.03 ± 0.01 [b]	1.20	0.13 ± 0.01 [a]	5.20
Geraniol	0.02 ± 0.00 [a]	0.67	0.02 ± 0.00 [a]	0.67
Nerol	0.04 ± 0.00 [a]	2.67	0.02 ± 0.01 [b]	1.33
NORISOPRENOIDS				
β-damascenone	0.02 ± 0.01 [a]	0.00	0.03 ± 0.01 [a]	0.00
HIGHER ALCOHOLS				
Hexanol	0.02 ± 0.00 [a]	0.00	0.02 ± 0.00 [a]	0.00
β-Phenyl ethanol	96.2 ± 1.12 [a]	0.69	113.7 ± 0.68 [a]	0.81
n-propanol	18.70 ± 1.80 [a]	0.06	17.26 ± 1.36 [a]	0.06
Isobutanol	16.41 ± 0.95 [b]	0.41	45.93 ± 4.79 [a]	1.15
Amyl alcohol	71.08 ± 0.58 [a]	1.11	44.00 ± 4.20 [b]	0.69
Isoamyl alcohol	158.97 ± 1.04 [a]	2.65	141.00 ± 11.48 [a]	2.35

3.4. Sensorial Analysis

The wines underwent sensory analysis, and the data reported in Figure 3 showed significant differences for all the wines analyzed for some of the aromatic notes. In particular, the olfactory analysis highlighted a significant increase in aromatic herbs and citrus notes in wine carried out with *S. bombicola*. Moreover, in regard to taste analysis, the same wines showed a significant increase in balance, structure, and softness. In general, the results highlighted the positive judgment of testers regarding each wine, characterized by specific aromatic notes and without defects.

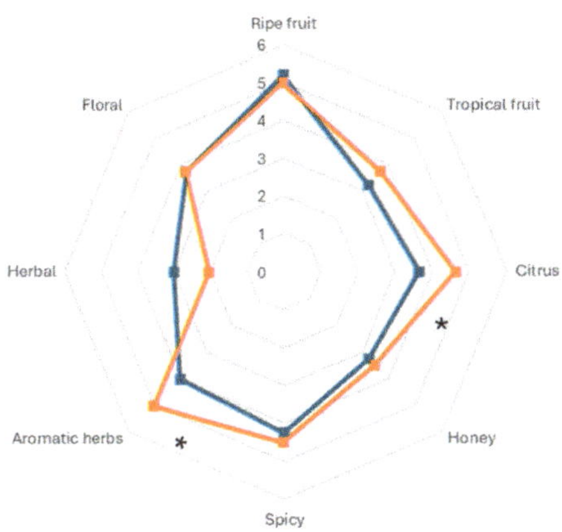

(a)

Figure 3. *Cont.*

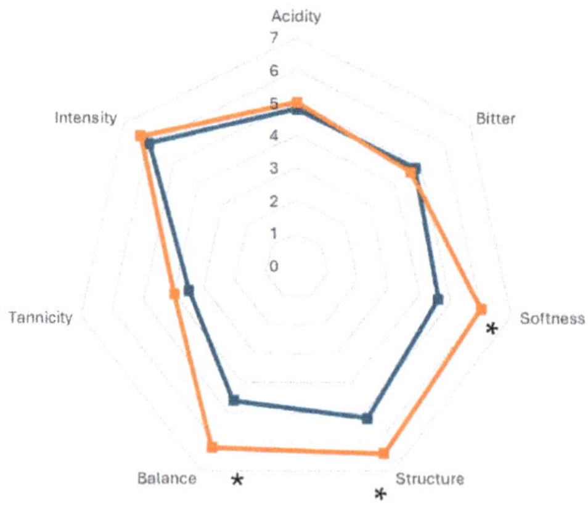

(b)

Figure 3. Sensory analysis of wines produced in the pure and mixed fermentations: (**a**) olfactory analysis and (**b**) taste analysis. *S. cerevisiae* (—■—), *S. bombicola*/*S. cerevisiae* (—■—). *, Significantly different (Fisher ANOVA; *p*-value 0.05).

4. Discussion

In the last two decades, there has been an average increase of two degrees of alcohol in wine due to climate change and the technology used and required by consumer demands. Overall, the combination of health, economic, and quality issues associated with high-alcohol wines has led to significant interest in the development of technologies for producing wines with reduced ethanol concentrations. Several approaches have been proposed to reduce the ethanol content in wine. Among biotechnological approaches, the use of non-*Saccharomyces* yeasts in different fermentation conditions can lead to the achievement of this goal [8,15,18,19,23,43–45]. A valuable fermentation strategy is the addition of air during the first stages of fermentation to allow the consumption of sugar through yeast respiration [46,47]. Among non-*Saccharomyces* yeasts, *S. bombicola* has been extensively studied in different fermentation conditions and grape varieties to achieve the goal of improving some specific analytical compounds [48,49].

In this investigation, the use of *S. bombicola*/*S. cerevisiae* under limited aeration conditions (20 mL/L min) at pilot scale sequential fermentation was evaluated to reduce the ethanol content and increase the aroma complexity in Montepulciano wine production. The use of selected non-*Saccharomyces* yeasts in sequential fermentation with *S. cerevisiae* to reduce the ethanol content was well investigated at the laboratory level but is poorly evaluated under winery conditions [43]. In previous studies, this strain of *S. bombicola* was used in different fermentation conditions to reduce the alcohol content in wine and increase the glycerol content [15,48,49]. *S. bombicola* was used in immobilized form to start fermentation, followed by inoculation of free *S. cerevisiae* cells. This led to wine with a 1.6% *v*/*v* reduction in the final ethanol content in comparison with *S. cerevisiae* starter strains [15]. In a subsequent study, *S. bombicola* bench-top fermentation conducted in Verdicchio grape juice supplemented with 20 mL/L/min of air during the first 72 h produced a wine with an ethanol reduction of 1.46% (*v*/*v*) [30]. Here, under winery conditions, *S. bombicola*/*S. cerevisiae* red winemaking fermentation, supplemented with air during the first 72 h, resulted in a reduction of ethanol of 0.80% (*v*/*v*). In comparison with the laboratory conditions and immobilized cell inoculum, there was a lower ethanol reduction, even if an 0.80% (*v*/*v*) reduction may be a valuable starting point in winery conditions. The scale-up conditions showed several concerns linked to winery conditions and difficulty in controlling the fermentation process with particular reference to mixed fermentations. Also, in this case, a relevant presence of wild yeasts (mainly *H. uvarum* and *S. bacillaris*) in the sequential trials was found. The presence of grape skins (fermentation with maceration) facilitates the presence of wild yeasts in the must and makes it more difficult for the inoculated *S. bombicola* strain to dominate the fermentation process. Better control of wild yeasts and an increase in the inoculation level of *S. bombicola* could enhance the competitivity. Regarding the structure and aromatic profile of wine, this fermentation strategy led to an increase in glycerol content of approximately 4.5 g/L. The ethanol reduction achieved in the present work could be, at least in part, explained by the relevant increase in glycerol, as previously reported by Ciani and Ferraro [50]. A similar result was obtained with *Candida zemplinina* (synonym *Starmerella bacillaris*, a closely related species with similar oenological features to *S. bombicola*), which was widely investigated to produce wine with fewer ethanol levels and higher glycerol content [51]. Oxygen supplementation influences the formation of some volatile compounds such as esters, higher alcohols, ethyl esters, and acetate esters [52–56]. In the present study, the air supplementation increased ethyl acetate and phenyl ethyl acetate in *S. bombicola*/*S. cerevisiae* trials in regard to the ester compounds and only isobutanol in regard to higher alcohols. Sensory analysis confirmed the analytical profile, highlighting citrus notes, balance, structure, and softness in inoculated wines with *S. bombicola* and determining the preference of the tasting panel. In summary, our results confirm the ability of the *S. bombicola* selected strain in sequential fermentation under aeration conditions to reduce the ethanol content in wine and increase the glycerol content in pilot scale winery conditions. Moreover, *S. bombicola* significantly influenced the aroma composition of wine, leading to a more balanced wine with lower ethanol concentration and a pleasant sensory profile.

5. Conclusions

For the first time, a sequential fermentation of *S. bombicola*/*S. cerevisiae* in partial aeration conditions (three days) was conducted in a winery at the pilot scale level. The results previously obtained at the laboratory scale were substantially confirmed. Here, an ethanol reduction of close to 1% and an enhancement of glycerol of 50% were obtained. The analytical profile and sensory evaluation of the wines revealed that *S. bombicola*/*S. cerevisiae* sequential fermentation displayed better overall characteristics than pure fermentation with *S. cerevisiae*. On the other hand, the relevant presence and development of wild yeasts in the inoculum of *S. bombicola* indicates that better control of the fermentation process is necessary. The results indicated that this biotechnological strategy could be used in favorable vintages where the control of wild yeasts is easier. Further studies should be carried out to confirm and improve these results under winery conditions through the enhancement of the starter inoculum and better control of the wild microbiota.

Author Contributions: Conceptualization, M.C., F.C., A.A., and L.C.; validation, M.C., F.C., A.A., L.C., S.G., and L.M.; formal analysis L.C., S.G., L.M., and A.A.; investigation L.C., S.G., A.A., and L.C.; resources, M.C.; data curation, M.C., F.C., A.A., and L.C.; writing, L.C., F.C., A.A., and M.C.; writing—review and editing, M.C. and F.C.; visualization, M.C., F.C., A.A., S.G., L.M., and L.C.; supervision, M.C. and F.C.; project administration, M.C. All authors have read and agreed to the published version of the manuscript.

Funding: This research was funded by MISE Accordo di Innovazione Agrifood DM 02/08/2019—progetto n. 6-project n. F/250006/01-04/X50 "Nuove tecnologie di prodotto per il Food & Wine tipico della tradizione italiana MADE IN ITALY, processo intelligente, integrato, ed interconnesso nella logica di agricoltura di precisione ed industria 4.0, tracciabilità della supply chain, metodi e servizi in ottica blockchain ed in accordo con i criteri del biologico, sostenibilità, sicurezza, design, competitività e globalizzazione.

Institutional Review Board Statement: The study design adhered to the ethical principles of the Declaration of Helsinki and participant privacy and data security were strictly protected. All participants were informed about the study and participated voluntarily. The grape juice was the same used by the winery to prepare the commercial wine. The yeast strains used are GRAS and already used in wine fermentation. In addition, the final wine products were filtered and added with sulphur dioxide, which, together with the alcohol content, provided the microbiological safety of the wines.

Informed Consent Statement: Informed consent was obtained from all subjects involved in the study.

Data Availability Statement: The original contributions presented in the study are included in the article, further inquiries can be directed to the corresponding author.

Conflicts of Interest: The authors declare no conflicts of interest.

References

1. Jones, G.V.; White, M.A.; Cooper, O.R.; Storchmann, K. Climate change and global wine quality. *Clim. Chang.* **2005**, *73*, 319–343. [CrossRef]
2. Alston, J.M.; Fuller, K.B.; Lapsley, J.T.; Soleas, G. Too Much of a Good Thing? Causes and consequences of increases in sugar content of california wine grapes. *J. Wine Econ.* **2011**, *6*, 135–159. [CrossRef]
3. Mira de Orduña, R. Climate change associated effects on grape and wine quality and production. *Food Res. Int.* **2010**, *43*, 1844–1855. [CrossRef]
4. Martinez de Toda, F.; Balda, P. Quantifying the effect of temperature on decoupling anthocyanins and sugars of the grape (*Vitis vinifera* L.'Maturana Tinta de Navarrete'). *Vitis* **2015**, *54*, 117–120.
5. Leeuwen, C.; van Darriet, P. The impact of climate change on viticulture and wine quality. *J. Wine Econ.* **2016**, *11*, 150–167. [CrossRef]
6. Kumar, Y.; Ricci, A.; Parpinello, G.P.; Versari, A. Dealcoholized wine: A ccoping review of volatile and Non-volatile profiles, consumer perception, and health benefits. *Food Bioprocess. Technol.* **2024**, *17*, 3525–3545. [CrossRef]
7. Varela, J.; Varela, C. Microbiological strategies to produce beer and wine with reduced ethanol concentration. *Curr. Opin. Biotechnol.* **2019**, *56*, 88–96. [CrossRef]
8. Varela, C.; Dry, P.R.; Kutyna, D.R.; Francis, I.L.; Henschke, P.A.; Curtin, C.D.; Chambers, P.J. Strategies for reducing alcohol concentration in wine. *Aust. J. Grape Wine Res.* **2015**, *21*, 670–679. [CrossRef]
9. Gonzalez, R.; Guindal, A.M.; Tronchoni, J.; Morales, P. Biotechnological approaches to lowering the ethanol yield during wine fermentation. *Biomolecules* **2021**, *11*, 1569. [CrossRef]
10. Belisario-Sánchez, Y.Y.; Taboada-Rodríguez, A.; Marín-Iniesta, F.; López-Gómez, A. Dealcoholized wines by spinning cone column distillation: Phenolic compounds and antioxidant activity measured by the 1,1-Diphenyl-2-Picrylhydrazyl Method. *J. Agric. Food Chem.* **2009**, *57*, 6770–6778. [CrossRef]
11. Kutyna, D.R.; Varela, C.; Henschke, P.A.; Chambers, P.J.; Stanley, G.A. Microbiological approaches to lowering ethanol concentration in wine. *Trends Food Sci. Technol.* **2010**, *21*, 293–302. [CrossRef]
12. Stoll, M.; Lafontaine, M.; Schultz, H.R. Possibilities to reduce the velocity of berry maturation through various leaf area to fruit ratio modifications in *Vitis vinifera* L. Riesling. *Progrès Agric. Et Vitic.* **2010**, *127*, 68–71.
13. Schmidtke, L.M.; Blackman, J.W.; Agboola, S.O. Production technologies for reduced alcoholic wines. *J. Food Sci.* **2012**, *77*, R25–R41. [CrossRef]
14. Bindon, K.; Varela, C.; Kennedy, J.; Holt, H.; Herderich, M. Relationships between harvest time and wine composition in *Vitis vinifera* L. Cv. Cabernet Sauvignon. Grape and Wine Chemistry. *Food Chem.* **2013**, *138*, 1696–1705. [CrossRef]
15. Canonico, L.; Comitini, F.; Oro, L.; Ciani, M. Sequential Fermentation with Selected immobilized Non-*Saccharomyces* yeast for reduction of ethanol content in wine. *Front. Microbiol.* **2016**, *7*, 278. [CrossRef]
16. Longo, R.; Blackman, J.W.; Torley, P.J.; Rogiers, S.Y.; Schmidtke, L.M. Changes in volatile composition and sensory attributes of wines during alcohol content reduction. *J. Sci. Food Agric.* **2017**, *97*, 8–16. [CrossRef]
17. Guindal, A.M.; Morales, P.; Tronchoni, J.; Gonzalez, R. Reduction of ethanol content in wine with an improved combination of yeast strains and process conditions. *Food Microbiol.* **2023**, *115*, 104344. [CrossRef]
18. Ciani, M.; Capece, A.; Comitini, F.; Canonico, L.; Siesto, G.; Romano, P. Yeast interactions in inoculated wine fermentation. *Front. Microbiol.* **2016**, *7*, 555. [CrossRef]
19. Contreras, A.; Hidalgo, C.; Schmidt, S.; Henschke, P.A.; Curtin, C.; Varela, C. The application of Non-*Saccharomyces* yeast in fermentations with limited aeration as a strategy for the production of wine with reduced alcohol content. *Int. J. Food Microbiol.* **2015**, *205*, 7–15. [CrossRef]

20. Contreras, A.; Hidalgo, C.; Henschke, P.A.; Chambers, P.J.; Curtin, C.; Varela, C. Evaluation of non-*Saccharomyces* yeasts for the reduction of alcohol content in wine. *Appl. Environ. Microbiol.* **2014**, *80*, 1670–1678. [CrossRef]
21. Gobbi, M.; De Vero, L.; Solieri, L.; Comitini, F.; Oro, L.; Giudici, P.; Ciani, M. Fermentative aptitude of non-*Saccharomyces* wine yeast for reduction in the ethanol content in wine. *Eur. Food Res. Technol.* **2014**, *239*, 41–48. [CrossRef]
22. Loira, I.; Vejarano, R.; Bañuelos, M.A.; Morata, A.; Tesfaye, W.; Uthurry, C.; Villa, A.; Cintora, I.; Suárez-Lepe, J.A. Influence of sequential fermentation with *Torulaspora delbrueckii* and *Saccharomyces cerevisiae* on wine quality. *LWT-Food Sci. Technol.* **2014**, *59*, 915–922. [CrossRef]
23. Quirós, M.; Rojas, V.; Gonzalez, R.; Morales, P. Selection of Non-*Saccharomyces* yeast strains for reducing alcohol levels in wine by sugar respiration. *Int. J. Food Microbiol.* **2014**, *181*, 85–91. [CrossRef]
24. Morales, P.; Rojas, V.; Quirós, M.; Gonzalez, R. The impact of oxygen on the final alcohol content of wine fermented by a mixed starter culture. *Appl. Microbiol. Biotechnol.* **2015**, *99*, 3993–4003. [CrossRef]
25. Gonzalez, R.; Quirós, M.; Morales, P. Yeast respiration of sugars by non-*Saccharomyces* yeast species: A promising and barely explored approach to lowering alcohol content of wines. *Trends Food Sci. Technol.* **2013**, *29*, 55–61. [CrossRef]
26. Guedes, J.P.; Cardoso, T.V.; Fernandes, T.; Mendes, F.; Baleiras-Couto, M.M.; Duarte, F.L.; Sousa, M.J.; Franco-Duarte, R.; Chaves, S.R.; Côrte-Real, M. Exploring wine yeast natural biodiversity to select strains with enological traits adapted to climate change. *Heliyon* **2024**, *10*, e36975. [CrossRef]
27. Ciani, M.; Maccarelli, F. Oenological properties of non-*saccharomyces* yeasts associated with winemaking. *World J. Microbiol. Biotechnol.* **1997**, *14*, 199–203. [CrossRef]
28. Comitini, F.; Agarbati, A.; Canonico, L.; Ciani, M. Yeast interactions and molecular mechanisms in wine fermentation: A comprehensive review. *IJMS* **2021**, *22*, 7754. [CrossRef]
29. Canonico, L.; Solomon, M.; Comitini, F.; Ciani, M.; Varela, C. Volatile profile of reduced alcohol wines fermented with selected non-*Saccharomyces* yeasts under different aeration conditions. *Food Microbiol.* **2019**, *84*, 103247. [CrossRef]
30. Canonico, L.; Galli, E.; Agarbati, A.; Comitini, F.; Ciani, M. *Starmerella bombicola* and *Saccharomyces cerevisiae* in wine sequential fermentation in aeration condition: Evaluation of ethanol reduction and analytical profile. *Foods* **2021**, *10*, 1047. [CrossRef]
31. Aplin, J.J.; Paup, V.D.; Ross, C.F.; Edwards, C.G. Chemical and sensory profiles of Merlot wines produced by sequential inoculation of *Metschnikowia pulcherrima* or *Meyerzyma guilliermondii*. *Fermentation* **2021**, *7*, 126. [CrossRef]
32. Jolly, N.; Mehlomakulu, N.N.; Nortje, S.; Beukes, L.; Hoff, J.; Booyse, M.; Erten, H. Non-*Saccharomyces* yeast for lowering wine alcohol levels: Partial aeration versus standard conditions. *FEMS Yeast Res.* **2022**, *22*, foac002. [CrossRef] [PubMed]
33. Ivit, N.N.; Longo, R.; Kemp, B. The effect of non-*Saccharomyces* and *Saccharomyces* non-*cerevisiae* yeasts on ethanol and glycerol levels in wine. *Fermentation* **2020**, *6*, 77. [CrossRef]
34. Ciani, M.; Ferraro, L. Role of oxygen on acetic acid production by *Brettanomyces/Dekkera* in winemaking. *J. Sci. Food Agric.* **1997**, *75*, 489–495. [CrossRef]
35. Agarbati, A.; Canonico, L.; Ciani, M.; Comitini, F. The impact of fungicide treatments on yeast biota of Verdicchio and Montepulciano grape varieties. *PLoS ONE* **2019**, *14*, e0217385. [CrossRef]
36. Total Acidity (Type-I) | OIV. Available online: https://www.oiv.int/standards/annex-a-methods-of-analysis-of-wines-and-musts/section-3-chemical-analysis/section-3-1-organic-compounds/section-3-1-3-acids/total-acidity-(type-i) (accessed on 5 June 2024).
37. OIV-MA-AS313-15. Available online: https://www.oiv.int/public/medias/7907/oiv-vol1-compendium-of-international-methods-of-analysis.pdf (accessed on 21 November 2024).
38. Volatile Acidity (Type-I) | OIV. Available online: https://www.oiv.int/standards/annex-a-methods-of-analysis-of-wines-and-musts/section-3-chemical-analysis/section-3-1-organic-compounds/section-3-1-3-acids/volatile-acidity-(type-i) (accessed on 5 June 2024).
39. pH (Type-I) | OIV. Available online: https://www.oiv.int/standards/annex-a-methods-of-analysis-of-wines-and-musts/section-3-chemical-analysis/section-3-1-organic-compounds/section-3-1-3-acids/ph-(type-i) (accessed on 5 June 2024).
40. Determination of Total Ethanol in Wine by High-Performance Liquid Chromatography (Type-IV) | OIV. Available online: https://www.oiv.int/standards/annex-a-methods-of-analysis-of-wines-and-musts/section-3-chemical-analysis/section-3-1-organic-compounds/section-3-1-3-acids/determination-of-total-ethanol-in-wine-by-high-performance-liquid-chromatography-(type-iv) (accessed on 5 June 2024).
41. OIV-MA-AS311-01A. Available online: https://www.oiv.int/public/medias/7907/oiv-vol1-compendium-of-international-methods-of-analysis.pdf (accessed on 21 November 2024).
42. Sulphur Dioxide (Type-IV) | OIV. Available online: https://www.oiv.int/standards/compendium-of-international-methods-of-wine-and-must-analysis/annex-f/annex-f-specific-methods-for-the-analysis-of-grape-sugar/sulphur-dioxide-(type-iv) (accessed on 5 June 2024).

43. Rolle, L.; Englezos, V.; Torchio, F.; Cravero, F.; Río Segade, S.; Rantsiou, K.; Giacosa, S.; Gambuti, A.; Gerbi, V.; Cocolin, L. Alcohol reduction in red wines by technological and microbiological approaches: A comparative study. *Aust. J. Grape Wine Res.* **2018**, *24*, 62–74. [CrossRef]
44. Mestre Furlani, M.V.; Maturano, Y.P.; Combina, M.; Mercado, L.A.; Toro, M.E.; Vazquez, F. Selection of non-*Saccharomyces* yeasts to be used in grape musts with high alcoholic potential: A strategy to obtain wines with reduced ethanol content. *FEMS Yeast Res.* **2017**, *17*, fox010. [CrossRef]
45. Alonso-del-Real, J.; Contreras-Ruiz, A.; Castiglioni, G.L.; Barrio, E.; Querol, A. The use of mixed populations of *Saccharomyces cerevisiae* and *S. Kudriavzevii* to reduce ethanol content in wine: Limited aeration, inoculum proportions, and sequential inoculation. *Front. Microbiol.* **2017**, *8*, 2087. [CrossRef]
46. Röcker, J.; Schmitt, M.; Pasch, L.; Ebert, K.; Grossmann, M. The use of glucose oxidase and catalase for the enzymatic reduction of the potential ethanol content in wine. *Food Chem.* **2016**, *210*, 660–670. [CrossRef]
47. Xynas, B.; Barnes, C. Yeast or water: Producing wine with lower alcohol levels in a warming climate: A review. *J. Sci. Food Agric.* **2023**, *103*, 3249–3260. [CrossRef]
48. Soden, A.; Francis, I.L.; Oakey, H.; Henschke, P.A. Effects of co-fermentation with *Candida stellata* and *Saccharomyces cerevisiae* on the aroma and composition of Chardonnay wine. *Aust. J. Grape Wine Res.* **2000**, *6*, 21–30. [CrossRef]
49. Ferraro, L.; Fatichenti, F.; Ciani, M. Pilot scale vinification process using immobilized *Candida stellata* cells and *Saccharomyces cerevisiae*. *Process Biochem.* **2000**, *35*, 1125–1129. [CrossRef]
50. Ciani, M.; Ferraro, L. Combined use of immobilized *Candida stellata* cells and *Saccharomyces cerevisiae* to improve the quality of wines. *J. Appl. Microbiol.* **1998**, *85*, 247–254. [CrossRef]
51. Ciani, M.; Comitini, F. Use of non-*Saccharomyces* yeasts in red winemaking. In *Red Wine Technology*; Morata, A., Ed.; Academic Press: Cambridge, MA, USA, 2019; pp. 51–68. ISBN 978-0-12-814399-5.
52. Shekhawat, K.; Bauer, F.F.; Setati, M.E. Impact of oxygenation on the performance of three non-*Saccharomyces* yeasts in co-Fermentation with *Saccharomyces cerevisiae*. *Appl. Microbiol. Biotechnol.* **2017**, *101*, 2479–2491. [CrossRef]
53. Valero, E.; Moyano, L.; Millan, M.C.; Medina, M.; Ortega, J.M. Higher alcohols and esters production by *Saccharomyces cerevisiae*. Influence of the initial oxygenation of the grape must. *Food Chem.* **2002**, *78*, 57–61. [CrossRef]
54. Evaluation of Gene Modification Strategies for the Development of Low-Alcohol-Wine Yeasts | Applied and Environmental Microbiology. Available online: https://journals.asm.org/doi/full/10.1128/aem.01279-12 (accessed on 21 November 2024).
55. Bertrand, A.; Torres-Alegre, V. Incidence de l'action de l'oxygène sur la formation des produits secondaires de la fermentation alcoolique du moût de raisin. *Sci. Des Aliment.* **1984**, *4*, 45–64.
56. Fariña, L.; Medina, K.; Urruty, M.; Boido, E.; Dellacassa, E.; Carrau, F. Redox Effect on Volatile Compound Formation in Wine during Fermentation by *Saccharomyces cerevisiae*. *Food Chem.* **2012**, *134*, 933–939. [CrossRef]

Disclaimer/Publisher's Note: The statements, opinions and data contained in all publications are solely those of the individual author(s) and contributor(s) and not of MDPI and/or the editor(s). MDPI and/or the editor(s) disclaim responsibility for any injury to people or property resulting from any ideas, methods, instructions or products referred to in the content.

foods

MDPI

Article

^{1}H NMR Spectroscopy Primitivo Red Wine Screening After Grape Pomace Repassage for Possible Toxin Contamination Removal

Laura Del Coco [1], Chiara Roberta Girelli [1], Lucia Gambacorta [2], Michele Solfrizzo [2] and Francesco Paolo Fanizzi [1,*]

[1] Department of Biological and Environmental Science and Technology, University of Salento, Via Lecce-Monteroni, 73100 Lecce, Italy; chiara.girelli@unisalento.it (C.R.G.)

[2] Institute of Sciences of Food Production (ISPA), National Research Council (CNR), Via Amendola 122/O, 70126 Bari, Italy; lucia.gambacorta@ispa.cnr.it (L.G.); michele.solfrizzo@ispa.cnr.it (M.S.)

* Correspondence: fp.fanizzi@unisalento.it

Abstract: Food safety and quality awareness have reached significant relevance as consumers are more interested in authentic foods and drinks with specific organoleptic values. Among foodstuffs, grape products can be contaminated by Ochratoxin A (OTA), a mycotoxin that can occur in red grape after infection with *Aspergillus carbonarius.* The high affinity of grape pomace with OTA makes its use advantageous as an adsorbing/decontaminating material whether the pomace is fresh, has undergone pressing, or has undergone a stabilizing process. The effects of different grape repassage treatments on wine metabolic profiles were studied by ^{1}H NMR spectroscopy coupled with metabolomics. The relative quantification of discriminating metabolites for activated-carbon-treated samples revealed higher levels of ethyl acetate and succinate than for the grape-pomace-repassed wine samples. On the contrary, the latter exhibited a relatively high content of glycerol, lactate, tartaric, isobutanol, isopentanol, and polyphenols. Although a specific decrease in aromatic compounds such as gallic acid, tyrosine, and tyrosol was also observed compared with the controls, for the pomace-based processes, the activated carbon treatment led to a marked general impoverishment of the metabolomic profiles, with a reduction in organic acids and glycerol. The repassage of wine over the grape pomace did not significantly affect the quality attributes of the wine, offering an alternative natural adsorbing/decontaminating material for the removal of OTA.

Keywords: wine; Ochratoxin A; grape pomace repassage; ^{1}H-NMR; metabolomics

1. Introduction

Among foodstuffs, grape products cover a wide group of foods such as grape berries and related processing derivatives, including wine, grape juice, distillates, vinegar, and jellies. All these products can be contaminated by harmful metabolites, in particular mycotoxin produced by specific fungi [1]. Among these, Ochratoxin A (OTA) is a naturally occurring secondary metabolite produced by different species of *Aspergillus* and *Penicillium* fungi genera [2]. Following cereals, wine is currently recognized as the second most significant source of OTA human exposure [1]. This mycotoxin has been officially classified as a potential human carcinogen (group 2B) by the International Agency for Research on Cancer (IARC) [3,4]. The molecular mechanisms of the oxidative pathway in OTA-mediated cytotoxicity (ROS accumulation, DNA damage, G1 arrest) in human immune cells have been recently described [5]. The OTA nephrotoxic, neurotoxic, teratogenic, and immunotoxic effects on human and animal health are also well described in

https://doi.org/10.3390/foods14050734

the literature [4]. It was reported to have a correlation with the human Balkan endemic nephropathy (Ben), a chronic renal disease, and human urinary tract cancers [6]. After European Food Safety Authority (EFSA) OTA toxicity evaluations (CONTAM Panel) [7], the EU Commission (EC) introduced a maximum tolerable value of 2 μg/L in must and wine [8]. Being frequently present on the grapes surface, the skins are responsible for OTA contamination of wine [6]. The transfer of mycotoxin from grapes to the final products may occur at any step of the winemaking operation, from the berry crushing, when the toxin passes into the liquid, to the maceration process, in which the OTA level increases [6]. Thus, although the fermentation and clarification stages induce OTA reduction, the vinification process strictly affects the toxin content, thereby leading to the development of several OTA concentration reduction strategies into the final product [1,6,9,10]. Biological (yeast, bacteria, fungi, enzymes) and physical (heating, UV light) strategies are currently available to remove OTA from contaminated wines [1,6]. Biological methods use different microorganisms (yeast, bacteria, fungi) to degrade the mycotoxins in wine [1,6,11]. Physical methods include filtration, radiation, and thermal treatments [1,6]. Chemical methods involve the use of fining agents to reduce the OTA level in wine [12]. Among these, activated carbon showed a high affinity with OTA but the organoleptic properties and the quality parameters could be strongly affected [4,6,13]. Indeed, the efficacy of oenological charcoal in OTA removal was reported to be directly related to a polyphenol content decrease, with a resulting negative impact on wine quality [14,15]. Moreover, among the physical treatment proposed to reduce mycotoxins in wine, an innovative, environmentally friendly, and efficient decontamination procedure using grape pomace repassage was developed by Solfrizzo et al. [13]. Pomace treatment is not uncommon in wineries, since repassage over Recioto or Amarone pomace is largely used in the case of Venetian high-quality wines, like "Valpolicella Ripasso", to give more body and sweetness aromas to Valpolicella red wines [13]. The repassage of contaminated wines over grape pomaces of the same variety showed a strong affinity for OTA, even after multiple uses (up to four times). Moreover, the mitigation effect using pomace repassage is a fast process and it can be observed after only 24 h after the repassage, reaching a 50–65% OTA reduction [13]. Recently, a winery prototype and a specific repassage process have been developed, patented, and used with excellent results (OTA removal of 70–80% from 1000 L of wine in just 5 h) [16]. Unlike other oenological fining agents, the high affinity of grape pomaces for OTA efficiently reduced its concentration in repassed wine without affecting its quality parameters such as color intensity and, above all, healthy compound content [13]. Although the role of grape pomace as a wine fining and clarification agent has been studied since 2013 [17], to the best of our knowledge, very few works have focused on the effect of repassage on wine quality parameters [18–20], and to date, no metabolomic studies are available. It should be noted that metabolomics coupled with NMR spectroscopy is widely used in food science since it is proven to be a reliable tool in the chemical characterization of food matrices for quality control, authenticity, and geographical origin assessment [21–25]. Specifically, NMR-based metabolomics is particularly suitable for wine screening analysis [26] and has been successfully applied to grape variety discrimination [27,28], PGI classification [29,30], adulteration detection [31], and wine authentication [32]. Thus, in this work, we characterized, for the first time, by a ^{1}H-NMR- metabolomic method, the chemical profiles of Apulian high-quality Primitivo red wine [33], which is quite susceptible to OTA contamination [13], and was submitted to different grape pomace repassage. In particular, the ^{1}H-NMR metabolic profiles of the untreated wine (control) samples were compared with activated carbon, fresh/stabilized Primitivo and fresh Aglianico grape pomace repassage treatments to investigate the possible variation in the wine metabolome and related quality parameters after the decontamination treatments.

2. Materials and Methods

2.1. Chemicals and Reagents

All chemical reagents for analysis were of analytical grade. Deuterium oxide (99.9 atom %D) containing 0.05% wt 3-(trimethylsilyl)propionic-2,2,3,3 d4 acid sodium salt (TSP) potassium phosphate monobasic was purchased from Armar Chemicals (Döttingen, Switzerland). Sodium azide was purchased by J.T. Baker (Phillipsburg, NJ, USA).

2.2. Wine Sampling

Primitivo is one of the most important and representative Apulian red grape varieties in Southern Italy, from which high-quality wine is produced, which is quite susceptible to OTA contamination [13,33]. In order to focus on possible metabolomic alteration due to decontamination procedures, minimizing any other variability source, Primitivo di Gioia cv OTA-free biological red wine samples from a 3000 L bulk tank representative of the commercial product were subjected to four different treatments (each treatment in triplicate or duplicate for a total of 11 different biological samples) and compared with untreated samples (3 controls). Therefore, in order to assess the differences produced in the metabolic profile of the same bulk product according to the specific performed treatment, the following samples were considered:

1. Controls, namely, untreated OTA-free Primitivo di Gioia red wine: three samples.
2. Repassage over Primitivo di Gioia fresh pomace: two different biological samples.
3. Repassage over Primitivo di Gioia stabilized pomace, in duplicate: three different biological samples.
4. Repassage over Aglianico fresh pomace: three different biological samples.
5. Enological activated carbon treatment: a total of 5 L of Primitivo stored in a steel tank containing 1.5 g of enological activated carbon, which was stirred for few seconds and left in static conditions for 24 h: three biological samples.

Thus, a total of 14 Primitivo red wine biological samples, as listed in Table 1, were obtained and stored in a cellar until analysis.

For the stabilized grape pomace, a total of 150 kg of Primitivo di Gioia grape pomaces were stored at room temperature for two months in 50 L steel tanks before being used for the repassage experiments. For the fresh grape pomace, a total of 250 kg of Primitivo and 250 kg of Aglianico pomaces were used fresh for the repassage experiments. The Primitivo di Gioia wine and grape pomace were purchased from a winery in Gioia del Colle (Apulia region, Italy). The Aglianico grape pomace was purchased in a winery in Nova Siri (Lucania region, Italy).

The determination of the total polyphenols (gallic acid content) and anthocyanins was carried out according to the International Organization of Vine and Wine (OIV) methods [34] and is reported in Supplementary Table S1. In particular, according to the OIV-MA-AS2-10 method, the spectrophotometric Folin–Ciocalteu assay was used to estimate the total polyphenol content (TPC) [35]. First, 1 mL of the wine, previously diluted 1/5, 50 mL of distilled water, 5 mL of Folin-Ciocalteu reagent, and 20 mL of sodium carbonate solution were introduced into a 100 mL flask. Distilled water was used to bring the volume up to 100 mL as a final volume. After mixing to dissolve all the contents, the solution was left for 30 min for the reaction to stabilize. The absorbance at 750 nm was determined through a path length of 1 cm compared with a blank prepared with distilled water in place of the wine. Gallic acid equivalents, or mg GAE, were used to express the results per liter of wine. For the anthocyanin content, the OIV-MA-AS315-11 method, involving the HPLC determination of nine major anthocyanins in red and rosé wine (type-II method) was used. The analysis of the wine was performed by direct separation by HPLC with the reverse phase column, gradient elution by water/formic acid/acetonitrile, and detection at 518 nm. The HPLC analysis was performed according to the following conditions: injection volume: 50 μL (red wine); flow: 0.8 mL/min; temperature: 40 °C; run time: 45 min; post run time: 5 min; detection: 518 nm. Malvidin chloride content (mg/L) values were used to express the results.

Table 1. List of analyzed control/OTA-removal-treated wine samples and related initial and reference adjusted pH values.

NMR ID	Treatment	Wine	Initial pH	Final pH * t
A	Control (absence of OTA)	Primitivo di Gioia	3.74	3.74
A	Control (absence of OTA)	Primitivo di Gioia	3.66	3.74
A	Control (absence of OTA)	Primitivo di Gioia	3.62	3.72
B	Repassage over fresh Primitivo pomace	Primitivo di Gioia	3.69	3.72
B	Repassage over fresh Primitivo pomace	Primitivo di Gioia	3.68	3.74
B	Repassage over fresh Primitivo pomace	Primitivo di Gioia	3.68	3.73
C	Repassage over stabilized Primitivo pomace	Primitivo di Gioia	3.67	3.73
C	Repassage over stabilized Primitivo pomace	Primitivo di Gioia	3.69	3.75
D	Repassage over fresh Aglianico pomace	Blend Primitivo di Gioia and Primitivo di Manduria	3.55	3.75
D	Repassage over fresh Aglianico pomace	Blend Primitivo di Gioia and Primitivo di Manduria	3.59	3.74
D	Repassage over fresh Aglianico pomace	Blend Primitivo di Gioia and Primitivo di Manduria	3.61	3.74
E	Activated carbon	Primitivo di Gioia	3.83	3.74
E	Activated carbon	Primitivo di Gioia	3.70	3.74
E	Activated carbon	Primitivo di Gioia	3.90	3.74

* After adjustment to pH wine reference (±0.02).

2.3. *Repassage of Red Wines over Grape Pomace from Different Varieties*

The prototype (Figure S1) and the process used in this study for the repassage experiments were those previously developed and optimized to remove OTA from contaminated wine by repassing the contaminated wine over uncontaminated grape pomaces [13,16]. The prototype [16] mainly consists of a 1000 L steel tank containing the wine to be repassed over the pomaces, 3 steel tanks, each filled with 73 kg of pomace, and a system of circulating pumps, sensors, and a control unit that makes the whole repassage process automated. The process used by the prototype was originally developed to maximize the removal of OTA during the repassage in the shortest possible time [16]. In particular, an aliquot of wine is pumped into the first tank until it covers the pomaces, left for 10 min, pumped in the second tank, left for 10 min, pumped in the third tank, left for 10 min, and then pumped into a collecting tank. As soon as the wine is transferred from the first tank to the second tank, a new amount of wine is added to the first tank and so on to continue the process. The process is completed when all the wine (~1000 L) has been repassed. The treatment with the enological activated carbon was performed in static conditions in a steel tank containing 5 L of Primitivo and 1.5 g of enological activated carbon, stirred for few seconds, and left in static conditions for 24 h.

2.4. *NMR Measurements*

The protocol applied for the wine samples' preparation and NMR acquisition and analysis was according to the Bruker standard procedure established for the NMR screening of wine (Bruker BioSpin GmbH, Rheinstetten, Germany). The sample preparation and measurement protocol facilitates and optimizes the quantitative determination of metabolites and statistical analysis [24,36]. For each sample, 900 μL of wine was added to 100 μL of buffer (KH_2PO_4, NaN_3 in D_2O + 0.03% v/v 3-trimethylsilyl-propionic-2,2,3,3-d4 acid sodium salt, TSP). In order to minimize possible variations in the chemical shifts of signals among the samples, the pH values for the wines were adjusted (±0.02) to a wine reference value measured at 3.74, as reported in Table 1. This value is intermediate between the measured pH obtained for all the samples studied (ranging from 3.9 to 3.5) and is close to the reported value found for Primitivo wines [37]. Then, a total of 600 μL of the resulting solution was transferred into a 5 mm NMR tube for spectral acquisition. All the NMR spectra were recorded at a temperature of 300.0 K (±0.05), after 5 min for thermal equilibration, on a Bruker Avance III spectrometer (Bruker, Karlsruhe, Germany), operating at 400.13 MHz for ^1H observation. Automated tuning and matching, locking, shimming, and calibration of the 90° hard pulse P (90°), including adjustment of the 25 Hz presaturation pulse, was performed for each sample using the standard Bruker routines (ATMA, LOCK, TOPSHIM, and PULSECAL, respectively) to optimize the NMR conditions [36]. Two ^1H NMR experiments were performed for each sample in the automation procedure (zgpr.mod and noesygppsld.comp1 Bruker pulse sequences). The measurements were repeated once in random order after the completion of the first entire set. The automated noesygppsld.comp1 pulse sequence was used to enhance the minor compound signals [26,38–40]. The AQ parameters were as follows: 32 scans, 64K data points, a spectral width of 8223.685 Hz, an acquisition time of 3.98 s, a relaxation delay of 4 s, and a mixing time of 10 ms. The FIDs were multiplied by an exponential weighting function corresponding to a line broadening of 0.3 Hz before Fourier transformation, phasing, and base line correction. The metabolites were assigned on the basis of analysis of 2D NMR spectra (2D ^1H Jres, ^1H COSY, ^1H-^{13}C HSQC, and HMBC, which were also acquired for assignment purposes) and compared with the published data [22,29,36,38–40].

2.5. Multivariate Data Analysis

The NMR spectra were processed using Topspin 3.6.5 and visually inspected using Amix 3.9.13 (Bruker, Biospin, Italy). The ^1H NMR (noesygppsld.comp1 Bruker pulse sequence) spectra were segmented into rectangular buckets of a fixed 0.04 ppm width and integrated. In order to reduce the number of variables (assigned/not assigned NMR signals) and to compensate for small shifts in the positions of the peaks, the NMR spectra were converted into data matrices through a bucketing procedure. By this method, the integration of the NMR signals into small spectral regions, called "buckets" or "bins", was performed. Simple equidistant binning is the most commonly used method, and it has already been described as a robust procedure in metabolomic fingerprinting for sample classification [41,42]. The spectral regions between 4.90 and 4.75, 3.70 and 3.60, and 1.22 and 1.15 ppm were discarded because of the residual peaks of water and ethanol signals (in particular, the quartet at 3.66 ppm, the -CH_2 group and the triplet at 1.18 ppm, and the-CH_3 group of ethanol). The resulting data sets consisted of the 228 variables (^1H NMR spectra bucketed values, in columns) measured for each wine sample (row). The description of the statistical analyses refers to Pareto scaled data (performed by dividing the mean-centered data by the square root of the standard deviation) [43]. The data table generated from all 16 spectra was considered for multivariate data analysis (MVA) by using Simca-P version 14 (Sartorius Stedim Biotech, Umea, Sweden). In particular, unsupervised (principal component analysis (PCA)) and supervised (partial least squares discriminant analysis (PLS-DA) and orthogonal partial least squares discriminant analysis (OPLS-DA)) methods were performed to analyze the intrinsic variation in the data [44–46]. The models validation was performed by using the internal cross-validation default method (7-fold) and permutation test (400 permutations). The quality of the models was evaluated by the R^2 and Q^2 parameters. The first (R^2) is a cross-validation parameter defined as the data variance portion explained by the model and specifies goodness-of-fit. The second (Q^2) describes the portion of data variance predictable by the model [44,46–48]. The S-line plot for the OPLS-DA models visualizes the centered loading vector p (ctr), colored according to the correlation loading absolute value, p(corr). The loading line plot for the PLS-DA model displays the correlation loading as the structure between X and Y (w*c) [45].

3. Results

3.1. ^1H NMR Spectroscopy

The relative expansions of the regions of the representative studied Primitivo wine ^{1}H-NMR spectrum and the peak assignments of the relevant metabolites carried out by a comparison of literature data [22,36,38–40] and bi-dimensional experiments are reported in Figure 1. Besides the highest peaks, assigned to the -CH_3 ethanol signal, the presence of other alcohols (isobutanol, isopentanol, 2,3 butanediol) as well as aliphatic groups of amino acid (alanine, proline, GABA) and organic acid (lactic, acetic, succinic) resonances were observed in the low-frequency field region of the spectrum (0.5–3.00 ppm). Intense peaks assigned to glycerol and resonances ascribable to tartrate, sucrose, and methanol were observed in the middle frequency region of the spectrum (3–5 ppm). Protons of sugars such as glucose, sucrose, xylose, arabinose, and galacturonic acid were also observed in the 5–5.5 ppm spectral region. Finally, in the aromatic spectral region (5.5–9.5 ppm), proton resonances of caffeic acid, tyrosol, gallic acid, tyrosine, phenethyl alcohol, formic acid, and trigonelline were identified. A summary of the identified metabolites is reported in Table S2. The whole set of 14 NMR spectra for all the studied samples is reported in Figure S2.

3.2. Multivariate Analysis

The first level of investigation was carried out using the supervised PLS-DA (Figure 2) methods, on the whole bucket-reduced spectra in the 0.5–10 ppm spectral range [49]. The application of these methods to the spectral dataset allowed us to obtain information on both the general trends of the data and the discriminating metabolites responsible for the class grouping [50]. The samples obtained from the repassage over the stabilized and fresh Primitivo pomace were further considered as a single class. As shown in the t [1]/t [2] PLS-DA score plot (two components, $R^2X = 0.59$, $R^2Y = 0.53$, $Q^2 = 0.20$) in Figure 2a, a certain degree of separation was found among the sample classes. In particular, the wines treated with the activated carbon were clearly separated along the first component t [1] from all the other repassed wine classes (repassage over fresh/stabilized Primitivo pomace and repassage over fresh Aglianico pomace) and the control samples. This could suggest that the metabolic profiles of the activated-carbon-treated samples exhibited qualitative chemical composition different from the grape-repassed and untreated samples. Moreover, it should be noted that the control wine samples (Primitivo wine not repassed over pomace) results show that it is clearly separated on the second component t [2] from the grape-repassed ones. The metabolites responsible for the observed separation of the activated-carbon-treated samples compared with the other wine classes (both controls and repassed over Primitivo and/or Aglianico pomace) were seen in the S-line plot for the model, as reported in Figure 2b. Relative higher levels of specific metabolites such as ethyl acetate and succinate were seen in the activated-carbon-treated samples, while relatively higher levels of glycerol, lactate, tartaric, isobutanol, isopentanol, and polyphenol content were found in the remaining samples.

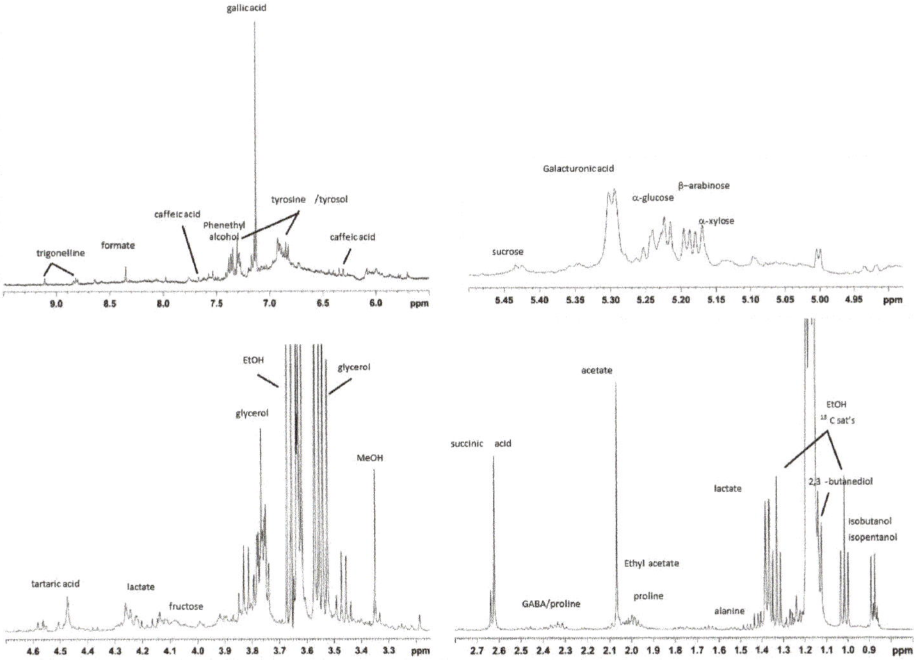

Figure 1. ^{1}H NMR relative expansions of significant spectral regions (400 MHz, D_2O) of a specimen of analyzed wine spectra, acquired with automated noesygppsld.comp1 Bruker sequence.

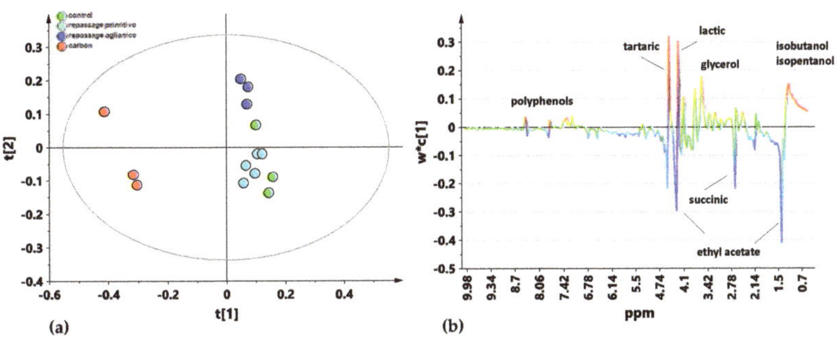

Figure 2. (**a**) PLS-DA t [1]/t [2] score plot (two components, $R^2X = 0.59$, $R^2Y = 0.53$, $Q^2 = 0.20$) for the whole wine sample data set, classified according to the performed OTA decontamination treatment: Primitivo grape pomace repassage (sky-blue circle), Aglianico grape pomace repassage (blue circle), activated carbon (red circle) and control samples (green circles). (**b**). Loading S-line plot for the PLS-DA model, colored according to the correlation scaled coefficient. The *x*-axis indicates binned signals (ppm) in the ^{1}H NMR spectrum.

In order to further analyze the wine metabolic response due to the repassage over the pomace using different grape varieties (Primitivo and Aglianico), the metabolic profiles of the treated and untreated classes were studied by supervised OPLS-DA pairwise comparisons. Two different OPLS-DA models were built, comparing separately, against the control group, the wines with repassage over fresh/stabilized Primitivo (Figure 3a,b) and over fresh Aglianico pomace (Figure 3c,d), respectively. This was carried out in order to ascertain whether the grape variety could have a different effect on the processed wine samples' characteristics. The OPLS-DA models were both obtained by using one predictive and two orthogonal components (1 + 2 + 0). Good model quality parameters were also found with $R^2X = 0.82$, $R^2Y = 0.98$ and $Q^2 = 0.83$ for the controls vs. samples treated with Primitivo pomace repassage (Figure 3a) and $R^2X = 0.84$, $R^2Y = 0.99$ and $Q^2 = 0.91$ for the controls vs. samples treated with Aglianico pomace repassage (Figure 3c) OPLS-DA, respectively. As shown by the limited dispersion along the first orthogonal component (Figure 3a), the samples obtained from repassage over the Primitivo stabilized pomace did not differ significantly from the samples with repassage over the fresh Primitivo pomace, confirming the original PLS-DA results. This aspect could be of interest for wineries, since stabilized pomace can be recovered over time and reused for repassage with efficacy [13]. A clear difference was also observed in the OPLS-DA model shown in Figure 3c, comparing the samples obtained by repassage over the Aglianico pomace with the control group. From the corresponding S-line plots for both the OPLS-DA models, the molecular components, distinctive for each class and responsible for differentiation with the controls, were identified (Figure 3b,d). Interestingly, the discriminating metabolites between the controls vs. Primitivo and Aglianico grape-pomace-repassed samples appear to be similar (Figure 3b,d). Thus, in both pairwise comparison, the untreated control wines were shown to be characterized by a higher relative content of organic acids, such as tartaric and succinic acids, and containing isopentanol as well as isobutanol. On the other hand, the Aglianico and Primitivo grape-pomace-treated samples showed a higher relative content of sugars and lactic acid. Moreover, differently from the wines repassed over the Primitivo pomace, the Aglianico repassed samples were also characterized by a higher relative amount of glycerol and acetic acid. This could suggest that the usage of different pomace grape varieties may lead to a certain degree of specific quality modification in the final product.

Finally, a selected bucketing from the ^{1}H 1D-NOESY spectra was further considered, focusing only on the aromatic spectral region (6.00–10.00 ppm). From the PLS-DA t [1]/t [2] score plot visual inspection (five components, $R^2X = 0.92$, $R^2Y = 0.85$, $Q^2 = 0.56$, Figure 4a), the control samples were well separated from all the other samples along the first component t [1]. The activated-carbon-treated samples were shown to be clearly separated on the second component t [2], both from the controls and from all the wines obtained by repassage over the different varieties of grape pomace. The loading line plot of the model (Figure 4b) revealed the binned signals assigned to the functional groups of the aromatic molecules responsible for class discrimination. The phenol content was shown to be generally higher in the controls than in the other treated classes. Specifically, a higher content of gallic and caffeic acids, phenethyl alcohol, and tyrosol could be seen for the control samples compared with the grape-repassed and carbon-treated samples.

A supervised pairwise OPLS-DA analysis was then performed, obtained by focusing on the aromatic spectral region (Figure 5), to further investigate the metabolites responsible for the partition of the different classes. A clear partition between the controls vs. Aglianico (Figure 5a) grape-repassed, Primitivo (Figure 5b) grape-repassed, and activated-carbon-treated (Figure 5c) samples can be observed in the related score plot. Moreover, the S-line plot for the models showed that a higher relative content of phenolic compounds characterized the control wine samples, in comparison with both the carbon-treated (Figure 5d), Primitivo and Aglianico (Figure 5e) grape-pomace-repassed wines (Figure 5e) and Aglianico (Figure 5f) pomace samples. In particular, the molecules responsible for discrimination among the samples were shown to be tyrosol (bins at 6.94 and 7.18 ppm), caffeic acid (bin at 6.46 ppm), and phenethyl alcohol (bins at 7.30 and 7.38 ppm) together with gallic acid (bin at 7.14 ppm). A lower relative content of gallic acid was found in particular for the samples with repassage over the Aglianico pomace compared with the controls (Figure 5f). These results for the metabolite contents are consistent with previously obtained classical data and are reported in Supplementary as Table S1.

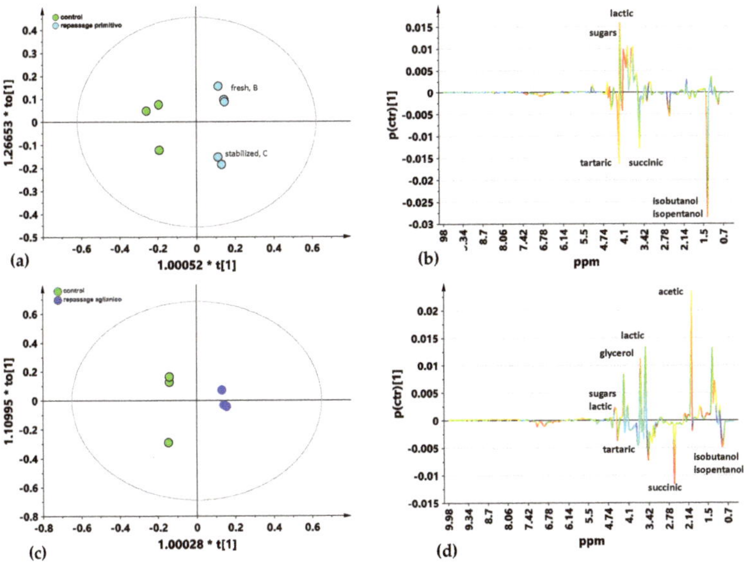

Figure 3. Pairwise OPLS-DA t [1]/t [2] score plots for untreated (controls) vs. (**a**) Primitivo (1 + 2 + 0; $R^2X = 0.82$, $R^2Y = 0.98$ and $Q^2 = 0.83$) and (**c**) Aglianico (1 + 2 + 0; $R^2X = 0.84$, $R^2Y = 0.99$ and $Q^2 = 0.91$) repassed wine samples. Sample symbols are colored according to the different OTA decontamination treatment: control, green circle; Aglianico, blue circles; and Primitivo, green circles. Metabolites responsible for the class separation can be observed in the model-related S-line plots, (**b**,**d**) colored according to the correlation scaled coefficient. The *x*-axis indicates binned signals (ppm) in the ^{1}H NMR spectrum.

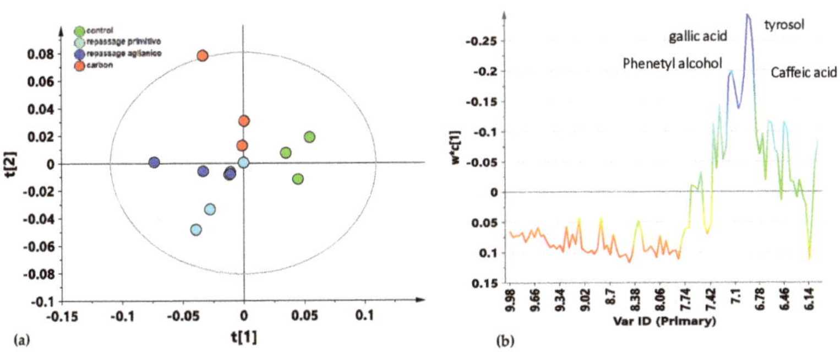

Figure 4. (**a**) Focused aromatic spectral region PLS-DA t [1]/t [2] score plot (five components, $R^2X = 0.92$, $R^2Y = 0.85$, $Q^2 = 0.56$) for the wine samples, classified according to the performed OTA decontamination treatment: Primitivo grape pomace repassage (sky-blue circle), Aglianico grape pomace repassage (blue circle), activated carbon (red circle) and control samples (green circles). (**b**) Metabolites responsible for the class separation could be observed in the model-related loading line plots, colored according to the correlation scaled coefficient. The *x*-axis indicates binned signals (ppm) in the ^{1}H NMR spectrum. Quality parameters (Correct Classification Rate), CCR; area under the Receiver Operating Characteristic (ROC) curve (AUC); intercepts of R^2 and Q^2 values on the *y*-axis; Fisher's probability index) of the model are reported in Table S3.

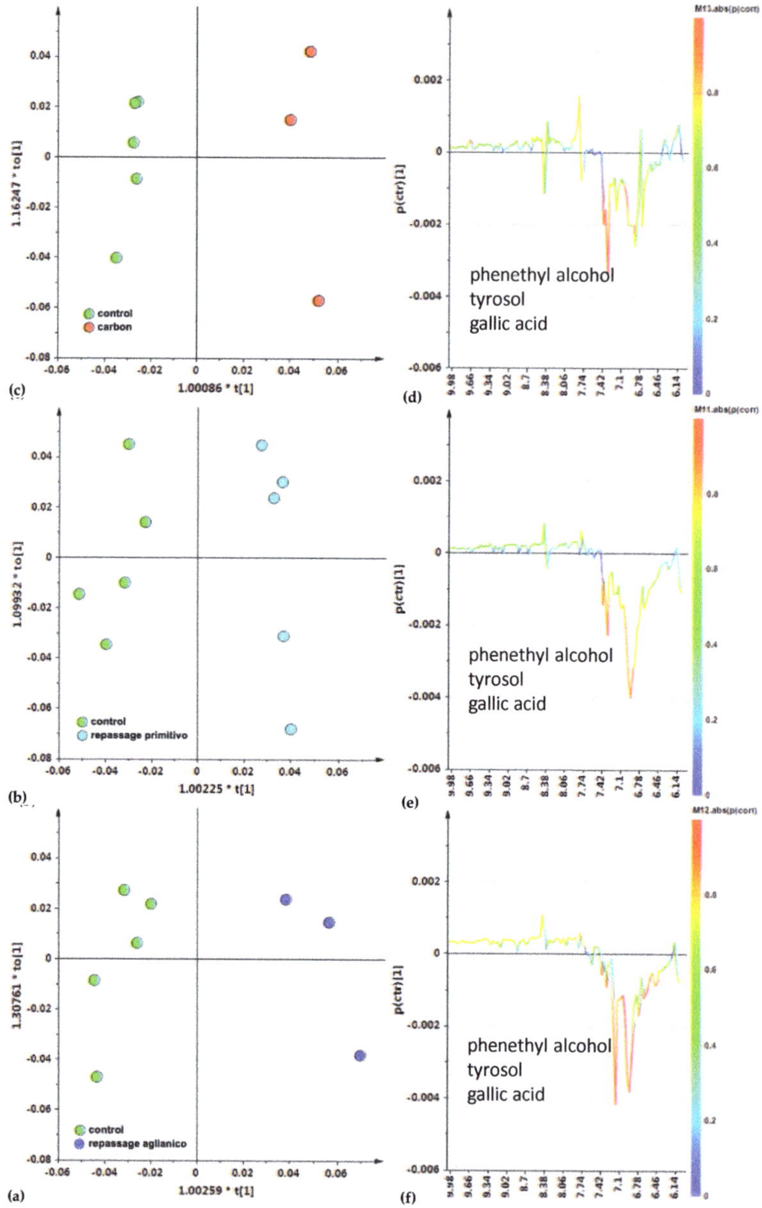

Figure 5. Focused aromatic spectral region OPLS-DA t [1]/to [1] score plot, for the control vs. (**a**) activated carbon (1 + 1 + 0; $R^2X = 0.69$, $R^2Y = 0.99$, $Q^2 = 0.96$); (**b**) Primitivo pomace (1 + 1 + 0; $R^2X = 0.79$, $R^2Y = 0.96$, $Q^2 = 0.90$) and (**c**) Aglianico pomace (1 + 1 + 0; $R^2X = 0.78$, $R^2Y = 0.94$, $Q^2 = 0.83$) wine samples. Sample symbols are colored according to the different OTA decontamination treatment: control, green circle; Aglianico, blue circles; and Primitivo green circles. Metabolites responsible for the class separation can be observed in the OPLS-DA model-related loading S-line plots (**d**–**f**), colored according to the correlation scaled coefficient. The *x*-axis indicates binned signals (ppm) in the ^{1}H NMR spectrum.

4. Discussion

Multivariate statistical analyses were applied to the NMR-based metabolomics data of the Primitivo wine samples. These resulted from OTA decontamination treatments based on repassage over Primitivo and/or Aglianico pomace and activated carbon treatments compared with untreated Primitivo wines (controls). The analyses allowed the characterization of the metabolic response to the decontamination treatment in the processed products. Relatively higher levels of specific metabolites, such as ethyl acetate and succinate, were found in the activated-carbon-treated samples compared with all the otherwise-processed samples, including the controls. Ethyl acetate, one of the main contributors to volatile acidity, is the most abundant ester in wine and it also constitutes an oxidation indicator, since it is responsible for acescency, a typical alteration of sensory properties [51]. However, at concentrations below 80 mg/L, it could also positively influence the wine's aroma [52–54]. Succinic acid is one of the main organic acids, formed during alcoholic fermentation, and its level varies between grape cultivars, with higher concentrations in red grape [55]. Succinic acid is one of the most important non-volatile acids in wine, and it was reported to be responsible for the largest part of the increase in titratable acidity [56,57]. Although its production can be affected by alcohol concentration, the observed succinic acid content in wines is very stable, since it does not change over aging. The organoleptic character of succinic acid has been described as sour with a salty, bitter taste. Due to its bitter, salty flavor, winemakers watch out for the levels of succinic acid in wine [57]. On the contrary, in all the grape-pomace-repassed wine samples, relatively higher contents of glycerol, lactate, tartaric, isobutanol, isopentanol, and phenolic compounds were found compared with the activated-carbon-treated samples. Like succinic acid, tartaric acid is one of the major contributors to wine acidity and it influences the perception of a tart taste in wine [58]. The level of tartaric acid is strictly related to the cultivar, ripening stage, and maturity, and it may be used as a biomarker for characterizing grape cultivars. Nevertheless, it should be also considered that its content can be significantly altered depending on the levels of potassium bitartrate and calcium tartrate formation, which precipitate in wines [59]. In addition, the precipitation process is also dependent on the winemaking conditions, including temperature, pH, and the concentration of calcium or potassium [60]. Also, lactic acid contributes to the wine total acidity and, unlike malic and tartaric acid, being a softer and milder acid, its level is associated with a creamier mouthfeel of wine [57]. Glycerol is one of the most abundant components of wine and its content has been associated with several attributes of taste, such as oiliness, persistence, and mellowness in the mouth [61]. Thus, glycerol plays a significant role in the organoleptic properties of wine, although an overproduction of this compound is usually linked with an acetic acid accumulation [60]. Finally, a relatively high content of isobutanol and isopentanol was selectively found in the control wine samples compared with the repassed wine samples. These molecules (isobutanol and isopentanol) were generally identified as wine higher alcohols and contribute to assessing the aromatic sensory perception of red wine aroma (fruity, flowery notes and aroma complexity) [62,63]. Finally, the multivariate analysis focused on the aromatic spectral region showed that, consistent with the literature data, carbon treatment induces a marked decrease in the polyphenol content of wine [64], thus affecting the flavor, color, and bouquet of the final product [65,66]. In fact, as described in the literature, the removal of OTA through the use of fining agents such as carbon is based on a non-specific adsorption mechanism, which causes a decrease in the content of phenols, flavonoids, total anthocyanins, and polymeric pigments [64]. Moreover, the presence of relatively higher levels of glycerol, lactate, tartaric, isobutanol, isopentanol, and phenols, observed in all the otherwise-processed samples, compared with the activated carbon treatment samples, strongly supports the possible use of pomace for OTA decontamination without significantly affecting the organoleptic properties of wine. Indeed, the ^{1}H NMR profile-based data analyses suggest that treatment with activated carbon is definitely more invasive compared with those samples treated with pomace for OTA removal. Although some decrease in specific metabolites was also observed compared with the controls for the pomace-based processes, the activated carbon treatment produced a general impoverishment of the metabolomic profiles of the studied wines. Moreover, as already reported in the literature, despite high efficiency, the OTA detoxification of wine, by the use of activated carbon fining adsorbents, causes a modification of organoleptic properties even when used at the recommended dosage range [4,64]. The alternative technique of repassage over pomace could be therefore be a successful method for the removal of OTA. Interestingly, although a certain degree of differentiation in the final product was observed, the wine samples obtained by repassage over the pomace from different grape varieties did not show remarkable differences. At the same time, it should be also highlighted that the usage of stabilized compared with fresh pomace did not show major differences, suggesting the possible use of stored material as a useful tool for modern wineries.

5. Conclusions

In this work, a ^{1}H-NMR-based metabolomic method was applied to a wine data set with the aim of identifying any possible variation in the chemical profiles after different OTA removal treatments. In particular, the untreated (control) Primitivo wine sample metabolic profiles were compared with Primitivo fresh/stabilized and Aglianico grape-pomace-repassed and activated-carbon-treated wines. As shown in the results of the statistical analysis, the metabolic profiles of the activated-carbon-treated wine samples exhibited a decrease in important quality-related metabolites such as glycerol, lactic, and tartaric acids, isobutanol/pentanol, and polyphenols compared with the untreated and grape-pomace-repassed wine samples. Moreover, although a lower content of phenolic compounds was observed in the pomace-based processed wines compared with the control samples, our findings suggest the repassage over pomace could be a successful method for efficient OTA removal without significantly altering the chemical profile and related oenological characteristics. Thus, the alternative technique of repassage over pomace may be proposed as a natural adsorbing/decontaminating material for the removal of OTA.

Supplementary Materials: The following supporting information can be downloaded at: https://www.mdpi.com/article/10.3390/foods14050734/s1, Figure S1. Prototype scheme (a,b) photo of must/wine decontamination. Figure S2. ^{1}H NMR spectrum (noesygppsld.compl pulse program) of (a–c) control; (d–f) Primitivo fresh grape pomace; (g,h) Primitivo stabilized grape pomace; (i–k) Aglianico fresh grape pomace; (l–n) Activated-carbon-treated wine samples. Table S1. Mean and standard deviation of total polyphenols (gallic acid content) and anthocyanins (malvin chloride) of wine samples (mg/L). Table S2. ^{1}H-NMR signals (chemical shifts) of assigned metabolites in wine spectra. Table S3. Performance and validation results to assess accuracy and robustness of PLS-DA model (Figure 4) for discrimination of untreated and treated wine samples, focusing on aromatic spectral regions.

Author Contributions: Conceptualization, M.S. and F.P.F.; methodology, L.D.C. and F.P.F.; software, L.D.C. and F.P.F.; validation, L.G., M.S. and F.P.F.; formal analysis, L.D.C. and L.G.; investigation, L.D.C., L.G. and F.P.F.; data curation M.S, L.G., C.R.G., F.P.F. and L.D.C.; writing—original draft preparation, L.D.C. and F.P.F.; writing—review and editing, C.R.G., M.S. and F.P.F.; visualization, C.R.G. and F.P.F.; supervision, M.S. and F.P.F.; project administration, M.S. All authors have read and agreed to the published version of the manuscript.

Funding: This research received no external funding.

Institutional Review Board Statement: Not applicable.

Informed Consent Statement: Not applicable.

Data Availability Statement: The original contributions presented in this study are included in the article/Supplementary Materials. Further inquiries can be directed to the corresponding author.

Acknowledgments: C.R.G. and F.P.F. thank the WINPROAGE project Wine by-products metabolomic profiling in Natural chemoPRevention Of vascular AGEings—(P2022TKYNN); PRIN 2022 PNRR-Call (D.D. 1409 of 14/09/2022) of the Italian Ministry of University and Research (MUR).

Conflicts of Interest: The authors declare no conflicts of interest.

References

1. Cosme, F.; Ribeiro, M.; Filipe-Ribeiro, L.; Nunes, M.F. Review of Mycotoxins in Grapes and Grape Products. In *Grapes and Grape Products—Chemistry and Technology*; IntechOpen: Rijeka, Croatia, 2024.
2. Pfohl-Leszkowicz, A.; Manderville, R.A. An Update on Direct Genotoxicity as a Molecular Mechanism of Ochratoxin a Carcinogenicity. *Chem. Res. Toxicol.* **2012**, *25*, 252–262. [CrossRef] [PubMed]
3. IARC Working Group on the Evaluation of Carcinogenic Risks to Humans, International Agency for Research on Cancer, and World Health Organization. *Some Naturally Occurring Substances: Food Items and Constituents, Heterocyclic Aromatic Amines and Mycotoxins*; International Agency for Research on Cancer, Ed.; IARC monographs on the evaluation of the carcinogenic risk of chemicals to humans; IARC: Lyon, France, 1993; ISBN 978-92-832-1256-0.
4. Gil-Serna, J.; Vázquez, C.; González-Jaén, M.; Patiño, B. Wine Contamination with Ochratoxins: A Review. *Beverages* **2018**, *4*, 6. [CrossRef]
5. Liu, J.; Wang, Y.; Cui, J.; Xing, L.; Shen, H.; Wu, S.; Lian, H.; Wang, J.; Yan, X.; Zhang, X. Ochratoxin A Induces Oxidative DNA Damage and G1 Phase Arrest in Human Peripheral Blood Mononuclear Cells in Vitro. *Toxicol. Lett.* **2012**, *211*, 164–171. [CrossRef] [PubMed]
6. La Placa, L.; Tsitsigiannis, D.; Camardo Leggieri, M.; Battilani, P. From Grapes to Wine: Impact of the Vinification Process on Ochratoxin A Contamination. *Foods* **2023**, *12*, 260. [CrossRef]
7. Opinion of the Scientific Panel on Contaminants in the Food Chain [CONTAM] Related to Ochratoxin A in Food | EFSA. Available online: https://www.efsa.europa.eu/en/efsajournal/pub/365 (accessed on 5 July 2024).
8. *Commission Regulation (EU) 2023/915 of 25 April 2023 on Maximum Levels for Certain Contaminants in Food and Repealing Regulation (EC) No 1881/2006 (Text with EEA Relevance)Text with EEA Relevance*; European Union: Brussels, Belgium, 2024.
9. Ding, L.; Han, M.; Wang, X.; Guo, Y. Ochratoxin A: Overview of Prevention, Removal, and Detoxification Methods. *Toxins* **2023**, *15*, 565. [CrossRef]
10. Wang, L.; Wang, Q.; Wang, S.; Cai, R.; Yuan, Y.; Yue, T.; Wang, Z. Bio-Control on the Contamination of Ochratoxin A in Food: Current Research and Future Prospects. *Curr. Res. Food Sci.* **2022**, *5*, 1539–1549. [CrossRef]
11. Petruzzi, L.; Sinigaglia, M.; Corbo, M.R.; Campaniello, D.; Speranza, B.; Bevilacqua, A. Decontamination of Ochratoxin A by Yeasts: Possible Approaches and Factors Leading to Toxin Removal in Wine. *Appl. Microbiol. Biotechnol.* **2014**, *98*, 6555–6567. [CrossRef]
12. Behfar, M.; Heshmati, A.; Mehri, F.; Khaneghah, A.M. Removal of Ochratoxin A from Grape Juice by Clarification: A Response Surface Methodology Study. *Foods* **2022**, *11*, 1432. [CrossRef]
13. Solfrizzo, M.; Avantaggiato, G.; Panzarini, G.; Visconti, A. Removal of Ochratoxin A from Contaminated Red Wines by Repassage over Grape Pomaces. *J. Agric. Food Chem.* **2010**, *58*, 317–323. [CrossRef]
14. Visconti, A.; Perrone, G.; Cozzi, G.; Solfrizzo, M. Managing Ochratoxin A Risk in the Grape-Wine Food Chain. *Food Addit. Contam. Part A* **2008**, *25*, 193–202. [CrossRef]
15. Filipe-Ribeiro, L.; Milheiro, J.; Matos, C.C.; Cosme, F.; Nunes, F.M. Reduction of 4-ethylphenol and 4-ethylguaiacol in Red Wine by Activated Carbons with Different Physicochemical Characteristics: Impact on Wine Quality. *Food Chem.* **2017**, *229*, 242–251. [CrossRef]
16. Solfrizzo, M.; Perrone, G.; Piemontese, L.; Zivoli, R.; Ferrara, M.; Gambacorta, L.; Magistà, D.; Fracchiolla, R.; Leo, R. *Process and Plant for the Treatment of Musts and Wines*; Consiglio Nazionale delle Ricerche e Industrie Fracchiolla SpA: Bari, Italy, 2016.
17. Guerrero, R.F.; Smith, P.; Bindon, K.A. Application of Insoluble Fibers in the Fining of Wine Phenolics. *J. Agric. Food Chem.* **2013**, *61*, 4424–4432. [CrossRef]
18. Jiménez-Martínez, M.D.; Gómez-Plaza, E.; Molero, N.; Bautista-Ortín, A.B. Fining of Red Wines with Pomace Cell Wall Material: Effect on Wine Phenolic Composition. *Food Bioprocess Technol.* **2017**, *10*, 1531–1539. [CrossRef]
19. Jiménez-Martínez, M.D.; Gil-Muñoz, R.; Gómez-Plaza, E.; Bautista-Ortín, A.B. Performance of Purified Grape Pomace as a Fining Agent to Reduce the Levels of Some Contaminants from Wine. *Food Addit. Contam. Part A* **2018**, *35*, 1061–1070. [CrossRef] [PubMed]

20. Niculescu, V.-C.; Ionete, R.-E. An Overview on Management and Valorisation of Winery Wastes. *Appl. Sci.* **2023**, *13*, 5063. [CrossRef]
21. Calò, F.; Girelli, C.R.; Wang, S.C.; Fanizzi, F.P. Geographical Origin Assessment of Extra Virgin Olive Oil via NMR and MS Combined with Chemometrics as Analytical Approaches. *Foods* **2022**, *11*, 113. [CrossRef] [PubMed]
22. Colì, C.S.; Girelli, C.R.; Cesari, G.; Hussain, M.; Fanizzi, F.P. Biodynamic, Organic and Integrated Agriculture Effects on cv. Italia Table Grapes Juice, over a 3-Year Period Experiment: An ^{1}H NMR Spectroscopy-Based Metabolomics Study. *Chem. Biol. Technol. Agric.* **2024**, *11*, 35. [CrossRef]
23. Cevallos-Cevallos, J.M.; Reyes-De-Corcuera, J.I.; Etxeberria, E.; Danyluk, M.D.; Rodrick, G.E. Metabolomic Analysis in Food Science: A Review. *Trends Food Sci. Technol.* **2009**, *20*, 557–566. [CrossRef]
24. Spraul, M.; Schütz, B.; Humpfer, E.; Mörtter, M.; Schäfer, H.; Koswig, S.; Rinke, P. Mixture analysis by NMR as Applied to Fruit Juice Quality Control. *Magn. Reson. Chem.* **2009**, *47*, S130–S137. [CrossRef]
25. Girelli, C.R.; Papadia, P.; Pagano, F.; Miglietta, P.P.; Fanizzi, F.P.; Cardinale, M.; Rustioni, L. Metabolomic NMR Analysis and Organoleptic Perceptions of Pomegranate Wines: Influence of Cultivar and Yeast on the Product Characteristics. *Heliyon* **2023**, *9*, e16774. [CrossRef]
26. Mao, I.L.; Costa, G.D.; Richard, T. ^{1}H-NMR Metabolomics for Wine Screening and Analysis. *OENO One* **2023**, *57*, 15–31. [CrossRef]
27. Mascellani, A.; Hoca, G.; Babisz, M.; Krska, P.; Kloucek, P.; Havlik, J. 1H NMR Chemometric Models for Classification of Czech Wine Type and Variety. *Food Chem.* **2021**, *339*, 127852. [CrossRef] [PubMed]
28. Bambina, P.; Spinella, A.; Lo Papa, G.; Chillura Martino, D.F.; Lo Meo, P.; Cinquanta, L.; Conte, P. 1H-NMR Spectroscopy Coupled with Chemometrics to Classify Wines According to Different Grape Varieties and Different Terroirs. *Agriculture* **2024**, *14*, 749. [CrossRef]
29. Serio, F.; Girelli, C.R.; Acito, M.; Imbriani, G.; Sabella, E.; Moretti, M.; Fanizzi, F.P.; Valacchi, G. Preliminary Characterization of "Salice Salentino" PDO Wines from Salento (South Italy) Negroamaro Grapes: NMR-Based Metabolomic and Biotoxicological Analyses. *Foods* **2024**, *13*, 3554. [CrossRef] [PubMed]
30. Gerginova, D.; Simova, S. Chemical Profiling of Wines Produced in Bulgaria and Distinction from International Grape Varieties. *ACS Omega* **2023**, *8*, 18702–18713. [CrossRef]
31. Viskić, M.; Bandić, L.M.; Korenika, A.-M.J.; Jeromel, A. NMR in the Service of Wine Differentiation. *Foods* **2021**, *10*, 120. [CrossRef]
32. Solovyev, P.A.; Fauhl-Hassek, C.; Riedl, J.; Esslinger, S.; Bontempo, L.; Camin, F. NMR Spectroscopy in Wine Authentication: An Official Control Perspective. *Compr. Rev. Food Sci. Food Saf.* **2021**, *20*, 2040–2062. [CrossRef]
33. Pittari, E.; Moio, L.; Piombino, P. Screening of the Volatile Composition and Olfactory Properties of Aglianico and Primitivo, Two Southern Italian Red Wines. *Appl. Sci.* **2023**, *13*, 2165. [CrossRef]
34. International Code of Oenological Practices | OIV. Available online: https://www.oiv.int/standards/international-code-of-oenological-practices (accessed on 9 July 2024).
35. Singleton, V.L.; Rossi, J.A. Colorimetry of Total Phenolics with Phosphomolybdic-Phosphotungstic Acid Reagents. *Am. J. Enol. Vitic.* **1965**, *16*, 144–158. [CrossRef]
36. Amargianitaki, M.; Spyros, A. NMR-Based Metabolomics in Wine Quality Control and Authentication. *Chem. Biol. Technol. Agric.* **2017**, *4*, 9. [CrossRef]
37. Losciale, P.; Conti, L.; Seripierri, S.; Alba, V.; Mazzone, F.; Rustioni, L.; di Leo, G.; Tarricone, F.; Tarricone, L. Effect of Different Deficit Irrigation Regimes on Vine Performance, Grape Composition and Wine Quality of the "Primitivo" Variety Under Mediterranean Conditions. *Irrig. Sci.* **2024**, *42*, 877–890. [CrossRef]
38. Godelmann, R.; Fang, F.; Humpfer, E.; Schütz, B.; Bansbach, M.; Schäfer, H.; Spraul, M. Targeted and Nontargeted Wine Analysis by 1H NMR Spectroscopy Combined with Multivariate Statistical Analysis. Differentiation of Important Parameters: Grape Variety, Geographical Origin, Year of Vintage. *J. Agric. Food Chem.* **2013**, *61*, 5610–5619. [CrossRef] [PubMed]
39. Monakhova, Y.B.; Schäfer, H.; Humpfer, E.; Spraul, M.; Kuballa, T.; Lachenmeier, D.W. Application of Automated Eightfold Suppression of Water and Ethanol Signals in ^{1}H NMR to Provide Sensitivity for Analyzing Alcoholic Beverages. *Magn. Reson. Chem. MRC* **2011**, *49*, 734–739. [CrossRef] [PubMed]
40. López-Rituerto, E.; Savorani, F.; Avenoza, A.; Busto, J.H.; Peregrina, J.M.; Engelsen, S.B. Investigations of La Rioja Terroir for Wine Production Using ^{1}H NMR Metabolomics. *J. Agric. Food Chem.* **2012**, *60*, 3452–3461. [CrossRef]
41. Mulder, F.A.A.; Tenori, L.; Licari, C.; Luchinat, C. Practical Considerations for Rapid and Quantitative NMR-Based Metabolomics. *J. Magn. Reson.* **2023**, *352*, 107462. [CrossRef]
42. Vignoli, A.; Ghini, V.; Meoni, G.; Licari, C.; Takis, P.G.; Tenori, L.; Turano, P.; Luchinat, C. High-Throughput Metabolomics by 1D NMR. *Angew. Chem. Int. Ed.* **2019**, *58*, 968–994. [CrossRef]
43. van den Berg, R.A.; Hoefsloot, H.C.; Westerhuis, J.A.; Smilde, A.K.; van der Werf, M.J. Centering, Scaling, and Transformations: Improving the Biological Information Content of Metabolomics Data. *BMC Genom.* **2006**, *7*, 142. [CrossRef]
44. Trygg, J.; Wold, S. Orthogonal projections to latent structures (O-PLS). *J. Chemom.* **2002**, *16*, 119–128. [CrossRef]

45. Eriksson, L.; Byrne, T.; Johansson, E.; Trygg, J.; Vikström, C. *Multi-and Megavariate Data Analysis Basic Principles and Applications*; Umetrics Academy: Umeå, Sweden, 2013; Volume 1, ISBN 91-973730-5-2.
46. Wheelock, Å.M.; Wheelock, C.E. Trials and Tribulations of 'Omics Data Analysis: Assessing Quality of SIMCA-Based Multivariate Models Using Examples from Pulmonary Medicine. *Mol. Biosyst.* **2013**, *9*, 2589. [CrossRef]
47. Triba, M.N.; Le Moyec, L.; Amathieu, R.; Goossens, C.; Bouchemal, N.; Nahon, P.; Rutledge, D.N.; Savarin, P. PLS/OPLS Models in Metabolomics: The Impact of Permutation of Dataset Rows on the K-Fold Cross-Validation Quality Parameters. *Mol. Biosyst.* **2015**, *11*, 13–19. [CrossRef]
48. Girelli, C.R.; Del Coco, L.; Papadia, P.; De Pascali, S.A.; Fanizzi, F.P. Harvest Year Effects on Apulian EVOOs Evaluated by ^{1}H NMR Based Metabolomics. *PeerJ* **2016**, *4*, e2740. [CrossRef] [PubMed]
49. De Pascali, S.A.; Coletta, A.; Del Coco, L.; Basile, T.; Gambacorta, G.; Fanizzi, F.P. Viticultural Practice and Winemaking Effects on Metabolic Profile of Negroamaro. *Food Chem.* **2014**, *161*, 112–119. [CrossRef] [PubMed]
50. Worley, B.; Powers, R. PCA as a Practical Indicator of OPLS-DA Model Reliability. *Curr. Metabolomics* **2016**, *4*, 97–103. [CrossRef] [PubMed]
51. Plata, C.; Millán, C.; Mauricio, J.C.; Ortega, J.M. Formation of Ethyl Acetate and Isoamyl Acetate by Various Species of Wine Yeasts. *Food Microbiol.* **2003**, *20*, 217–224. [CrossRef]
52. Hopfer, H.; Nelson, J.; Ebeler, S.E.; Heymann, H. Correlating Wine Quality Indicators to Chemical and Sensory Measurements. *Molecules* **2015**, *20*, 8453–8483. [CrossRef]
53. Paolini, M.; Tonidandel, L.; Larcher, R. Development, Validation and Application of a Fast GC-FID Method for the Analysis of Volatile Compounds in Spirit Drinks and Wine. *Food Control* **2022**, *136*, 108873. [CrossRef]
54. *Wine Flavours, Faults and Taints*; The Australian Wine Research Institute: Adelaide, Australia. Available online: https://www.awri.com.au/industry_support/winemaking_resources/sensory_assessment/recognition-of-wine-faults-and-taints/wine_faults/ (accessed on 5 July 2024).
55. Ali, K.; Maltese, F.; Toepfer, R.; Choi, Y.H.; Verpoorte, R. Metabolic Characterization of Palatinate German White Wines According to Sensory Attributes, Varieties, and Vintages using NMR Spectroscopy and Multivariate Data Analyses. *J. Biomol. NMR* **2011**, *49*, 255–266. [CrossRef]
56. Coulter, A.D. Succinic Acid-How It Is Formed, What Is Its Effect on Titratable Acidity, and What Factors Influence Its Concentration in Wine? *Wine Ind. J.* **2004**, *19*, 16–25.
57. Chidi, B.S.; Bauer, F.F.; Rossouw, D. Organic Acid Metabolism and the Impact of Fermentation Practices on Wine Acidity: A Review. *S. Afr. J. Enol. Vitic.* **2018**, *39*, 1–15. [CrossRef]
58. Volschenk, H.; van Vuuren, H.J.J.; Viljoen-Bloom, M. Malic Acid in Wine: Origin, Function and Metabolism during Vinification. *S. Afr. J. Enol. Vitic.* **2006**, *27*, 123–136. [CrossRef]
59. Hu, B.; Yue, Y.; Zhu, Y.; Wen, W.; Zhang, F.; Hardie, J.W. Proton Nuclear Magnetic Resonance-Spectroscopic Discrimination of Wines Reflects Genetic Homology of Several Different Grape (*V. vinifera* L.) Cultivars. *PLoS ONE* **2015**, *10*, e0142840. [CrossRef] [PubMed]
60. Son, H.-S.; Hwang, G.-S.; Ahn, H.-J.; Park, W.-M.; Lee, C.-H.; Hong, Y.-S. Characterization of Wines from Grape Varieties Through Multivariate Statistical Analysis of ^{1}H NMR Spectroscopic Data. *Food Res. Int.* **2009**, *42*, 1483–1491. [CrossRef]
61. Lubbers, S.; Verret, C.; Voilley, A. The Effect of Glycerol on the Perceived Aroma of a Model Wine and a White Wine. *LWT Food Sci. Technol.* **2001**, *34*, 262–265. [CrossRef]
62. Sun, Y.; Zhang, T.; Lü, H.; Yu, Z.; Li, X. Effect of Added Sulphur Dioxide Levels on the Fermentation Characteristics of Strawberry Wine. *J. Inst. Brew.* **2016**, *122*, 446–451. [CrossRef]
63. Zhang, L.; Tao, Y.S.; Wen, Y.; Wang, H. Aroma Evaluation of Young Chinese Merlot Wines with Denomination of Origin. *S. Afr. J. Enol. Vitic.* **2013**, *34*, 46–53. [CrossRef]
64. Cosme, F.; Inês, A.; Silva, D.; Filipe-Ribeiro, L.; Abrunhosa, L.; Nunes, F.M. Elimination of Ochratoxin A from White and Red Wines: Critical Characteristics of Activated Carbons and Impact on Wine Quality. *LWT* **2021**, *140*, 110838. [CrossRef]
65. Varga, J.; Kocsubé, S.; Péteri, Z.; Vágvölgyi, C.; Tóth, B. Chemical, Physical and Biological Approaches to Prevent Ochratoxin Induced Toxicoses in Humans and Animals. *Toxins* **2010**, *2*, 1718–1750. [CrossRef]
66. Quintela, S.; Villarán, M.C.; López de Armentia, I.; Elejalde, E. Ochratoxin A removal in wine: A review. *Food Control* **2013**, *30*, 439–445. [CrossRef]

Disclaimer/Publisher's Note: The statements, opinions and data contained in all publications are solely those of the individual author(s) and contributor(s) and not of MDPI and/or the editor(s). MDPI and/or the editor(s) disclaim responsibility for any injury to people or property resulting from any ideas, methods, instructions or products referred to in the content.

foods

MDPI

Article

Relationship Between Health Benefit Perception Moderate Wine Consumption, Wine Label and Healthy Behaviour

Anita Silvana Ilak Peršurić [1], Ana Težak Damijanić [2] and Sanja Radeka [3,*]

[1] Department Economics and Agricultural Development, Institute of Agriculture and Tourism, 52440 Poreč, Croatia; anita@iptpo.hr
[2] Department Tourism, Institute of Agriculture and Tourism, 52440 Poreč, Croatia
[3] Department Agriculture and Nutrition, Institute of Agriculture and Tourism, 52440 Poreč, Croatia
* Correspondence: sanja@iptpo.hr; Tel.: +385-052-408339

Abstract: Moderate wine consumption is, generally, the focus of various medical studies, while consumer behaviour research does not specifically centres on moderation in wine consumption. Wine consumption in moderation is an important part of various healthy diets; still, consumers need to make informed choices when purchasing wine and the information printed on wine labels partially contributes to this process. Therefore, the main aims of this paper were to develop a scale for measuring perceptions of the health benefits associated with moderate wine consumption, and to test the effect of dietary habits and non-obligatory wine label information on the perception of the health benefits associated with moderate wine consumption. The data were collected on a sample of wine consumers who participated in an interdisciplinary experiment regarding the impact of moderate wine consumption on human health. Univariate, bivariate, and multivariate statistics were used. The consumers' socio-demographic characteristics were used as a starting point in the analysis because they influence wine consumption. Gender was identified as a consistently important variable in predicting the perception of health benefits associated with moderate wine consumption. Health behaviour was a significant predictor along with gender, but after introducing non-obligatory wine label information, its significance in explaining the dependent variable was diminished. The results suggest that a consumer's perception of the scale of moderate wine consumption is a unidimensional construct. Furthermore, the non-obligatory information on wine labels was identified and classified as either wine-related warnings or wine-related health benefits.

Keywords: consumers; wine; wine label; food; health benefits; Croatia

1. Introduction

Wine consumption has been an important part of different healthy diets [1–3], like the Mediterranean diet, for centuries [4], and it has been a particularly popular choice of beverage with a meal [5,6]. However, since wine is an alcoholic beverage and recent health-related studies have stressed both its negative and positive impacts on consumers' health, its consumption may be regarded with caution [7,8]. Nevertheless, moderate wine consumption is still considered a crucial part of the Mediterranean diet and other diets considered healthy [6,9–12], with the emphasis on the word "moderate". Although the literature does not strictly define what moderate wine consumption means [7,13,14], it generally means limiting wine consumption to a maximum of 300 mL per day. Moderate wine consumption may help in dealing with different diseases [13,15–20]. These benefits could prompt consumers to consume wine, and the information that wine producers put on wine labels may help them in their purchasing process. However, consumers are often confused about the positive influence that moderate wine consumption can have on their health [7]. Therefore, appropriate information should be provided for them, and this information should include information about the health benefits of wine. Furthermore, for consumers to make an informed choice, they also need to know the potential side effects, which may be considered to be wine-related warnings.

Current research regarding moderate wine consumption generally focuses on various medical studies [13,15–20], while research regarding healthy diets mostly mentions that moderate wine consumption is a part of a particular diet [9–12]. Furthermore, non-obligatory wine label information is rarely investigated [21–24]. Therefore, there is a lack of research focusing on how consumers perceive the health benefits of moderate wine consumption, especially in consumer research. Consequently, there is also a lack of research in the consumer research field about moderate wine consumption and healthy diets, and moderate wine consumption and non-obligatory wine label information. Therefore, the aims of this paper were to develop scales for measuring the perception of health benefits associated with moderate wine consumption and non-obligatory wine label information, and to test the effect of dietary habits and non-obligatory wine label information on the perception of health benefits associated with moderate wine consumption.

2. Literature Review

2.1. Moderate Wine Consumption and Health Benefits

Wine consumption is often a part of the human diet, especially in traditionally wine-producing countries [1–3], and it is usually consumed as a part of a meal [6]. However, excess alcohol intake has a negative impact on human health [13,22–24]; therefore, wine should be consumed in limited quantities, because moderate wine consumption is generally perceived as beneficial to one's health [23,25]. However, there are slight differences in descriptions of what moderate wine consumption means.

Moderate wine consumption, according to [13], refers to the intake of less than 15.0 g of alcohol per day, which is equivalent to about one drink. Contrary to their methodology, [7] accept the definition of moderate wine consumption as the consumption of 200 mL of wine for women and 300 mL of wine for men, which equates to around an average of 20 g or 30 g of pure alcohol, respectively. The latter definition might be considered a kind of compromise between the strict limitation provided by [13] and the findings of [7] that have determined that wine consumers generally consider one glass of wine moderate, but 38% of Italian wine consumers and 22% of Spanish wine consumers believed that two to three glasses are still moderate. This definition of "moderate" wine consumption is also supported by [14] who identified that 40% of Italian, 20% of Spanish, and 12% of French participants considered two glasses a day to be moderate wine consumption.

Although there are still some disagreements about what moderate wine consumption includes, health research focusing on wine's benefits has identified its different positive effects on human health [13,15–20]. Those effects are related to cognitive, metabolic, and cardiovascular diseases. Both men and women drinking light to moderate amounts of wine daily scored better on all cognitive tests compared to those respondents drinking other alcoholic beverages [15]. Moderate wine drinkers, especially women, demonstrated better cognitive scores compared to non-drinkers [13] and male wine drinkers had reduced a risk of Alzheimer's disease [16,26]. Additionally, medical research has identified the benefits of moderate wine consumption regarding different neurological problems [17,22–24], especially in protecting neuronal cells from damage, preventing neurodegenerative diseases, facilitating the treatment of depression, improving cognitive function, and increasing memory [27]. One study, conducted during the COVID-19 pandemic, based on the research performed by [28], identified that the anti-microbe and anti-bacterial properties of certain wine components have a positive impact on COVID-19-related health issues [29]. Furthermore, the positive effects of wine on cardiovascular diseases were confirmed due to the presence of polyphenols, mainly resveratrol, in wine [18]. Finally, some authors reported that moderate wine consumption may grant people a longer life expectancy, compared to consuming wine in excess or not consuming wine at all [19,20].

Various medical studies, to a certain degree, influence public opinion about wine consumption. Positive aspects are usually the basis for a perception of health benefits. The health benefits that consumers generally link to moderate wine consumption include different aspects such as lowering cholesterol levels and reducing the risk of cardiovascular diseases [7,30]. Furthermore, wine consumers believe that wine has a positive effect on atherosclerosis, hypertension, cancer, type 2 diabetes, and neurological disorders [8]. However, wine consumers had a slightly negative perception of how wine influences sleep [7].

2.1.1. Wine Consumption and Diet

The moderate consumption of wine is often a part of a healthy diet; however, the main question is what is a healthy diet. Different sources stress diets like the Mediterranean, DASH, MIND, and traditional Asian diets as healthy diets [9–12]. Some of them are characteristic of certain regions like the Mediterranean region and are well known worldwide. The primary prerequisites of all healthy diets are variety in the intake of food and the limitation of the quantities consumed. This often results in the minimisation of fats, sugar, processed and readymade foods, etc., and the encouragement of legume intake, fruit and vegetable intake, etc., in order to improve one's health [31–34]. Consumers practising one of these healthy diets should experience different health-related improvements by diminishing the risks related to obesity, high blood cholesterol, coronary diseases, and type two diabetes [31,35]. Healthy diets often include moderate wine consumption because moderate wine consumption contributes by a certain degree to the achievement of different health benefits [7,30]. For instance, when wine is a part of a healthy diet, it is consumed as a part of a meal [4,6,36], so its negative effects on human health are diminished [37].

2.1.2. Moderate Wine Consumption and Label Information

Consumers' food purchase-related choices are often based on information provided on product labels. Information on product labels is not unified worldwide [38]; however, all product labels must contain information that is mandatory, as defined by different countries [14,38]. The information presented on wine labels has a certain impact on consumers' intention to purchase and consume wine [30,39]. Wine labels often include information about the wine's extrinsic characteristics, i.e., the grape variety, region, country, vintage [40], and this information could be regarded as obligatory information; however, other aspects related to wine, like warnings about the harmful consequences of wine consumption [23] or stating the health benefits that moderate wine consumption can have on consumers health [25], are generally not mentioned as information on wine labels. Furthermore, there is no agreement worldwide about the information that should be stated on wine labels [23,41]. Nevertheless, the information presented on wine labels is important to consumers, although its importance and content might vary due to different consumer characteristics, like their socio-demographic characteristics [23,41,42]. However, nowadays consumers might have a problem understanding what is written on the label. Some authors [43] identified that consumers were confused about food labels in general; namely, they often had difficulties identifying whether some label content (especially nutritional components) is important or valuable to the overall population, but they had even more problems with food labels containing health claims.

2.1.3. Wine Consumption and Socio-Demographic Characteristics

In general, consumers' socio-demographic characteristics influence their food purchasing decision-making process, and the same applies to wine consumption [8,14,39,44]. The socio-demographic characteristics of consumers were identified as important in relation to their nutrition [14], their perception of wine as a product [25], wine label information [22–24,39], and wine consumption habits [45]. Regarding wine consumption, the sociodemographic characteristics of consumers are generally used as segmentation criteria [14,44,46]. Wine-related research has identified gender [8,14,24,25,41,44,45,47], age [8,23,25,30,39,42,45,48,49], education [8,24], occupation [8], and income [8,45] as the most relevant consumer characteristics. Considering consumers' gender, [23] determined that women are more interested in information on wine labels compared to men, while [25] identified that women are more likely to believe that sulphites in wine may give some people headaches and that men were prone to believing that white and rosé are healthier wine types. Age was also identified as an important demographic characteristic in relation to wine consumption; namely, [39] determined that Generation X reported higher levels of subjective wine knowledge compared to Generation Y, while the findings of [25] suggest that millennials were most likely to consider white, rosé, and sparkling wines as healthier wine types compared to the other age groups. Furthermore, [30] determined that those wine consumers who identified a few wine-related health benefits tended to be older. Income was a significant variable in predicting consumers' spending on wine [45], while occupation and income were differentiating factors among wine consumer segments identified by [8].

In summary, consumers' socio-demographic characteristics, their health behaviour, and the information on wine labels can influence consumers' wine consumption; therefore, this paper poses the following hypotheses:

H1: *Gender has a positive influence on moderate wine consumption.*

H2: *Age has a positive influence on moderate wine consumption.*

H3: *Education has a positive influence on moderate wine consumption.*

H4: *Income has a positive influence on moderate wine consumption.*

H5: *Profession has a positive influence on moderate wine consumption.*

H6: *Health behaviour has a positive influence on moderate wine consumption.*

H7: *Wine warnings have a negative influence on moderate wine consumption.*

H8: *The perception of wine-related health benefits has a positive influence on moderate wine consumption.*

3. Materials and Methods

3.1. Research Design

The research presented in this paper is part of a wider research project conducted as a part of the Vinum Sanum project financed by the Croatian Science Foundation (HRZZ), aiming to examine moderate wine consumption in relation to human health. The Vinum

Sanum project is a multidisciplinary scientific project that includes experts from the life sciences (agronomy and medicine) and social sciences and humanities (economy and psychology). Therefore, in the sample design process, the main characteristics related to both research fields were considered. The data collected from participants included variables regarding the participants' health (medicine), the results of a wine consumption experiment (agronomy), and their consumer behaviour (economy and psychology). The research was conducted in Istria County and Primorsko-goranska County in Croatia, and it included the local population. Data were collected during 2019 and 2020 by researchers. For each year, a minimum sample size of 200 participants was planned.

The selection of participants was influenced by medical and agronomy research requirements; namely, the following categories were excluded: people with physical and mental health problems, people with addiction problems, and pregnant and lactating women. Furthermore, the participants' health-related variables were collected before and after the application of the wine consumption experiment, and no interventions regarding the participants' diet or physical activity were carried out. The participants were recruited based on public announcements published in different printed and online media. They were asked to submit their application, and the first 200 applicants satisfying the previously mentioned requirements were selected. The researchers invited the selected participants, and the following tests were performed before they were accepted as research participants: a routine medical exam, blood tests, and an interview with a psychologist. Only healthy individuals were included in the research and placed in a control group or wine consumer group. They all signed a letter of consent after being informed about the research. The Ethical Committee approved this research. The participants consumed 200 mL of wine daily during mealtimes for six weeks. All participants were asked to maintain their usual dietary habits during the consumption period, and to abstain from other alcoholic beverages except for wines provided by the project team. During the medical part of the project, medical examinations and laboratory tests were conducted for all study participants. The medical examination included measuring the participant's weight, height, and waistline, and their hip width and BMI. Heart rate and systolic and diastolic blood pressure were measured, and an electrocardiogram (ECG) was obtained. Data regarding health, therapy, smoking, and alcohol consumption habits were recorded. Laboratory tests included routine biochemistry and haematology analysis and the determination of serotonin and dopamine concentrations.

3.2. Consumer Behaviour Research

The data presented in this paper were collected as part of consumer behaviour research in the field of economics, and they included only participants who regularly drank moderate amounts of wine and were older than 18 years of age, namely the wine consumer group. The participants were asked to fill in the questionnaire only once; that is, at the same as when the researchers conducted the previously mentioned medical exams and the interview with a psychologist.

The questionnaire containing 23 questions was divided into three sections—wine consumption-related questions (wine consumption motivation, wine consumption perception, wine-related information, and wine attributes), dietary habits, and the participants' socio-demographic characteristics. All the scales used in this research were constructed based on previous research (see Appendix A).

The moderate wine consumption scale measured the consumers' perception of the benefits that limited wine consumption has for their health. It was operationalized as a unidimensional construct and the items for measuring the health benefits of moderate wine consumption were generated based on research proposed by [7,12,13,15,16,50,51]. Initially, 10 items were generated. A Likert-type scale ranging from 1 to 5 (1—Not important at all,

2—Not important, 3—Neutral, 4—Important, 5—Very important) was used. Dietary habits were measured as a unidimensional construct, and the items were adopted from [10,12,51–55]. To measure dietary habits, a total of 17 items were used. A Likert-type scale ranging from 1 (totally not agree) to 5 (totally agree) was used. Questions about non-obligatory wine label information were drawn from [7,8,22–24,56–58], and they were measured using a scale from 1 to 5 (1—not important, 2—minimally important, 3—medium importance, 4—very important, 5—exceptionally important). This construct was operationalised as a two-dimensional construct; namely, four items measuring wine's health benefits and four items measuring wine-related warnings were generated. All constructs were first examined by a psychologist and then by three experts (agronomy, medicine, and economy) to achieve content adequacy, and later they were tested on 10 volunteers working at the Institute of Agriculture and Tourism, who were not directly related to the project, to examine the clarity of each statement [59].

3.3. Data Analysis

Data were processed using IBM SPSS software version 21 and included descriptive statistics (to provide a general description of the sample), bivariate statistics (correlation), and multivariate statistics (exploratory factor analysis to determine the dimensions of the constructs and regression analysis to test the relationships between the perception of health benefits associated with moderate wine consumption and dietary habits and non-obligatory wine label information). Prior to data processing, the individual items measuring the health benefits of moderate wine consumption, wine-related information, and dietary habits were examined by checking data accuracy, missing data, and distribution. Missing values were replaced using the Markov chain Monte Carlo (MCMC) method for item imputation. To identify potential dimensions or to determine the unidimensionality of constructs, exploratory factor analysis was applied, using the maximum likelihood method and Promax rotation with an eigenvalue of 1.00 or more to identify potential factors. Internal reliability was determined by computing Cronbach's alpha. The factors of the perception of health benefits associated with moderate wine consumption, non-obligatory wine label information, and dietary habits were calculated as a mean value for each respondent [60–62]. To test the influence of dietary habits and non-obligatory wine label information, a hierarchical regression was conducted. Correlation between independent variables was calculated prior to the performance of hierarchical regression. Hierarchical regression was chosen to examine the effect of dietary habits and non-obligatory wine label information on the perception of health benefits associated with moderate wine consumption [63]. Since socio-demographic characteristics were identified as factors influencing the wine consumers' behaviour, they were used as control variables. Dietary habits were introduced in the second step to determine the proportion of variance explained by the respondent's diet, and then non-obligatory wine label information variables were introduced to test their influence on the perception of health benefits associated with moderate wine consumption. To test for multicollinearity, the variance inflation factors (VIFs) were calculated.

4. Results

4.1. Sample Characteristics

The sample consisted of 357 moderate wine consumers identified as adequate to participate in the research. Regarding the participants' gender, the proportion of the female respondents was higher than that of male respondents (Table 1). The majority of the participants were 41 years of age or older. Based on their education, the participants were divided into two segments, elementary/secondary and tertiary (since less than 0.7% had only elementary education). The participants were divided into either an employee category or others (this category included students (7%), retirees (5%), and those not in employment (4.6%)). Income was measured in the former Croatian currency (kuna) and was later calculated in euros (therefore, the range is not a round number). The majority of the respondents had a personal monthly income below EUR 929.2.

Table 1. Socio-demographic features of participants.

N = 357				
Gender	**Age (Years)**	**Education**	**Occupation**	**Monthly Personal Income**
Female (61.7%) Male (38.3%)	≤40 (42.2%) ≥41 (57.8%)	Elementary and Secondary (25.6%) University (74.4%)	Employee (83.4%) Other (16.6%)	≤€929.1 (54.1%) ≥€929.2 (45.9%)

The relationships between wine consumption and food and diets (Table 2) were evaluated by participants with five grades that variated from the most negative (not at all) to the most positive (very important). About ¾ of the survey participants consumed wines mostly at home. Our findings on consumption frequency were similar to the pattern found in Spain by [44], where the home consumption pattern was the most frequent one (places of consumption: at home, restaurants, bars, and other places). The preferred wine type for the majority of participants was dry wines. A minority of participants preferred special wine types, e.g., sparkling and liqueur wines. Considering the frequency of wine consumption, every fifth participant consumed wine daily and one out of three consumed wine several times a week.

Table 2. Participants' wine consumption characteristics.

Wine Consumption					
Preferred wine type	Dry 56.8%	Semi-dry 22.8%	Semi-sweet 17.2%	Sparkling 2.4%	Liquor 0.8%
Consumption place	At home 74.2%	Restaurant 21.8%	While travelling 0.3%	In bars, nightclubs 3.5%	Other places 0.2%
Consumption frequency	Special Occasions 16.3%	Several times a year 4.1%	Several times a month 25.5%	Several times a week 32.2%	Daily 21.9%

Wine was evaluated in a positive sense (Table 3), as a beverage with positive health aspects; namely, as good for one's health, as a natural beverage, and as part of a healthy diet. In over sixty percent of cases, this evaluation was important or very important. In the evaluation of their diet (Table 3), the participants expressed their serious concerns about their food choices and particular food components. Their awareness about the influence of food on their health was evident and taking care of their own health was validated as highly important (six out of ten participants found it was important or very important). Consequently, seven out of ten participants evaluated the impacts of food on their health as very important. Their food choices revealed preferences for fresh food in their daily diet, consequently fresh food consumption was very important or important to three-quarters of the participants. In accordance with their fresh food choices, their cooking practices also revealed a healthy preference for cooking at home from scratch. Consequently, less than ten percent of participants regularly consumed prepared/readymade meals or opted for fast food in their diet.

Table 3. Participants' evaluation of food, food compounds, and wine as part of their diet.

Diet/Food Evaluation	I Do Not Agree at All (%)	I Do Not Agree (%)	I'm Neutral (%)	I Agree (%)	I Totally Agree (%)
Wine is a natural beverage	2.8	2.8	18.6%	40.0	35.8
Wine is good for one's health	1.9	4.4	23.3%	41.9	28.3
I consume wine as part of healthy diet	6.9	9.4	24.0	37.7	21.9
Wine has positive impact on health	2.8	7.5	23.6	40.3	25.8
I think food affects health	0.6	2.5	5.8	20.6	70.6
I take care of my health	3.1	7.8	28.3	44.5	16.2
I take care of daily calories intake in my diet	11.7	22.6	31.8	26.5	7.3
I take care of daily fat content in my diet	7.8	20.1	32.6	28.7	10.9
I take care of daily sugar content in my diet	6.1	14.8	30.4	33.0	15.6
I take care of daily salt intake in my diet	5.9	18.5	33.6	28.6	13.4
I take care of daily fibre content in my diet	5.6	16.9	40.7	27.5	9.3
I take care of daily vitamin content in my diet	5.9	13.8	34.0	33.1	13.2
I take care of daily additives content in my diet	3.9	10.1	27.0	30.6	28.4
I eat 3–5 meals a day	7.8	12.6	24.1	24.9	30.5
I eat whatever suits me in my diet	18.6	27.8	31.4	13.6	8.3
I prefer fresh food in my diet	2.8	2.8	18.6	40.0	35.8
I prefer to cook or myself	4.4	8.9	12.8	27.2	46.7
I eat often fast-food	58.3	25.3	9.7	5.6	1.1
I prefer pre-cooked/readymade food in my diet	48.9	27.5	13.1	6.4	4.2

Regarding particular compounds in food, such as the content of fats, sugar, salt, vitamins, and additives in food, the participants were divided, with half of them stating that these food compounds should be counted and controlled, about one-third finding this to be neither important nor unimportant, and the rest finding it to be of low or no importance.

The evaluation of diet, food components, and wine as part of a daily diet showed that the survey participants were highly aware of food, beverages, and their impact on health. The encompassed data also indicate that the sample was divided into the half who evaluated all statements highly, about a third who were indecisive, and the ones who found it to be of low importance. Consequently, it was presumed that certain socio-demographic features of the participants may have affected this division of the sample.

4.2. Scale Analysis

Exploratory factor analysis was performed to identify the dimensions of the perception of health benefits associated with moderate wine consumption and dietary habits and non-obligatory wine label information (Table 4). Items with a loading below 0.5 and cross-loadings were deleted, resulting in the retention of 26 items—nine items measuring the health benefits of moderate wine consumption (a unidimensional construct), eight items measuring dietary habits (a unidimensional construct), and nine items on non-obligatory wine label information (a two-dimensional construct—five items measuring wine-related warnings and five items measuring health benefit claims).

Table 4. Exploratory factor analysis.

Items	Mean	SD	F 1	F 2	F 3	F 4
MCW—positive influence on memory	3.3	1.14	0.909			
MCW—positive influence on heart health	3.6	1.07	0.891			
MCW—positive influence on blood vessels	3.7	1.04	0.875			
MCW—positive influence on cholesterol levels	3.3	1.14	0.869			
MCW—positive influence on blood sugar	3.2	1.16	0.815			
MCW—positive influence on weight	2.9	1.24	0.786			
MCW—positive influence on energy level	3.0	1.25	0.760			
MCW—positive influence on metabolism	3.2	1.16	0.712			
MCW—positive influence on neurovegetative diseases	3.3	1.16	0.635			
Pays attention to fat intake	3.1	1.12		0.856		
Pays attention to sugar intake	3.4	1.09		0.826		
Pays attention to fibre intake	3.1	0.99		0.782		
Pays attention to calorie intake	2.9	1.14		0.778		
Pays attention to salt intake	3.2	1.09		0.759		
Pays attention to vitamin intake	3.3	1.06		0.734		
Pays attention to additive intake	3.7	1.08		0.612		
Being health-conscious	3.6	0.95		0.607		
Label: Lower risk of heart disease	3.5	1.26			0.989	
Label: Reduces cholesterol levels	3.7	1.24			0.978	
Label: Positive influence on health	3.7	1.20			0.633	
Label: Reduces vascular disease	3.5	1.20			0.603	
Label: Contains antioxidative components	3.8	1.22			0.406	
Label: Do not consume with medicine	4.2	1.21				0.860
Label: Do not drink and drive	4.4	1.19				0.813
Label: Not for younger than 18	4.2	1.08				0.728
Label: Consume in moderate amounts	4.1	1.17				0.705
Label: Do not consume during pregnancy and breastfeeding	3.9	1.24				0.600
Eigenvalues			7.993	3.233	3.423	2.597
% variance			29.604	11.973	12.678	9.620
% cumulative variance			29.604	41.577	54.255	63.874
Cronbach's α			0.946	0.909	0.917	0.853

The first factor, F1, measured the perception of health benefits associated with moderate wine consumption. It included different questions regarding how moderate wine consumption can have a positive impact on human health. The items on the perception of health benefits associated with moderate wine consumption varied from moderate consumption having a positive influence on weight (2.9) to moderate consumption having a positive influence on blood vessels (3.7). The second factor, F2, measured dietary habits, and showed an active approach to food choices in daily diet, concerning both the harmful effects of excessive fats (3.1), salt (3.2), sugar (3.4), and additives (3.7), and the possible positive effects of vitamins (3.3) and fibre (3.1). Non-obligatory wine label information was divided into two dimensions. The third factor F3, included items that emphasised the positive impacts of wine consumption on human health, and it was labelled as health benefit claims, while the second factor, F4, contained items related to warnings, and was labelled as wine warnings. Survey participants evaluated the health benefit claims as generally important; namely, they varied from wine contains antioxidative components

(3.8) to wine reduces vascular diseases and lowers the risk of heart disease (3.5). However, respondents found wine warnings to be slightly more important; specifically, they rated the warning Do not drink and drive (4.4) as the most important, while Consume in moderate amounts (4.1) was the least important warning. Jointly, all factors accounted for 64.12% of the accumulated variance, and most of the factor loadings were greater than 0.60. Cronbach's alpha coefficients were between 0.850 and 0.952. The KMO of the data was 0.897, while the significance level of Bartlett's test was 0.000, which means the data were suitable for factor analysis. All of the items in the scales fall into corresponding factors. Cronbach's alpha coefficients were above 0.6, which showed the great reliability of all of the scales. Most of the factor loadings exceeded 0.6, suggesting that all of the scales had a high convergent validity. The correlation between an item and the total score (Table 5) was acceptable for all four scales [64].

Table 5. Internal correlation coefficient of test scale.

Variable	Measurement	Correlation Coefficient
Moderate wine consumption (MCW)	MCW—positive influence on memory	0.860 ***
	MCW—positive influence on heart health	0.849 ***
	MCW—positive influence on blood vessels	0.847 ***
	MCW—positive influence on cholesterol levels	0.883 ***
	MCW—positive influence on blood sugar	0.865 ***
	MCW—positive influence on weight	0.846 ***
	MCW—positive influence on energy level	0.829 ***
	MCW—positive influence on metabolism	0.815 ***
	MCW—positive influence on neurovegetative diseases	0.871 ***
Health behaviour (HB)	Pays attention to fat intake	0.850 ***
	Pays attention to sugar intake	0.816 ***
	Pays attention to fibre intake	0.824 ***
	Pays attention to calorie intake	0.800 ***
	Pays attention to salt intake	0.807 ***
	Pays attention to vitamin intake	0.826 ***
	Pays attention to additive intake	0.716 ***
	Being health-conscious	0.691 ***
Label—health benefit claims (LHBC)	Label: Lower risk of heart disease	0.881 ***
	Label: Reduces cholesterol levels	0.917 ***
	Label: Positive influence on health	0.865 ***
	Label: Reduces vascular disease	0.911 ***
	Label: Contains antioxidative components	0.838 ***
Label—warnings (LW)	Label: Do not consume with medicine	0.827 ***
	Label: Do not drink and drive	0.809 ***
	Label: Not for younger than 18	0.819 ***
	Label: Consume in moderate amounts	0.775 ***
	Label: Do not consume during pregnancy and breastfeeding	0.742 ***

Note: *** significant at $\alpha = 0.001$.

4.3. Model Analysis

To test the effect of dietary habits and non-obligatory wine label information on the perception of health benefits associated with moderate wine consumption, hierarchical regression analysis was performed. Prior to this analysis, descriptive statistics for all four constructs were generated and Pearson's correlation coefficients were calculated to check if the independent variables were highly correlated. Non-obligatory wine label warnings were the most important aspect (Table 6), but they were slightly more important than non-obligatory wine label health benefit claims. Additionally, the perception of health benefits associated with moderate wine consumption and dietary habits were also important, but they were less important to consumers than non-obligatory wine label information. All constructs used as independent variables were statistically significantly related, but the correlation was relatively low.

Table 6. The descriptive statistics of the constructs.

Variable	Mean	SD	Correlation			
			MCW	HB	LW	LHBC
MCW	3.4	0.95	1	0.189 ***	−0.001	0.612 ***
HB	3.4	0.81		1	0.127 *	0.306 ***
LW	3.7	1.06			1	0.280 ***
LHBC	4.2	0.89				1

Note: * significant at $\alpha = 0.05$. *** significant at $\alpha = 0.001$.

Hierarchical regression analysis was used to test the research hypothesis (Table 7). Variance inflation factors varied across the models, but they were between 1.033 and 1.315. Model 1 included control variables only; namely, consumers' socio-demographic characteristics that explained 4.5% of the variance of the perception of health benefits associated with moderate wine consumption. Gender was the only variable significant in predicting the perception of health benefits associated with moderate wine consumption, and the model was significant at the 0.01 level of significance. In the next step, health behaviour was included.

Table 7. Results of the regression analysis.

Variable	Model 1	Model 2	Model 3
Constant	2.888	2.448	2.018
Age	0.096	0.034	−0.046
Gender	−0.173 **	−0.196 ***	−0.134 **
Education	0.094	0.056	0.049
Income	0.007	0.006	−0.025
Profession	0.000	0.014	0.005
HB		0.199 ***	0.039
LHBC			−0.173 ***
LW			0.651 ***
R^2	0.045	0.080	0.428
Adjusted R^2	0.031	0.064	0.414
F	3.153 **	4.838 ***	30.862 ***
R^2 Change	0.045 **	0.035 ***	0.348 ***
F Change	3.153	12.708	100.256

Note: ** significant at $\alpha = 0.01$. *** significant at $\alpha = 0.001$.

Although the F test was significant, and health behaviour was also a significant variable, besides gender, the proportion of explained variance increased by 3.5%, increasing

R^2 to only 8%. However, adding health behaviour into this relationship increased the overall level of model significance. In the next step, non-obligatory wine label information was entered; namely, wine warnings and health benefit claims. These two factors increased the proportion of explained variance by 34.8%, increasing R^2 to 42.8%. After adding these two factors to the regression model, respondents' gender remained a significant predictor of the perception of health benefits associated with moderate wine consumption. However, health behaviour lost its significance in explaining the dependent variable.

Gender had a negative impact on the perception of health benefits associated with moderate wine consumption, suggesting that women were likely to perceive wine consumption, even if moderate, as not beneficial to human health. Since gender was identified as an important variable in relation to moderate wine consumption, its relationship with the independent variables was tested. Pearson's correlation coefficients indicated no statistically significant relationships between gender and health behaviour (0.096, sig. 0.077), label warnings (0.080, sig. 0.136), and label health benefits claims (−0.064, sig. 239). Health behaviour had a positive impact on the perception of health benefits associated with moderate wine consumption, suggesting that respondents who pay attention to their diet are more likely to perceive moderate wine consumption to be beneficial to human health. Despite the initial impact of health behaviour in model 2, after adding non-obligatory wine label information, health behaviour lost its significance. Both additional factors were significant; however, wine warnings had a negative impact on the perception of health benefits associated with moderate wine consumption, while health benefit claims had a positive impact on the perception of health benefits associated with moderate wine consumption.

5. Discussion

This paper explored wine the relationship between wine consumers' perceptions of moderate wine consumption, and its relationship to consumers' dietary habits and the importance they placed on non-obligatory wine label information. First, it assesses the applicability of four measurement scales (moderate wine consumption, health behaviour, wine warnings, and wine health benefit claims). Second, it empirically tests the impact of three different constructs of moderate wine consumption.

All ten items measuring the consumers' perception of moderate wine consumption loaded significantly onto the factor measuring moderate wine consumption, supporting the findings of [7,12,13,15,16,50,51]. Therefore, this study suggests the scale for measuring consumers' perception of the benefits of moderate wine consumption includes a comprehensive number of aspects related to the positive influence of wine on human health. Furthermore, the items measuring health behaviour and non-obligatory wine label information loaded significantly onto their factors, confirming the finding related to health behaviour [10,12,51–54,65] and non-obligatory wine label information [7,8,22–24,56–58].

By testing the effects of socio-demographic characteristics in relation to moderate wine consumption, this study revealed a strong influence of gender on perceptions of moderate wine consumption, confirming the findings of [8,14,24,25,41,44,45,47]. Namely, men had better opinions about the health benefits of moderate wine consumption compared to women. Health behaviour was a significant predictor along with gender; however, after introducing non-obligatory wine label information, its significance was diminished. These findings suggest that moderate wine consumption is considered an important part of healthy diets, supporting the findings of [9–12]. However, when non-obligatory wine label information was added, health behaviour was no longer a significant predictor of moderate wine consumption. Both factors representing non-obligatory wine label information were significant in terms of predicting moderate wine consumption, suggesting that non-obligatory wine label information is crucial in relation to consumers' perception of moderate wine consumption. These results may point to the fact that this information is used in the process forming an opinion on how wine influences human health. Therefore, communicating information on the package or label, especially any health benefits, was found to be important in relation to moderate wine consumption, confirming the findings of [22–25]. Contrary to the findings of [43], wine consumers were generally well informed about non-obligatory wine label information, recognising both positive and negative wine effects.

6. Conclusions

Moderate wine consumption is a part of various healthy diets, but wine consumers need to make informed choices and non-obligatory information on wine labels may help in this process. This paper explored the relationship between wine consumers' perception of moderate wine consumption, health benefits, socio-demographic characteristics, health behaviour, and wine labels. Wine has various health benefits, but only if consumed in moderation. Consumers in this study recognised the health benefits of moderate wine consumption and generally expressed their healthy behaviour. They also considered both non-obligatory wine label elements, namely health benefit claims and health warnings, to be important, although they placed a higher importance on health warnings. Gender played an important part in relation to consumers' perception of moderate wine consumption. On the other hand, healthy diet behaviour was an important factor when examined in relation to socio-demographic characteristics. However, when the non-obligatory wine label information variables were added to the equation, health behaviour lost its significance, suggesting that wine-related information plays a greater role in wine consumption compared to diet habits.

This study makes several contributions to the extant literature.

First, by considering the research proposed by [7,12,13,15,16,50,51], this paper proposes a scale for measuring consumers' perception of moderate wine consumption. Secondly, by considering the non-obligatory wine label information suggested by [7,8,22–24,56–58], a scale for measuring wine warnings and wine health benefits was developed. Thirdly, dietary habits adopted from [10,12,51–54,65] were tested and items measuring them were identified as unidimensional constructs. Finally, the influence of socio-demographic characteristics, dietary habits, and non-obligatory wine label information on moderate wine consumption were evaluated. The findings suggest that gender is a vital predictor of moderate wine consumption. Furthermore, dietary habits exhibit marginal importance in relation to moderate wine consumption, while non-obligatory wine label information plays a major role in predicting perceptions of the health benefits of moderate wine consumption.

This study's findings bring forth some implications for policymakers and wine producers. Moderate wine consumption is often a part of different diets generally considered to be healthy diets, like the Mediterranean diet. However, consumers must be better informed about what moderate consumption includes in terms of the daily recommended quantity. Therefore, promoting moderate wine consumption as a part of a healthy diet and its benefits on human health is recommended. Although participants in this research were well aware of the health benefit claims and wine-related warnings that can be a part of wine labels, the general population lacks knowledge in this area. Therefore, additional effort should be placed on raising the awareness of the general population in terms of the health benefits of moderate wine consumption, emphasising the word "moderate". At the same time, they should also be informed about the negative aspects of excessive wine consumption. Non-obligatory wine label information often includes the positive and negative effects of wine on human health. The positive effects of moderate wine consumption could be used to promote different types of functional wines, i.e., wines with increased levels of bioactive compounds. These functional wines could address issues like lowering certain health risks by emphasising different compounds present in wines, like various bioactive compounds with high antioxidative activity. For wine producers, the production of new types of functional wines with additional value enables the diversification of products which enhances their economic status and enforces their long-term competitiveness on the wine market. Raising awareness of the effects of healthy choices of foods and beverages that provide good health, diminish health risks, and prevent diseases is becoming increasingly significant for the general public and also in a scientific sense.

There are a few limitations to this study. The sample consisted of only healthy participants, i.e., neither had any cardiovascular or neurovegetative disease. Future research could include participants with various health issues. The participants were recruited through public announcements published in different media; therefore, the results cannot be generalised to all wine consumers. Future research could test the scales proposed in this study (moderate wine consumption, dietary habits, and non-obligatory wine label information) on wine consumers. This research examined the influence of dietary habits and non-obligatory wine label information on the health benefits of moderate wine consumption. Dietary habits were identified as an important factor in relation to the health benefits of moderate wine consumption. However, they lost their importance when non-obligatory wine label information was added. Since the dietary habits of the participants were not modified for this study, future research could examine in detail the role of dietary habits in general, along with wine's health benefits and wine-related warnings. Due to the sample design limitations imposed by the multidisciplinary nature of the project, the scales for measuring moderate wine consumption and non-obligatory wine label information were not verified through confirmatory factor analysis. Future research should focus on fixing this issue by testing them on bigger and if possible cross-national samples. Furthermore, future research can centre on applying the findings presented in this paper by extending various consumer behaviour-related theories.

Author Contributions: A.S.I.P.; conceptualization, investigation, data curation, writing—original draft preparation, writing—review and editing; A.T.D.; software, validation, formal analysis, writing—original draft preparation, writing—review and editing. S.R.; project resources, project management, writing—review. All authors have read and agreed to the published version of the manuscript.

Funding: This work has been supported by the Croatian Science Foundation under the project "Influence of different vinification technologies on the qualitative characteristics of wines from Croatian autochthonous varieties: the role of Wine in human diet VINUM SANUM IP-2018-01-5049.

Institutional Review Board Statement: The study was carried out in accordance with the Declaration of Helsinki developed by the World Medical Association and was approved by the Ethics Committee of the Clinical Hospital Centre Rijeka (Croatia). The study was conducted in accordance with the Declaration of Helsinki, approval of Ethics Committee of KBC Rijeka, nr. (003-05/18-1/13 on 6 February 2018).

Informed Consent Statement: A written informed consent was obtained from all subjects involved in the study.

Data Availability Statement: The original contributions presented in this study are included in the article. Further inquiries can be directed to the corresponding author.

Acknowledgments: We are grateful to all the participants for investing their time and efforts during the survey.

Conflicts of Interest: The authors declare no conflicts of interest.

Appendix A

Table A1. Dimensions of the perception of health benefits perception of associated with moderate wine consumption and dietary habits and non-obligatory wine label information.

Variables	Description	Author	Year
MCW—positive influence on memory	Better cognitive tests for wine consumption Red wine as protective for women	[15] [16]	2011 2018
MCW—positive influence on heart health	Med, Dash diet, and MWC of red wine as protective factors Resveratrol and heart failure Effects on cardiovascular diseases	[50] [7]	2017 2017
MCW—positive influence on blood vessels	Resveratrol and vascular inflammation Effects on cardiovascular diseases	[50] [7]	2017 2015
MCW—positive influence on cholesterol levels	Resveratrol and oxidative stress Resveratrol treatments	[50]	2017
MCW—positive influence on blood sugar	Med, Dash diet, and MWC of red wine as protective factors Limitation of sugar in diet	[12]	2015
MCW—positive influence on weight	Flavonoid polyphenols in metabolic diseases and inflammation Antioxidant properties of polyphenols have also been shown to be effective in metabolic diseases Effects on obesity	[50] [7]	2017 2017
MCW—positive influence on energy level	Functional foods make people more energetic	[51]	2020
MCW—positive influence on metabolism	Consuming polyphenols from fruit and red wine	[12]	2015
MCW—positive influence on neurovegetative diseases	Better cognitive tests for wine consumption Women, cognitive scores, impairment, decline Red wine as protective for women Red wine reduced the risk of AD in men	[15]. [13] [16]	2011 2005 1999
Pays attention to fat intake	Limitation of fats, e.g., saturated Non-saturated fat intake (olive oil)	[10]	1999
Pays attention to sugar intake	Limitation of sugar in diet, Med diet positive for diabetes	[12]	2015
Pays attention to fibre intake	Adding fibre into everyday diet, eating a balanced diet	[54]	2019
Pays attention to calorie intake	Food choice questionnaire Diet proposals and nutrient substitutions	[55] [52]	1998 2020
Pays attention to salt intake	Food choice questionnaire Reduction of sodium Sodium and hypertension	[55] [52]	1998 2020
Pays attention to vitamin intake	Eating a balanced diet Vitamins as food supplements	[54] [66]	2019 2014
Pays attention to additive intake	Other substances use Xylitol Eating food without additives	[52] [51]	2020 2020
Being health-conscious	Conscious choices of foods in diet, Mediterranean, DASH, and AHEI diets as choices for healthy life	[12] [53]	2015 2017
Label: Lower risk of heart disease	Health impact with eco/environmental label	[7]	2017
Label: Reduces cholesterol levels	Better willingness to pay for wine with health aspects	[3]	2006
Label: Positive influence on health	Health aspects and WTP	[56]	2008
Label: Reduces vascular disease	Wine and health Olive oil and heart health	[8] [67]	2016 2014
Label: Contains antioxidative components	Wine and health	[8]	2016
Label: Do not consume with medicine	Opinion on label—do not consume with medicine	[21]	2017
Label: Do not drink and drive	Opinion—after drinking do not drive	[21]	2017
Label: Not for younger than 18	Ban for children under 18, concern of overconsumption	[21] [58]	2017 1999
Label: Consume in moderate amounts	Trilateral survey highlighted what MWC is	[7] [58]	2017 1999
Label: Do not consume during pregnancy and breastfeeding	In France, health warning on labels is obligatory	[7]	2017

References

1. Trevisan, M.; Schisterman, E.; Mennotti, A.; Farchi, G.; Conti, S. Drinking Pattern and Mortality: The Italian Risk Factor and Life Expectancy Pooling Project. *Ann. Epidemiol.* **2001**, *11*, 312–319. [CrossRef] [PubMed]
2. Rossi, A.; Fusco, F. Wine Index of Salubrity and Health (WISH): An Evidence-Based Instrument to Evaluate the Impact of Good Wine on Well-Being. *Int. J. Wine Res.* **2019**, *11*, 23–37. [CrossRef]
3. Bell, S.; Daskalopoulou, M.; Rapsomaniki, E.; George, J.; Britton, A.; Bobak, M.; Casas, J.P.; Dale, C.E.; Denaxas, S.; Shah, A.D.; et al. Association between Clinically Recorded Alcohol Consumption and Initial Presentation of 12 Cardiovascular Diseases: Population Based Cohort Study Using Linked Health Records. *BMJ* **2017**, *356*, j909. [CrossRef]
4. Minzer, S.; Estruch, R.; Casas, R.; Ding, C.; O'Neill, D.; Bell, S.; Stamatakis, E.; Britton, A.; Zatońska, K.; Psikus, P.; et al. Wine Intake in the Framework of a Mediterranean Diet and Chronic Non-Communicable Diseases: A Short Literature Review of the Last 5 Years. *Molecules* **2020**, *25*, 5045. [CrossRef] [PubMed]
5. Brunner, T.A.; Siegrist, M. A Consumer-oriented Segmentation Study in the Swiss Wine Market. *Br. Food J.* **2011**, *113*, 353–373. [CrossRef]
6. Fiore, M.; Alaimo, L.S.; Chkhartishvil, N. The Amazing Bond among Wine Consumption, Health and Hedonistic Well-Being. *Br. Food J.* **2020**, *122*, 2707–2723. [CrossRef]
7. Vecchio, R.; Decordi, G.; Grésillon, L.; Gugenberger, C.; Mahéo, M.; Jourjon, F. European Consumers' Perception of Moderate Wine Consumption on Health. *Wine Econ. Policy* **2017**, *6*, 14–22. [CrossRef]
8. Samoggia, A. Wine and Health: Faraway Concepts? *Br. Food J.* **2016**, *118*, 946–960. [CrossRef]
9. Giacosa, A.; Barale, R.; Bavaresco, L.; Faliva, M.A.; Gerbi, V.; La Vecchia, C.; Negri, E.; Opizzi, A.; Perna, S.; Pezzotti, M.; et al. Mediterranean Way of Drinking and Longevity. *Crit. Rev. Food Sci. Nutr.* **2014**, *56*, 635–640. [CrossRef]
10. Tjønneland, A.; Grønbæk, M.; Stripp, C.; Overvad, K. Wine Intake and Diet in a Random Sample of 48763 Danish Men and Women. *Am. J. Clin. Nutr.* **1999**, *69*, 49–54. [CrossRef]
11. Cena, H.; Calder, P.C. Defining a Healthy Diet: Evidence for the Role of Contemporary Dietary Patterns in Health and Disease. *Nutrients* **2020**, *12*, 334. [CrossRef] [PubMed]
12. Huhn, S.; Kharabian Masouleh, S.; Stumvoll, M.; Villringer, A.; Witte, A.V. Components of a Mediterranean Diet and Their Impact on Cognitive Functions in Aging. *Front. Aging Neurosci.* **2015**, *7*, 132. [CrossRef]
13. Stampfer, M.J.; Kang, J.H.; Chen, J.; Cherry, R.; Grodstein, F. Effects of Moderate Alcohol Consumption on Cognitive Function in Women. *N. Engl. J. Med.* **2005**, *352*, 245–253. [CrossRef] [PubMed]
14. Grunert, K.G.; Fernández-Celemín, L.; Wills, J.M.; Storcksdieck genannt Bonsmann, S.; Nureeva, L. Use and Understanding of Nutrition Information on Food Labels in Six European Countries. *J. Public Health* **2010**, *18*, 261–277. [CrossRef]
15. Arntzen, K.A.; Schirmer, H.; Wilsgaard, T.; Mathiesen, E.B. Moderate Wine Consumption Is Associated with Better Cognitive Test Results: A 7 Year Follow up of 5033 Subjects in the Tromsø Study. *Acta Neurol. Scand.* **2010**, *122*, 23–29. [CrossRef] [PubMed]
16. Fischer, K.; Melo van Lent, D.; Wolfsgruber, S.; Weinhold, L.; Kleineidam, L.; Bickel, H.; Scherer, M.; Eisele, M.; Van den Bussche, H.; Wiese, B.; et al. Prospective Associations between Single Foods, Alzheimer's Dementia and Memory Decline in the Elderly. *Nutrients* **2018**, *10*, 852. [CrossRef]
17. Radeka, S.; Rossi, S.; Bestulić, E.; Budić-Leto, I.; Kovačević Ganić, K.; Horvat, I.; Lukić, I.; Orbanić, F.; Zaninović Jurjević, T.; Dvornik, Š. Bioactive Compounds and Antioxidant Activity of Red and White Wines Produced from Autochthonous Croatian Varieties: Effect of Moderate Consumption on Human Health. *Foods* **2022**, *11*, 1804. [CrossRef]
18. Gutiérrez-Escobar, R.; Aliaño-González, M.J.; Cantos-Villar, E. Wine Polyphenol Content and Its Influence on Wine Quality and Properties: A Review. *Molecules* **2021**, *26*, 718. [CrossRef]
19. Plunk, A.D.; Syed-Mohammed, H.; Cavazos-Rehg, P.; Bierut, L.J.; Grucza, R.A. Alcohol Consumption, Heavy Drinking, and Mortality: Rethinking the J-Shaped Curve. *Alcohol. Clin. Exp. Res.* **2014**, *38*, 471–478. [CrossRef]
20. Ruf, J.C. Overview of Epidemiological Studies on Wine, Health and Mortality. *Drugs Under Exp. Clin. Res.* **2003**, *29*, 173–179.
21. Annunziata, A.; Vecchio, R.; Mariani, A. Alcohol Warnings and Moderate Drinking Patterns among Italian University Students: An Exploratory Study. *Nutrients* **2017**, *9*, 628. [CrossRef]
22. Annunziata, A.; Pomarici, E.; Vecchio, R.; Mariani, A. Nutritional Information and Health Warnings on Wine Labels: Exploring Consumer Interest and Preferences. *Appetite* **2016**, *106*, 58–69. [CrossRef] [PubMed]
23. Annunziata, A.; Pomarici, E.; Vecchio, R.; Mariani, A. Health Warnings on Wine: A Consumer Perspective. *Br. Food J.* **2016**, *118*, 647–659. [CrossRef]
24. Annunziata, A.; Pomarici, E.; Vecchio, R.; Mariani, A. Do Consumers Want More Nutritional and Health Information on Wine Labels? Insights from the EU and USA. *Nutrients* **2016**, *8*, 416. [CrossRef] [PubMed]
25. Chang, K.J.; Liz Thach, M.W.; Olsen, J. Wine and Health Perceptions: Exploring the Impact of Gender, Age and Ethnicity on Consumer Perceptions of Wine and Health. *Wine Econ. Policy* **2016**, *5*, 105–113. [CrossRef]
26. Arora, S.; Santiago, J.A.; Bernstein, M.; Potashkin, J.A. Diet and Lifestyle Impact the Development and Progression of Alzheimer's Dementia. *Front. Nutr.* **2023**, *10*, 1213223. [CrossRef] [PubMed]

27. Qi, C.-C.; Ge, J.-F.; Zhou, J.-N. Preliminary Evidence That Abscisic Acid Improves Spatial Memory in Rats. *Physiol. Behav.* **2015**, *139*, 231–239. [CrossRef]
28. Vaquero, M.J.R.; Alberto, M.R.; de Nadra, M.C.M. Antibacterial Effect of Phenolic Compounds from Different Wines. *Food Control.* **2007**, *18*, 93–101. [CrossRef]
29. Khalil, A.; Tazeddinova, D. The Upshot of Polyphenolic Compounds on Immunity amid COVID-19 Pandemic and Other Emerging Communicable Diseases: An Appraisal. *Nat. Prod. Bioprospecting* **2020**, *10*, 411–429. [CrossRef]
30. Higgins, L.M.; Llanos, E. A Healthy Indulgence? Wine Consumers and the Health Benefits of Wine. *Wine Econ. Policy* **2015**, *4*, 3–11. [CrossRef]
31. Champagne, C.M. Dietary Interventions on Blood Pressure: The Dietary Approaches to Stop Hypertension (DASH) Trials. *Nutr. Rev.* **2006**, *64*, S53–S56. [CrossRef]
32. Willcox, D.C.; Scapagnini, G.; Willcox, B.J. Healthy Aging Diets Other than the Mediterranean: A Focus on the Okinawan Diet. *Mech. Ageing Dev.* **2014**, *136–137*, 148–162. [CrossRef] [PubMed]
33. Fung, T.T.; Hu, F.B.; Wu, K.; Chiuve, S.E.; Fuchs, C.S.; Giovannucci, E. The Mediterranean and Dietary Approaches to Stop Hypertension (DASH) Diets and Colorectal Cancer. *Am. J. Clin. Nutr.* **2010**, *92*, 1429–1435. [CrossRef] [PubMed]
34. Giacosa, A.; Barale, R.; Bavaresco, L.; Gatenby, P.; Gerbi, V.; Janssens, J.; Johnston, B.; Kas, K.; La Vecchia, C.; Mainguet, P.; et al. Cancer Prevention in Europe. *Eur. J. Cancer Prev.* **2013**, *22*, 90–95. [CrossRef] [PubMed]
35. Shirani, F.; Salehi-Abargouei, A.; Azadbakht, L. Effects of Dietary Approaches to Stop Hypertension (DASH) Diet on Some Risk for Developing Type 2 Diabetes: A Systematic Review and Meta-Analysis on Controlled Clinical Trials. *Nutrition* **2013**, *29*, 939–947. [CrossRef]
36. Urquiaga, I.; Guasch, V.; Marshall, G.; San Martín, A.; Castillo, Ó.; Rozowski, J.; Leighton, F. Effect of Mediterranean and Occidental Diets, and Red Wine, on Plasma Fatty Acids in Humans: An Intervention Study. *Biol. Res.* **2004**, *37*, 253–261. [CrossRef]
37. Oksanen, A.; Kokkonen, H. Consumption of Wine with Meals and Subjective Well-Being: A Finnish Population-Based Study. *Alcohol Alcohol.* **2016**, *51*, 716–722. [CrossRef]
38. Lähteenmäki, L. Consumer Interpretation of Nutrition and Other Information on Food and Beverage Labels. In *Advances in Food and Beverage Labelling: Information and Regulations*; Elsevier: Amsterdam, The Netherlands, 2015; pp. 133–148. [CrossRef]
39. Barber, N.; Dodd, T.; Ghiselli, R. Capturing the Younger Wine Consumer. *J. Wine Res.* **2008**, *19*, 123–141. [CrossRef]
40. Lockshin, L.; Corsi, A.M. Consumer Behaviour for Wine 2.0: A Review since 2003 and Future Directions. *Wine Econ. Policy* **2012**, *1*, 2–23. [CrossRef]
41. Pabst, E.; Szolnoki, G.; Mueller Loose, S. The Effects of Mandatory Ingredient and Nutrition Labelling for Wine Consumers—A Qualitative Study. *Wine Econ. Policy* **2019**, *8*, 5–15. [CrossRef]
42. Pomarici, E.; Asioli, D.; Vecchio, R.; Næs, T. Young Consumers' Preferences for Water-Saving Wines: An Experimental Study. *Wine Econ. Policy* **2018**, *7*, 65–76. [CrossRef]
43. Corallo, A.; Latino, M.E.; Menegoli, M.; Spennato, A. A Survey to Discover Current Food Choice Behaviors. *Sustainability* **2019**, *11*, 5041. [CrossRef]
44. Rodríguez-Donate, M.C.; Romero-Rodríguez, M.E.; Cano-Fernández, V.J.; Guirao-Pérez, G. Gender and Wine Consumption: Sociodemographic Profiles. *Br. Food J.* **2019**, *122*, 242–257. [CrossRef]
45. Thach, L.; Olsen, J. Profiling the High Frequency Wine Consumer by Price Segmentation in the US Market. *Wine Econ. Policy* **2015**, *4*, 53–59. [CrossRef]
46. Ilak Peršurić, A.S.; Mann, S. What Distinguishes Connoisseurs from Spenders? A Case Study of Wine in Croatia. *Ekon. Poljopr.* **2019**, *66*, 929–940. [CrossRef]
47. Mora, M.; Urdaneta, E.; Chaya, C. Emotional Response to Wine: Sensory Properties, Age and Gender as Drivers of Consumers' Preferences. *Food Qual. Prefer.* **2018**, *66*, 19–28. [CrossRef]
48. Garcia, T.; Barrena, R.; Grande, I. The Wine Consumption Preferences of Young People: A Spanish Case Study. *Int. J. Wine Bus. Res.* **2013**, *25*, 94–107. [CrossRef]
49. Sogari, G.; Corbo, C.; Macconi, M.; Menozzi, D.; Mora, C. Consumer Attitude towards Sustainable-Labelled Wine: An Exploratory Approach. *Int. J. Wine Bus. Res.* **2015**, *27*, 312–328. [CrossRef]
50. Liberale, L.; Bonaventura, A.; Montecucco, F.; Dallegri, F.; Carbone, F. Impact of Red Wine Consumption on Cardiovascular Health. *Curr. Med. Chem.* **2019**, *26*, 3542–3566. [CrossRef]
51. Rizzo, G.; Borrello, M.; Dara Guccione, G.; Schifani, G.; Cembalo, L. Organic Food Consumption: The Relevance of the Health Attribute. *Sustainability* **2020**, *12*, 595. [CrossRef]
52. Domínguez Díaz, L.; Fernández-Ruiz, V.; Cámara, M. An International Regulatory Review of Food Health-Related Claims in Functional Food Products Labeling. *J. Funct. Foods* **2020**, *68*, 103896. [CrossRef]
53. Jannasch, F.; Kröger, J.; Schulze, M.B. Dietary Patterns and Type 2 Diabetes: A Systematic Literature Review and Meta-Analysis of Prospective Studies. *J. Nutr.* **2017**, *147*, 1174–1182. [CrossRef] [PubMed]

54. NHS. How to Get More Fibre into Your Diet. Available online: https://www.nhs.uk/live-well/eat-well/digestive-health/how-to-get-more-fibre-into-your-diet/ (accessed on 15 April 2019).
55. Pollard, T.M.; Steptoe, A.; Wardle, J. Motives Underlying Healthy Eating: Using the Food Choice Questionnaire to Explain Variation in Dietary Intake. *J. Biosoc. Sci.* **1998**, *30*, 165–179. [CrossRef]
56. Bazoche, P.; Deola, C.; Soler, L.G. An Experimental Study of Wine Consumers' Willingness to Pay for Environmental Characteristics. In Proceedings of the 12 th Congress of the European Association of Agricultural Economists-EAAE, Ghent, Belgium, 26–29 August 2008; pp. 1–8.
57. Sottomayor, M.; Souza Monteiro, D.M.; Teixeira, M.S. Valuing Nested Names in the Portuguese Olive Oil Market: An Exploratory Study. In Proceedings of the 116th Seminar, Parma, Italy, 27–30 October 2010.
58. Nohre, L.; MacKinnon, D.P.; Stacy, A.W.; Pentz, M.A. The Association between Adolescents' Receiver Characteristics and Exposure to the Alcohol Warning Label. *Psychol. Mark.* **1999**, *16*, 245–259. [CrossRef]
59. Hinkin, T.R.; Tracey, J.B.; Enz, C.A. Scale Construction: Developing Reliable and Valid Measurement Instruments. *J. Hosp. Tour. Res.* **1997**, *21*, 100–120. [CrossRef]
60. DiStefano, C.; Zhu, M.; Mindrilă, D. Understanding and Using Factor Scores: Considerations for the Applied Researcher. *Pract. Assess. Res. Eval.* **2009**, *14*, 20.
61. Brown, T.A. *Confirmatory Factor Analysis for Applied Research*, 2nd ed.; The Guilford Press: New York, NY, USA, 2015.
62. Hair, J.F.; Black, W.C.; Babin, B.J. *Multivariate Data Analysis. A Global Perspective*, 7th ed.; Pearson Education Inc.: Upper Saddle River, NJ, USA, 2009.
63. Lewis, M. Stepwise versus Hierarchical Regression: Pros and Cons. In Proceedings of the Annual Meeting of the Southwest Educational Research Association, Denton, TX, USA, 7–10 February 2007.
64. Kline, P. *A Handbook of Test Construction: Introduction to Psychometric Design*, 2nd ed.; Routledge: New York, NY, USA, 2015.
65. Steptoe, A.; Pollard, T.M.; Wardle, J. Development of a Measure of the Motives Underlying the Selection of Food: The Food Choice Questionnaire. *Appetite* **1995**, *25*, 267–284. [CrossRef]
66. European Comission. Regulation No 1924/2006 of the European Parliament and of the Council of 20 December 2006 on Nutrition and Health Claims Made on Foods. Available online: https://eur-lex.europa.eu/eli/reg/2006/1924/oj/eng (accessed on 2 May 2025).
67. Carrillo, E.; Fiszman, S.; Lähteenmäki, L.; Varela, P. Consumers' Perception of Symbols and Health Claims as Health-Related Label Messages. A Cross-Cultural Study. *Food Res. Int.* **2014**, *62*, 653–661. [CrossRef]

Disclaimer/Publisher's Note: The statements, opinions and data contained in all publications are solely those of the individual author(s) and contributor(s) and not of MDPI and/or the editor(s). MDPI and/or the editor(s) disclaim responsibility for any injury to people or property resulting from any ideas, methods, instructions or products referred to in the content.

MDPI AG
Grosspeteranlage 5
4052 Basel
Switzerland
Tel.: +41 61 683 77 34

Foods Editorial Office
E-mail: foods@mdpi.com
www.mdpi.com/journal/foods

Disclaimer/Publisher's Note: The title and front matter of this reprint are at the discretion of the Guest Editors. The publisher is not responsible for their content or any associated concerns. The statements, opinions and data contained in all individual articles are solely those of the individual Editors and contributors and not of MDPI. MDPI disclaims responsibility for any injury to people or property resulting from any ideas, methods, instructions or products referred to in the content.

www.ingramcontent.com/pod-product-compliance
Lightning Source LLC
Chambersburg PA
CBHW061507210926
51850CB00008B/1840
9783725862986